Dimitrios Tzovaras

Multimodal User Interfaces

From Signals to Interaction

Springer

Dr. Dimitrios Tzovaras
Informatics and Telematics Institute
Centre for Research and Technology Hellas
1st Km Thermi-Panorama Road
57001 (PO Box 361) Thermi-Thessaloniki
Greece

ISBN 978-3-540-78344-2   e-ISBN 978-3-540-78345-9

DOI 10.1007/978-3-540-78345-9

Springer Series on Signals and Communication Technology ISSN 1860-4862

Library of Congress Control Number: 2008921237

© 2008 Springer-Verlag Berlin Heidelberg

This work is subject to copyright. All rights are reserved, whether the whole or part of the material is concerned, specifically the rights of translation, reprinting, reuse of illustrations, recitation, broadcasting, reproduction on microfilm or in any other way, and storage in data banks. Duplication of this publication or parts thereof is permitted only under the provisions of the German Copyright Law of September 9, 1965, in its current version, and permission for use must always be obtained from Springer. Violations are liable for prosecution under the German Copyright Law.

The use of general descriptive names, registered names, trademarks, etc. in this publication does not imply, even in the absence of a specific statement, that such names are exempt from the relevant protective laws and regulations and therefore free for general use.

Typesetting and production: LE-TEX Jelonek, Schmidt & Vöckler GbR, Leipzig, Germany
Cover design: WMXDesign GmbH, Heidelberg

Printed on acid-free paper

9 8 7 6 5 4 3 2 1

springer.com

Springer Series on
SIGNALS AND COMMUNICATION TECHNOLOGY

Signals and Communication Technology

Passive Eye Monitoring
Algorithms, Applications and Experiments
R.I. Hammoud (Ed.) ISBN 978-3-540-75411-4

Multimodal User Interfaces
From Signals to Interaction
D. Tzovaras ISBN 978-3-540-78344-2

Human Factors and Voice Interactive Systems
D. Gardner-Bonneau, H.E. Blanchard (Eds.)
ISBN: 978-0-387-25482-1

Wireless Communications
2007 CNIT Thyrrenian Symposium
S. Pupolin (Ed.) ISBN: 978-0-387-73824-6

**Satellite Communications
and Navigation Systems**
E. Del Re, M. Ruggieri (Eds.)
ISBN: 978-0-387-47522-6

Digital Signal Processing
An Experimental Approach
S. Engelberg ISBN: 978-1-84800-118-3

**Digital Video and Audio
Broadcasting Technology**
A Practical Engineering Guide
W. Fischer ISBN: 978-3-540-76357-4

Three-Dimensional Television
Capture, Transmission, Display
H.M. Ozaktas, L. Onural (Eds.)
ISBN 978-3-540-72531-2

**Foundations and Applications
of Sensor Management**
A.O. Hero, D. Castañón, D. Cochran,
K. Kastella (Eds.) ISBN 978-0-387-27892-6

**Digital Signal Processing
with Field Programmable Gate Arrays**
U. Meyer-Baese ISBN 978-3-540-72612-8

Adaptive Nonlinear System Identification
The Volterra and Wiener Model Approaches
T. Ogunfunmi ISBN 978-0-387-26328-1

Continuous-Time Systems
Y.S. Shmaliy ISBN 978-1-4020-6271-1

Blind Speech Separation
S. Makino, T.-W. Lee, H. Sawada (Eds.)
ISBN 978-1-4020-6478-4

**Cognitive Radio, Software Defined Radio,
and Adaptive Wireless Systems**
H. Arslan (Ed.) ISBN 978-1-4020-5541-6

Wireless Network Security
Y. Xiao, D.-Z. Du, X. Shen
ISBN 978-0-387-28040-0

Terrestrial Trunked Radio – TETRA
A Global Security Tool
P. Stavroulakis ISBN 978-3-540-71190-2

Multirate Statistical Signal Processing
O.S. Jahromi ISBN 978-1-4020-5316-0

Wireless Ad Hoc and Sensor Networks
A Cross-Layer Design Perspective
R. Jurdak ISBN 978-0-387-39022-2

**Positive Trigonometric Polynomials
and Signal Processing Applications**
B. Dumitrescu ISBN 978-1-4020-5124-1

Face Biometrics for Personal Identification
Multi-Sensory Multi-Modal Systems
R.I. Hammoud, B.R. Abidi, M.A. Abidi (Eds.)
ISBN 978-3-540-49344-0

**Cryptographic Algorithms
on Reconfigurable Hardware**
F. Rodríguez-Henríquez
ISBN 978-0-387-33883-5

**Ad-Hoc Networking
Towards Seamless Communications**
L. Gavrilovska ISBN 978-1-4020-5065-7

Multimedia Database Retrieval
A Human-Centered Approach
P. Muneesawang, L. Guan
ISBN 978-0-387-25627-6

Broadband Fixed Wireless Access
A System Perspective
M. Engels, F. Petre
ISBN 978-0-387-33956-6

Acoustic MIMO Signal Processing
Y. Huang, J. Benesty, J. Chen
ISBN 978-3-540-37630-9

Algorithmic Information Theory
Mathematics of Digital Information
Processing
P. Seibt ISBN 978-3-540-33218-3

Continuous-Time Signals
Y.S. Shmaliy ISBN 978-1-4020-4817-3

Interactive Video
Algorithms and Technologies
R.I. Hammoud (Ed.) ISBN 978-3-540-33214-5

Content

1	**Introduction**..	1
2	**Multimodality Theory**...	5
	Niels Ole Bernsen	
	2.1 Introduction	5
	2.2 What is a Multimodal System?...........................	5
	2.2.1 An Enigma	5
	2.2.2 A Solution	6
	2.3 Which Modalities are There?..............................	10
	2.3.1 Deriving a Taxonomy of Input/Output Modalities...........	10
	2.3.2 Basic Concepts..	12
	2.3.3 Modality Taxonomy..................................	15
	2.3.4 Information Channels................................	19
	2.3.5 Interaction Devices	20
	2.3.6 Practical Uses of the Theory	20
	2.4 Multimodal Information Representation	24
	2.4.1 Advantages of Multimodality	24
	2.4.2 Constructing Multimodality from Unimodal Modalities............	25
	2.4.3 Linear Modality Addition and Replacement....................	25
	2.4.4 Non-linear Effects, Users, Design Detail, Purpose	28
3	**Information-Theoretic Framework for Multimodal Signal Processing**	31
	Jean-Philippe Thiran, Torsten Butz, and A. Murat Tekalp	
	3.1 Introduction	31
	3.2 Some Information Theoretic Concepts...............	32
	3.2.1 Stochastic Process and Error Probability	32
	3.2.2 Fano's Inequality and the Data Processing Inequality	33
	3.2.3 Information Theoretic Feature Extraction........	35

V

	3.3	From Error Probability to Multimodal Signal Processing	36
		3.3.1 Multimodal Stochastic Processes	36
		3.3.2 Objective Functions for Multimodal Signal Processing...	39
	3.4	Optimization	48
	3.5	Results	50
		3.5.1 Multimodal Medical Images	50
		3.5.2 Speech-Video Sequences	55
	3.6	Discussion	58
	3.7	Conclusions	58
4	**Multimodality for Plastic User Interfaces:** **Models, Methods, and Principles**		**61**
	Jean Vanderdonckt, Gaëlle Calvary, Joëlle Coutaz, *Adrian Stanciulescu*		
	4.1	Introduction	61
	4.2	Running Examples	62
		4.2.1 The Assisted Neuro-surgery System	62
		4.2.2 The Sedan-Bouillon Web Site	64
	4.3	Modality and Multimodality	66
		4.3.1 Definitions	66
		4.3.2 The CARE Properties	67
	4.4	The Problem Space of Plastic Multimodal UIs	70
		4.4.1 Two Adaptation Means: UI Re-molding and UI Re-distribution	71
		4.4.2 UI Components Granularity	72
		4.4.3 State Recovery Granularity	72
		4.4.4 UI Deployment	73
		4.4.5 Coverage of the Context of Use	73
		4.4.6 Coverage of Technological Spaces	74
		4.4.7 Existence of a Meta-UI	74
		4.4.8 UI Re-molding and Modalities	75
		4.4.9 UI Re-molding and Levels of Abstraction	76
		4.4.10 Summary	76
	4.5	Domain of Plasticity of a User Interface	77
	4.6	Three Principles for the Development of Plastic Multimodal UIs	81
		4.6.1 Blurring the Distinction between Design-time and Run-time	81
		4.6.2 Mixing Close and Open Adaptiveness	82
		4.6.3 Keeping Humans in the Loop	82
	4.7	Conclusion	83
	4.8	Acknowledgments	84

5   Face and Speech Interaction 85
*Mihai Gurban, Veronica Vilaplana, Jean-Philippe Thiran
and Ferran Marques*
- 5.1   Face and Facial Feature Detection 85
  - 5.1.1   Face Detection 86
  - 5.1.2   Facial Feature Detection 97
- 5.2   Interaction 98
  - 5.2.1   Multimodal Speaker Localization 99
  - 5.2.2   Audio-Visual Speech Recognition 104
- 5.3   Conclusions 117

6   Recognition of Emotional States in Natural Human-Computer Interaction 119
*R. Cowie1, E. Douglas-Cowie, K. Karpouzis, G. Caridakis,
M. Wallace and S. Kollias*
- 6.1   Introduction 119
- 6.2   Fundamentals 121
  - 6.2.1   Emotion Representation 121
  - 6.2.2   Methodology Outline 123
  - 6.2.3   Running Example 124
- 6.3   Feature Extraction 125
  - 6.3.1   Visual Modality 125
  - 6.3.2   Auditory Modality 138
- 6.4   Multimodal Expression Classification 140
  - 6.4.1   The Elman Net 140
  - 6.4.2   Classification 143
- 6.5   Experimental Results 146
  - 6.5.1   The Case for Naturalistic Data 146
  - 6.5.2   Statistical Results 147
  - 6.5.3   Quantitative Comparative Study 149
  - 6.5.4   Qualitative Comparative Study 151
- 6.6   Conclusions 152
- 6.7   Acknowledgements 153

7   Two SIMILAR Different Speech and Gestures Multimodal Interfaces 155
*Alexey Karpov, Sebastien Carbini, Andrey Ronzhin,
Jean Emmanuel Viallet*
- 7.1   Introduction and State-of-the-art 155
- 7.2   ICANDO Multimodal Interface 160
  - 7.2.1   Objectives 160
  - 7.2.2   System's Description 161

	7.3	Experimental Results	168
	7.4	MOWGLI Multimodal Interface	169
		7.4.1 Objectives	170
		7.4.2 System's Description	170
	7.5	Interfaces Comparison and Evaluation	180
	7.6	Main Future Challenges	183
	7.7	Acknowledgements	184

8 Multimodal User Interfaces in Ubiquitous Environments 185
Fabio Paternò, Carmen Santoro

8.1	Introduction	185
8.2	Related Work	186
8.3	Migratory User Interfaces	188
8.4	The Dimensions of Migration	188
8.5	An Architecture for a Migration Platform	192
8.6	Example Applications	196
	8.6.1 Domotic Scenario (mobile vocal+graphical->graphical desktop)	196
	8.6.2 Museum Scenario (pda->digital tv)	197
8.7	Conclusions	199

9 Software Engineering for Multimodal Interactive Systems 201
Laurence Nigay, Jullien Bouchet, David Juras, Benoit Mansoux, Michael Ortega, Marcos Serrano, Jean-Yves Lionel Lawson

9.1	Introduction	201
9.2	PAC-Amodeus: a Conceptual Architectural Solution	203
	9.2.1 Concurrent Processing of Data	204
	9.2.2 Data Fusion	204
9.3	Software Tools for Multimodality	207
	9.3.1 Existing Tools	207
	9.3.2 ICARE Platform	209
	9.3.3 ICARE Conceptual Model	210
	9.3.4 ICARE Graphical Editor	211
	9.3.5 OpenInterface Platform	213
9.4	Conclusion	216
9.5	Acknowledgments	218

10 Gestural Interfaces for Hearing-Impaired Communication 219
Oya Aran, Thomas Burger, Lale Akarun, Alice Caplier

10.1	Introduction	219
10.2	Modality Processing and Analysis	221
	10.2.1 Preprocessing	222
	10.2.2 Hand Shape	224
	10.2.3 Hand Location	229
	10.2.4 Hand Motion	230

		10.2.5 Facial Movements	233
		10.2.6 Lip Reading	234
		10.2.7 Facial Expressions	235
	10.3	Temporal Analysis	236
		10.3.1 Sign Language	236
		10.3.2 Cued Speech	237
	10.4	Multimodal Fusion	237
		10.4.1 Temporal Modelling	238
		10.4.2 Heterogenic Multiplexing	240
	10.5	Applications	246
		10.5.1 Sign Language Tutoring Tool	246
		10.5.2 Cued Speech Manual Gesture Interpreter	248
	10.6	Conclusion	249
11	**Modality Replacement Framework for Applications for the Disabled**		**251**
	Savvas Argyropoulos Konstantinos Moustakas		
	11.1	Introduction	251
	11.2	The Modality Replacement Concept	254
	11.3	Cued Speech	255
	11.4	Feature Extraction and Representation for the Cued Speech Language	257
		11.4.1 Audio Feature Extraction	257
		11.4.2 Lip Shape Feature Extraction	257
		11.4.3 Gesture Feature Extraction	258
	11.5	Coupled Hidden Markov Models	259
	11.6	Modality Reliability	260
	11.7	Modified Coupled Hidden Markov Model	261
		11.7.1 Training	261
	11.8	Evaluation of the Cued Speech Recognition System	262
	11.9	Multimodal Human-computer Interfaces	264
		11.9.1 Multimodal Collaborative Game	265
	11.10	Discussion	269
12	**A medical Component-based Framework for Image Guided Surgery**		**271**
	Daniela G. Trevisan, Vincent Nicolas, Benoit Macq, Luciana P. Nedel		
	12.1	Introduction	271
	12.2	MedicalStudio Framework	272
		12.2.1 Architecture	272
		12.2.2 Framework Implementation	273
	12.3	General Purpose Components	274
		12.3.1 Multimodal Registration	275
		12.3.2 Segmentation and 3D Reconstruction	276

	12.4	Applications	277
		12.4.1 3D Medical Assistant for Orthognatic Computer Surgery	277
		12.4.2 ACROGuide	278
	12.5	Conclusions and Future Works	279
	12.6	Acknowledgment	280

13 Multimodal Interfaces for Laparoscopic Training 281
Pablo Lamata, Carlos Alberola, Francisco Sánchez-Margallo,
Miguel Ángel Florido and Enrique J. Gómez.

	13.1	Introduction	281
	13.2	Functionality	282
	13.3	Technical Issues	282
		13.3.1 Simulator Architecture	283
		13.3.2 Collision Detection and Handling	285
		13.3.3 Design of Multimodal Interface Scenarios and Surgical Simulation Tasks	287
	13.4	Research Issues	288
		13.4.1 Study of Laparoscopic Sensory Interaction	288
		13.4.2 Laparoscopic Simulation Conceptual Framework	289
	13.5	Conclusion	289
	13.6	Acknowledgement	290

References 291

Chapter 1
Introduction

Human interfacing with the environment and with other humans is undoubtedly, fully multimodal. All human senses participate, even if some of then dominate, to the everyday human operations of perception, action and interaction. Interaction with the computer or computer-mediated interaction with others has been based for decades in a limited set of modalities and customised devices. Recent technological advancement in the areas of signal processing, computer vision and human-computer interaction, however, has made multimodal interfaces a reality.

Multimodal interaction is a multidisciplinary research area integrating efforts of various sciences such as cognition, psychology, computer science and engineering. This book reports one of the main outcomes of an attempt to bring closer experts from the HCI and the signal processing fields. Multimodal signal processing is fused with multimodal HCI to produce new knowledge in the field of multimodal interfaces. This knowledge goes all the way from signals to interaction and paves the way for a multiview exploration of the topic by scientists of both fields.

The book presents the most recent results of the SIMILAR EU funded Network of Excellence (with the contribution of one Chapter from groups participating in the HUMAINE Network of Excellence) on three main pillars in multimodal interfaces research i.e. theory, software development and application frameworks. The contributions in this book are fully towards the ten Grand Challenges identified by the SIMILAR NoE in the "SIMILAR Dreams" book published on 2005.

In Chapter 2, the notions of modality and multimodality are defined and discussed, followed by the presentation of a theory and taxonomy of all modalities of information representation in the media of graphics, acoustics and haptics. The final part of the chapter discusses practical uses of the theory, in particular with respect to the issue of modality choice and the idea of predictive construction of multimodal representation from unimodal modalities.

The main aim of Chapter 3, significantly inspired from [1], is the development of an information theoretic framework for multimodal signal processing, closely related to information theoretic feature extraction/selection. This important relationship indicates how multimodal medical image processing can be unified to a large extent, e. g. multi-channel segmentation and image registration, and extend information theoretic registration to other features than image intensities. The framework is not at all restricted to medical images though and this is illustrated by applying it to multimedia sequences as well.

In Chapter 4, the main results from the developments in plastic UIs and multimodal UIs are brought together using a theoretic and conceptual perspective as a unifying approach. It is aimed at defining models useful to support UI plasticity by relying on multimodality, at introducing and discussing basic principles that can drive the development of such UIs, and at describing some techniques as proof-of-concept of the aforementioned models and principles. In Chapter 4, the authors introduce running examples that serve as illustration throughout the discussion of the use of multimodality to support plasticity.

Face and speech combined analysis is the basis of a large number of human computer interfaces and services. Regardless of the final application of such interfaces, there are two aspects that are commonly required: detection of human faces (and, if necessary, of facial features) and combination of both sources of information. In Chapter 5, both aspects are reviewed, new alternatives are proposed and their behaviour is illustrated in the context of two specific applications: audio-visual speaker localization and audio-visual speech recognition. Authors have initially concentrated in face detection since most of facial feature detection algorithms rely on a previous face detection step. The various methods have been analysed from the perspective of the different models that they use to represent images and patterns. In this way, approaches have been classified into four different categories: pixel based, block based, transform coefficient based and region based techniques.

In Chapter 6, a multi-cue, dynamic approach to detect emotion in naturalistic video sequences is described, where input is taken from nearly real world situations, contrary to controlled recording conditions of audiovisual material. Recognition is performed via a recurrent neural network, whose short term memory and approximation capabilities cater for modelling dynamic events in facial and prosodic expressivity. This approach also differs from existing work in that it models user expressivity using a dimensional representation, instead of detecting discrete 'universal emotions', which are scarce in everyday human-machine interaction.

In Chapter 7, the main goal is to build multimodal interfaces that benefit from the advantages of both speech and gestures modalities. The objective is also to interact as easily and naturally as possible using a representation of the world with which we are familiar, where objects are described by their name and attributes and locations indicated with hand gesture or nose gesture in case of hands or arms disabilities. Multimodal information processing uses late-stage semantic architecture and two input modes are recognised in parallel and processed by understanding components. The results involve partial meaning representations that are fused by the multimodal fusion component, which also is influenced by the dialogue management and interpretation of current context. The best-ranked multimodal interpretation is transformed into control commands which are sent to a user application.

In Chapter 8, the state-of-the-art is initially discussed in the area of ubiquitous user interfaces followed by an introduction of migratory user interfaces and discussion of the relevant design dimensions. Then, an example architecture is briefly described, developed by the authors for supporting migratory interfaces and show a couple of example applications. Lastly, some conclusions are drawn along with indications for future evolutions in this area.

Chapter 9 focuses on the problem of software development of multimodal interfaces. The structure of the chapter is as follows: first the PAC-Amodeus software architectural model is presented as a generic conceptual solution for addressing the challenges of data fusion and concurrent processing of data of multimodal interfaces. The authors then focus on software tools for developing multimodal interfaces. After presenting several existing tools, the ICARE and OpenInterface platforms are then described and illustrated.

Chapter 10 includes a discussion and review of analysis techniques for the modalities that are used in hearing impaired communication. The authors concentrate on the individual modalities: hand, face, lips, expression and treat their detection, segmentation, and feature extraction. In the *Temporal analysis* section, the focus is on the temporal analysis of the modalities, specifically in sign languages and in CS. The next section presents temporal modelling and belief-based multimodal fusion techniques. In the last section, two example applications are given: a sign language tutoring tool and a cued speech interpreter.

Chapter 11 focuses on the effective processing of information conveyed in multiple modalities and their translation into signals that are more easily understood by impaired individuals. More specifically, the concept of modality replacement is introduced and two applications for the intercommunication of the disabled are thoroughly discussed. In particular this chapter discusses the development of a framework for the combination of all the incoming modalities from an individual, recognition of the transmitted message, and translation into another form that is perceivable by the receiver. Moreover the modality replacement concept and its applications are introduced. The basic idea is to exploit the correlation among modalities to enhance the information perceived by an impaired individual who can not access all incoming modalities. Finally a modified CHMM is employed to model the complex interaction and interdependencies among the modalities and combine them efficiently to recognize correctly the transmitted message.

Chapter 12 introduces MedicalStudio, a composable, open-source easily evolvable cross-platform framework that supports surgical planning and intra-operative guidance with augmented interactions. It is designed to integrate the whole computer aided surgery process to which both researchers and clinicians participate. A description of the framework architecture is provided as well as some medical components and applications already developed into this framework. Collaboration with several research centers and medical clinics has shown the versatility and promising dissemination of this medical framework.

Finally, Chapter 13 presents the SINERGIA simulator as an example that emphasizes the importance of introducing multimodal interfaces in laparoscopic training. This simulator is being developed by the SINERGIA Spanish cooperative network, which is composed by a consortium of clinical and technical partners from Spain. The Chapter describes the first prototype of the system, which has been built on the work of MedICLab developed by the Polytechnic University of Valencia.

Chapter 2
Multimodality Theory

Niels Ole Bernsen
Natural Interactive Systems Lab., University of Southern Denmark

2.1 Introduction

Since the early 2000s, interactive systems designers, developers, and evaluators have increasingly been focusing on tapping into the enormous potential of multimodality and multimodal systems. In this chapter, we discuss and define the notions of modality and multimodality followed by presentation of a theory and taxonomy of all modalities of information representation in the media of graphics, acoustics, and haptics. The final part of the chapter discusses practical uses of the theory, in particular with respect to the issue of modality choice and the idea of predictive construction of multimodal representation from unimodal modalities.

2.2 What is a Multimodal System?

2.2.1 An Enigma

Somewhat surprisingly, given the enormous attention to multimodal systems world-wide, there is still a certain amount of confusion about what multimodality actually is. To understand multimodality, it seems important to understand why this confusion persists.

The term 'modality' itself is not overly informative. One of its relevant senses is to be a *manner of something*; another, the so-called *sensory modalities* of psychology, i.e., vision, hearing, etc., but, as we shall see, the 'modalities' of 'multimodality' cannot be reduced to the sensory modalities of psychology.

Historically, the term 'modality' already appears in something close to its present-day meaning in Bolt's early paper on advantages of using combined speech and deictic gesture [2]. Another early appearance is in [3] who mention written text and beeps as different examples of forms of representing information as output from, or input to, computer systems.

Today, the following two definitions or explanations of multimodality are perhaps among the more widespread ones.

1. A multimodal system is a system which somehow involves several modalities. This is both trivially true and quite uninformative about what multimodality actually is. The definition does, however, put a nice focus on the question: what are *modalities*? It would seem rather evident that, if something – a system, interaction, whatever – is *multi*modal, there must be other things which are *uni*modal and which *combine* to make that something multimodal. However, another explanation often seems to be lurking in the background:
2. A multimodal system is a system which takes us beyond the ancient and soon-to-become-obsolete GUI (Graphical User Interfaces) paradigm for interactive systems. Multimodal systems represent a new and more advanced paradigm for interactive systems, or, quoting [4]: "Multimodal systems are radically different than standard GUIs".

   This view probably helps explain the sense of novelty and adventure shared by many researchers and developers of multimodal systems today. However, as an explanation of multimodality, and although still near-vacuous as to what multimodality is, this one is actually false and seriously misleading. Even though there are several more informative attempts at definitions and explanations of multimodality around, these all imply that GUI systems are *multimodal* systems. Furthermore, GUI-based interaction is far from obsolete it's a useful paradigm which is just far more familiar and better explored than most other kinds of multimodal interaction.

2.2.2 A Solution

To see why multimodality still has a lot of unexplored novelty to it, why the sensory modalities of psychology only far too insufficiently account for the modalities there are, and why GUIs are nevertheless multimodal, we need a theory of what's involved. The basic notions of that theory are interaction, media, modalities, and information channels, which we will now look at in turn.

2.2.2.1 Human-Computer Interaction and Media

Human-computer 'interaction' is, in fact, *exchange of information* with computer systems, and is of many different kinds which we need not go into here (for taxonomy, see [5]). Exchange of information is ultimately a physical process. We never exchange information in the abstract even if we are very much used to thinking and reasoning about information in abstract terms. When humans ex-

2.2 What is a Multimodal System?

change information, the information is *physically instantiated* in some way, such as in sound waves, light, or otherwise. In fact, humans are traditionally said to have five or six senses for physically capturing information, i.e., *sight, hearing, touch, smell, taste*, and, if this one is counted as well, *proprioception*. These are the sensory modalities of psychology.

Correspondingly, let us say that information, to be perceptibly communicated to humans, must be instantiated in one or more of the following six *physical media*, i.e.:

- light / vision / graphics;
- sound waves / hearing / acoustics;
- mechanical touch sensor contact / touch / haptics;
- molecular smell sensor contact / smell / olfaction;
- molecular taste sensor contact / taste / gustation; and
- proprioceptor stimulation (as when you sense that you are being turned upside down).

It is useful to define each medium through a *triplet* as above, the first element referring to the physical information carrier, the second to the perceptual sense needed for perceiving the information, and the third one to information presentation in that medium. Note that this entails a non-standard use of the term 'graphics' in English, because the graphics modalities come to include not only graphical images and the like, but also, in particular, ordinary text.

To the above we need to add a point about human *perceptual thresholds*. If a human or system is to succeed in getting dispatched physically instantiated information perceived by a *human* recipient, the instantiation must respect the limitations of the human sensors. For instance, the human eye can only perceive light within a certain band of electromagnetic frequency (approximately 380–780 Nm); the human ear can only perceive sound within a certain Hz band (approximately 18–20.000 Hz); touch information must be above a certain mechanical force threshold to be perceived and its perception also depends on the density of touch sensors in the part of human skin exposed to the touch; etc. In other words, issuing an ultrasonic command to a soldier will have no effect because no physically instantiated information will be perceived by the soldier.

2.2.2.2 Modalities

We can now define a 'modality' in a straightforward way:

3. A modality or, more explicitly, *a modality of information representation*, is a way of representing information in some physical medium. Thus, a modality is defined by its physical medium and its particular "way" of representation.

It follows from the definition that modalities do not have to be perceptible by humans. Even media do not have to be perceptible to humans. So modalities don't even have to be represented in physical media accessible to humans, since there are physical media other than the six media listed above, all of which are partly accessible to humans. In what follows, we focus on modalities perceptible to humans unless otherwise stated.

Now we need to look at those "ways" of information representation because these are the reason for having the notions of modalities and multimodality in the first place. The simple fact is that we need those "ways" because humans use *many* and *very different* modalities for representing information instantiated in *the same* physical medium and hence perceived by *the same* human sensory system. Consider, for example, *light* as physical medium and *vision* as the corresponding human sense. We use vision to perceive *language text, image graphics, facial expression, gesture,* and much more. These are *different* modalities of information representation instantiated in *the same* physical medium.

Although the above example might be rather convincing on its own, it is useful to ask *why* the mentioned modalities are considered to be different. There are two reasons. The first one is that all modalities differ in *expressiveness*, i.e., they are suited for representing different kinds of information. A photo-realistic image, for instance, is generally far better at expressing how a particular person looks than is a linguistic description. The second reason is to do with the properties of the *recipient* of the information represented, perceptual, cognitive, and otherwise. For instance, since the blind do not have access to the medium of light, we primarily use the acoustic and haptic media to represent information to the blind. But again, since different modalities differ in expressiveness, we need all the modalities we can implement for information representation for the blind, such as speech and Braille test for linguistic information, haptic images, etc. Even if two particular representations are completely *equivalent* in information content and instantiated in the *same* medium, the human perceptual and cognitive system works in such a way that each of the representations may be preferable to the other depending on the purpose of use. For instance, if we want a quick overview of trends in a dataset, we might use, e.g., a static graphics *bar chart*, but if we want to study the details of each data point, we might prefer to look at an informationally equivalent static graphics *table* showing each data point.

As the above examples suggest, we may actually have to reckon with a considerable number of different modalities.

2.2.2.3 Input and Output Modalities

Given the considerable number of different modalities there are, it is good practice to specify if some particular modality is an *input* modality or an *output* modality. For instance, what is a "spoken computer game"? This phrase does not reveal if the system takes speech input, produces speech output, or both, although these three possibilities are very different from one another and circumscribe very different classes of systems. By tradition, we say that, during interaction, the *user* produces *input modalities* to the system and the *system* produces *output modalities*

2.2 What is a Multimodal System?

to the user. And the point is that, for many multimodal systems, the set of input modalities is often different from the set of output modalities.

Another important point about input and output modalities is that we can form the abstract concepts of (i) the class of *all possible input modalities that can be generated by humans* and (ii) the class of *all possible output modalities that can be perceived by humans*. These two classes are *asymmetrical*. This follows from the limitations of the human sensory system as a whole both as regards the physical media it is sensitive to and its biologically determined sensory thresholds. Computers, on the other hand, can be made far more discriminative than humans on both counts. Computers can sense X-rays and other exotic "rays" (alpha, beta, gamma etc.), radar, infrared, ultraviolet, ultrasound, voltage, magnetic fields, and more; can sense mechanical impact better than humans do; and might become capable of sensing molecular-chemical stimuli better than humans do. This means that computers: (i) have more input modalities at their disposal than humans have; (ii) have or, in some cases, probably will get, far less restrictive sensory thresholds for perceiving information in some particular modalities than humans; and (iii) can output information that humans are incapable of perceiving. This is useful for interactive systems design because it allows us to think in terms of, e. g., human interaction with a magnetic field sensing application which no human could replace.

This point about input/output modality asymmetry raises many interesting issues which, however, we shall ignore in the following. Let us simply stipulate that we will only discuss multimodal interaction in *maximum symmetrical conditions*, i. e., we will discuss multimodal input/output interaction based on the physical media humans can perceive and to the extent that humans can perceive information instantiated in those media.

2.2.2.4 Unimodal and Multimodal Interactive Systems

We can now define a multimodal interactive system.

4. A *multimodal interactive system* is a system which uses at least two different modalities for input and/or output. Thus, IM1,OM2, IM1, IM2, OM1 and IM1, OM1, OM2 are some minimal examples of multimodal systems, *I* meaning input, *O* output, and *Mn* meaning a specific modality *n*.
   Correspondingly,
5. A *unimodal interactive system* is a system which uses the same single modality for input and output, i. e., IMn, OMn.

An over-the-phone spoken dialogue system is an example of a unimodal system: you speak to it, it talks back to you, and that's it. Other examples are a Braille text input/output dialogue or chat system for the blind, or a system in which an embodied agent moves as a function of the user's movements. There are lots more, of course, if we make creative use of all the modalities at our disposal. Still, the class of potential multimodal systems is exponentially larger than the class of potential unimodal systems. This is why we have to reckon with a quasi-unlimited number of new modality combinations compared to the GUI age.

2.2.2.5 Why GUIs are Multimodal

It is probably obvious by now why GUI systems are multimodal: standard GUI interfaces take *haptic* input and present *graphics* output. Moreover, both the haptic input and the graphics output involves a range of individually different modalities.

2.3 Which Modalities are There?

Given the definition of a *modality of information representation* as a way of representing information in a particular physical medium, and given the limitations of human perceptual discrimination we have adopted as a frame for the present discussion: which and how many input and output modalities are there? From a theoretical point of view, we would like to be able to: (i) identify all *unimodal* or elementary modalities which could be used to build multimodal interfaces and enable multimodal interaction; (ii) group modalities in one or several sensible ways, hierarchically or otherwise; and (iii) provide basic information on each of them.

From a practical point of view in design, development, and evaluation, we would like to have: (iv) a practical toolbox of all possible unimodal input/output modalities to choose from; (v) guidelines or something similar for which modality to use for a given development purpose; and (vi) some form of generalization or extension from unimodality to multimodality.

If possible, the contents of the theory and the toolbox should have various properties characteristic of scientific theory, such as being: transparently derived from unambiguous first principles, exhaustive, well-structured, and empirically validated. We do, in fact, have much of the above in *modality theory,* a first version of which was presented in [6]. Modality theory will be briefly presented in this section. Other fragments contributing to the desiderata listed above, as well as fragments missing, are discussed in the final section of this chapter.

2.3.1 Deriving a Taxonomy of Input/Output Modalities

Table 2.1 shows a taxonomy of input/output modalities. The scope of the taxonomy is this: it shows, in a particular way to be clarified shortly, all possible modalities in the three media of *graphics, acoustics* and *haptics*, which are currently the all-dominant media used for exchanging information with interactive computer systems. Following the principle of *symmetry* introduced above, the taxonomy only shows modalities that are perceptible to humans. The taxonomy is claimed to

2.3 Which Modalities are There?

Table 2.1 A taxonomy of input and output modalities

Super level	Generic Level	Atomic level	Sub-atomic level
Linguistic modalities	1. Sta. an. graphic elements		
	2. Sta-dyn an. acoustic elements		
	3. Sta-dyn an. haptic elements	4a. Sta.-dyn. gest. discourse	
		4b. Sta.-dyn. gest. lab.	5a1. Typed text
	4. Dyn. an. graphic	4c. Sta.-dyn. gest. notation	5a2. Hand-writ text
		5a. Written text	5b1. Typed lab.
	5. Sta. non-an. graphic	5b. Written lab.	5b2. Hand-writ lab.
		5c. Written notation	5c1. Typed not.
		6a. Spoken discourse	5c2. Hand-writ not.
	6. Sta.-dyn. non-an. acoustic	6b. Spoken lab.	
		6c. Spoken notation	Legend
		7a. Haptic text	an = analogue
	7. Sta.-dyn. non-an. haptic	7b. Haptic lab.	dyn = dynamic
		7c. Haptic notation	gest = gesture
		8a. Dyn. written text	lab = labels/
	8. Dyn. non-an. graphic	8b. Dyn. written lab.	keywords
		8c. Dyn. written notation	non-an = non-analogue
		8d. Sta.-dyn. spoken discourse	not = notation
		8e. Sta.-dyn. spoken lab.	sta = static
		8f. Sta.-dyn. spoken not.	writ = written
Analogue modalities	9. Static graphic	9a. Images	
		9b. Maps	
		9c. Compos. diagrams	
		9d. Graphs	
		9e. Conceptual diagrams	
	10. Sta.-dyn. acoustic	10a. Images	
		10b. Maps	
		10c. Compos. diagrams	
		10d. Graphs	
		10e. Conceptual diagrams	
	11. Sta.-dyn. haptic	11a. Images	
		11b. Maps	
		11c. Compos. diagrams	
		11d. Graphs	
		11e. Conceptual diagrams	
	12. Dynamic graphic	12a. Images	12a1. Facial expression
		12b. Maps	12a2. Gesture

Table. 2.1 Continued

Super level	Generic Level	Atomic level	Sub-atomic level
Arbitrary modalities	13. Static graphic	12c. Compositional diagrams	12a3. Body action
	14. Sta.-dyn. acoustic	12d. Graphs	
	15. Sta.-dyn. haptic	12e. Conceptual diagrams	
	16. Dynamic graphic		
Explicit structure modalities	17. Static graphic		
	18. Sta.-dyn. acoustic		
	19. Sta.-dyn. haptic		
	20. Dynamic graphic		

be *complete* in the sense that all possible modalities in those media are either shown in the taxonomy or can be generated from it by further extension downwards from the generic level. What this means will become clear as we proceed.

The taxonomy is represented as a tree graph with four hierarchical levels, called *super level, generic level, atomic level*, and *sub-atomic level*, respectively. The *second-highest* (generic) level is derived from basic principles or hypotheses. For the detailed derivation, see [7]. In what follows, we sketch the general ideas behind the derivation in the form of a meaning representation tree (Fig. 2.1) and then describe the taxonomy itself.

2.3.2 Basic Concepts

The taxonomy assumes that, within its scope, the meaning of physically instantiated information to be exchanged among humans or between humans and systems, can be categorised as belonging to one of the categories shown in Fig. 2.1.

Figure 2.1 says that, at this stage, modality theory addresses meaning represented in graphics, acoustics and haptics, and that meaning representation is either *standard*, in which case it is either linguistic, analogue or explicit structures, or *arbitrary*; or meaning is *non-standard*, in which case it can be viewed as a result of applying some function, such as the functions used to create metaphorical or metonymic meaning. We now briefly explain and illustrate each concept in the meaning representation tree.

Standard meaning is (i) shared meaning in some (sub-)culture. Shared meaning is basic to communicative interaction because it allows us to represent information in some modality in an already familiar way so that people in our (sub-)culture will understand. Secondly, (ii) standard meaning is opposed to (shared but) non-standard meaning as explained below. Words in the vocabulary in some language, for instance, have standard meanings which are explained in dictionaries and thesauri. An image of a computer would not have *that* meaning to Neanderthal man.

2.3 Which Modalities are There?

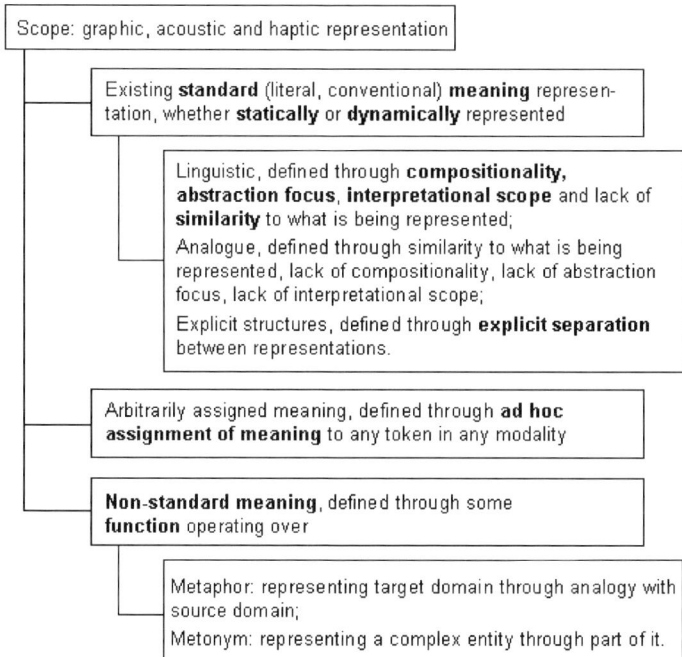

Fig. 2.1 Varieties of meaning representation

Static/dynamic: static representation is not defined in physical terms but, rather, as that which the recipient can perceptually inspect for as long as it takes, such as a GUI output screen, a blinking icon, or an acoustic alarm which must be switched off before it stops. Dynamic representations do not allow this freedom of perceptual inspection, such as a ringing telephone which may stop ringing at any moment.

Compositionality is a standard concept in linguistic analysis according to which linguistic meaning can be viewed, by way of approximation, at least, as built rule-by-rule from syntax to semantics [8]. For instance, the sentence "Mary loves John" is built in this way and systematically changes meaning if the word order is reversed, as in "John loves Mary".

Abstraction focus is the ability of language to focus meaning representation at any level of abstraction. If we write, e. g.: "A woman walks down the stairs", this is perfectly meaningful even though we are not told things like who she is, how she looks or walks, what the stairs and their surroundings look like, or whether or not the stairs go straight or turn right or left. Language can do that.

Interpretational scope: to continue the previous example, what we tend to do when reading the sentence is to construct our own (analogue, see below) representation. My representation may be very different from yours and none of the two are substantiated by what the declarative sentence about the woman walking down the stairs actually says. This is interpretational scope: we are both "right" but only as far as the standard meaning of the sentence goes. For more about interpretational scope and abstraction focus, see [9]. By contrast to these properties of linguistic representation.

Analogue representation is defined through similarity between a representation and what is being represented. A drawing of a cow more or less resembles a cow – if not, we have the right to ask if what is being represented in the drawing is really a cow. However, the *word* "cow" (German: Kuh, French: vache, Danish: ko) does not resemble a cow at all. Both the drawing and the word in some language are representations rather than the real thing, of course, the difference being that the drawing is an *analogue* representation whereas the second, linguistic representation is *non-analogue*.

Modality theory also deals with more tenuous analogue relationships than those of photo-realistic images to what they represent, such as between a line diagram and what *it* represents, or between the click-clicks of a Geiger counter, or the acoustic signatures of the sonar, and what *they* represent.

Explicit separation: this notion may not look much because it deals with what we often do when creating, e. g., tables or matrices. All we do is separate columns and rows using more or less straight lines. This is often very useful for separation and grouping purposes, however, and GUIs, for instance, are full of explicit structures – in windows using multi-layered explicit structures, in pull-down menus, etc. However, explicit structures are also useful in other modalities, such as when we use a beep to mark when the user can speak in non-barge-in spoken dialogue systems, or, in particular, in haptics for the blind because the blind do naturally group objects at-a-glance as the seeing do. Grey colour is used to support perceptual grouping in Table 2.1.

Ad hoc assignment of meaning: spies, for instance, always did that to avoid that anyone else could understand their communication; kids do it when playing the game of having to say "yes" when you mean "no" and vice versa until you flunk it; and we all do this when, e. g., using boldface, italics, font size, and other means to assign particular (non-shared) meanings to text items. If an ad hoc assigned meaning catches on, which is what happens when new phenomena get an accepted name in a language, it becomes standard meaning.

Non-standard meaning function: although we tend not to realize, human communication is shot through with non-standard meaning [10]. The reason we tend not to realize is that, e. g., metaphors which began their career as new creative inventions tend to become *dead metaphors* like when we speak of the "shoulder" of a mountain or the "head" of a noun phrase. Once dead, it takes a special focus to retrieve their origin and the words just behave as conventional standard meanings, like most Chinese characters which began their career as analogue signs. Typologies of non-standard meaning typically view each type of non-standard meaning as being created from standard meaning through application of some particular function, such as *metaphor* – e. g., "He blew the top" – using *analogy* with water cooling in a car; or *metonomy* – e. g., "The White House issued a statement saying..." – using the familiar physical entity of The White House as *part-representing-the-whole* for the executive branch of the US government).

2.3.3 Modality Taxonomy

The modality taxonomy is derived from the definitions and distinctions introduced in the previous paragraph. More specifically, the relationship between Fig. 2.1 and the taxonomy in Table 2.1 is as follows:

First, Fig. 2.1 introduces a set of orthogonal distinctions which are aimed to capture the core of what it is to represent information in the physical media scoped by the theory, i.e., graphics, acoustics and haptics.

Secondly, based on those distinctions, simple combinatorics mostly account for the derivation of the taxonomy's *generic level*. All other taxonomy levels are generated from the generic level. The qualification 'mostly' refers to the fact that, since the goal of derivation is to arrive at a practical toolbox of unimodal modalities which is reasonably intuitive to use by interactive systems developers, some *fusions* of otherwise separate derived categories have taken place [7]. However, these fusions are all reversible, and simply so, should future multimodal interactive systems development proceed in novel ways. A typical indicator in the taxonomy that such a *pragmatic fusion* has taken place is the definition of a modality as static/dynamic (**sta.-dyn.** in the taxonomy table), meaning that there is currently no useful point in maintaining separate modalities for static and dynamic representations of the kind specified.

Thirdly, the taxonomy does not distinguish between *standard meaning* and *non-standard meaning derivations* from standard meaning representations. In fact, it cannot because these are physically indistinguishable. The desktop metaphor representation, for instance, is a 2D static graphics plane with analogue icons, labels/keywords, explicit structures used for composing windows and other structures, etc. The fact that this representation is intended by its designers to serve as a metaphor is due to its designed similarity with an ordinary desktop. We might describe non-standard meaning, being derived from standard meaning, as a sort of third dimension relative to the 2D taxonomy. Had the taxonomy been 3D, you would have the desktop metaphor stretching out in the third dimension from the modality *analogue* static *graphic image* (9a).

2.3.3.1 Taxonomy Levels

Before we walk through the taxonomy levels, it is important to set one's mind to *unconventional* in order not to miss its richness. Consider that, among many other things, physically instantiated representations may be either *1D, 2D or 3D*, that analogue images, diagrams, graphs, etc., can be *acoustic* or *haptic* and not just *graphic*, and that *time*, as well as the presence or absence of *user control* of what is being represented, is essential to the distinction between static and dynamic representation. An animated interface character (or agent), for instance, is a (2D or 3D) analogue dynamic graphic (12) image (12a) whose modalities are facial expression (12a1), gesture (12a2) and body action (12a3). And what might an acoustic compositional diagram be used for (10c)?

Super level: since the taxonomy is generated from standard meaning at the *generic level*, the top *super level* modalities only represent one among several possible *classifications* of the derived generic-level modalities. The actual classification in the figure is in terms of *linguistic, analogue, arbitrary* and *explicit structure* modalities, respectively. However, it is perfectly possible and straightforward to re-classify the generic level modalities in terms of, e.g., the underlying media, getting a super level consisting of *graphics, acoustics* and *haptics*, respectively, or in terms of the static/dynamic distinction. The contents of the taxonomy will remain unchanged but the structure of the tree will be modified.

In the taxonomy shown above, super-level *linguistic modalities* represent information in some natural or formal language. *Analogue modalities* represent information by providing the representation with some amount of similarity with what it represents. *Arbitrary modalities* are representations which get their meaning assigned ad hoc when they are being introduced. *Explicit structure modalities* structure representations in space or time, as when information is structured and grouped by boxes-within-boxes in a GUI window.

Generic level: relative to the super level, the generic level expands the super level modalities by means of distinctions between static and dynamic modalities, and between the three physical media. Looking at the generic-level modalities, it is hard to avoid noticing that, in particular, many, if not all of the linguistic and analogue modalities are rather unfamiliar in the sense that we are not used to thinking in terms of them. This is true: they are theoretically derived abstractions at a level of abstraction which most of us visit quite infrequently. It's like trying to think and reason in terms of *furniture* instead of in the familiar terms tables, chairs and beds. In general, the atomic-level modalities are very different in this respect.

The reader may also note that the generic-level *arbitrary* modalities and *explicit structure* modalities are not expanded at the atomic level. The reason is that, at this point, at least, no additional distinctions seem to be needed by developers. If further distinction becomes needed, they can simply be added by expanding those modalities, i.e., one, several, or all of them, at the atomic level. Similarly, the first three numbered modalities at generic level, i.e., the graphic, acoustic, and haptic *analogue linguistic elements*, remain unexpanded. This is another example of pragmatic fusion. The modalities themselves cover possible languages in all three media which use analogue elements, like hieroglyphs or onomatopoetic, for linguistic representation. It would seem slightly contrived today to attempt to revive the analogue origins of the elements of many existing languages, which is why all discussion of languages in modality theory is done within the categories of non-analogue linguistic representation.

Sub-generic levels: there are potentially an unlimited number of sub-generic levels of which two are shown in the taxonomy in Table 2.1. The important point is that, in order to generate modalities at sub-generic levels, the definitions and distinctions in Fig. 2.1 are no longer sufficient. This means that all generation beyond generic level must proceed by establishing and validating new distinctions.

2.3 Which Modalities are There?

Atomic level: relative to the generic level, the atomic level expands parts of the taxonomy tree based on new sets of distinctions that can be easily read off from the taxonomy tree. The distinctions are defined in [7].

While the linguistic and analogue *generic-level* modalities are generally less familiar to our natural or prototypical conceptualizations [11] of information representation, their descendant modalities at atomic level are generally highly familiar. This is where we find, among many other modalities, GUI menu labels/keywords (5b), spoken discourse (6a), Braille text (7a), sign language (4a-c), and various static and dynamic analogue graphic representations, all thoroughly familiar. However, several of their corresponding acoustic and haptic sisters, though potentially very useful, are less familiar and may be viewed as products of the generative poser of the taxonomy.

Sub-atomic level: in the taxonomy in Table 2.1, only a few segments of the taxonomy tree have been expanded at sub-atomic level. The top right-hand corner shows expansion of *static written text*, *labels/keywords*, and *notation* (5a-c) into their typed and hand-written varieties, respectively. This is a rather trivial expansion which is there primarily to show the principle of downwards tree expansion onto the sub-atomic level. We are all familiar with the difference it makes to both users and the system if they have to read hand-written text as opposed to typed text.

In the lower right-hand corner, the *dynamic graphic image* modalities are expanded into the non-speech, natural interactive communication modalities of visible *facial expression*, (non-sign-language) *gesture*, and *body action* (12a1-3). This distinction is argued in [12]). If the animated interface agent also speaks, its multimodal information representation is increased even more by *acoustic speech* modalities (6a-c) and *visual speech* modalities, i.e., mouth and lip movements during speech (8d-f).

It seems clear that the sub-atomic-level expansions of the modality taxonomy tree shown above are incomplete in various ways, i.e., multimodal interactive system developers need additional modality expansions based on the atomic level. The important point is that everyone is free to expand the tree whenever there is a need. The author would very much appreciate information about any expansions made and as well as how they have been validated.

Below sub-atomic level: here are a couple of examples of highly desirable taxonomy tree expansions at sub-sub-atomic level. Although we still don't have a single complete, generally accepted or standard taxonomy of all different types of (non-sign language) *gesture* (12a2), there is considerable agreement in the literature about the existence of, at least, the following hand-arm gesture modalities, cf. [13]:

12a2a deictic gesture
12a2b iconic gesture
12a2c metaphoric gesture
12a2d emblems
12a2e beats (or batons)
12a2f other

We use the standard expedient of corpus annotators of adding an 'other' category for gestures which don't fit the five agreed-upon categories, thus explicitly marking the gesture modality scheme as being under development. Similarly, if there were a stable taxonomy for *facial expression of emotion*, it would constitute a set of sub-sub-atomic modalities expanding *facial expression* (12a3).

2.3.3.2 Some General Properties of the Taxonomy

In this and the following section, we list a number of properties of the taxonomy of input/output modalities and of the modalities themselves, which follow from the discussion above.

Unlimited downwards expansion. The taxonomy tree can be expanded downwards without limitation as needed, by analyzing, defining, and validating new distinctions that can serve as basis for downwards expansion, as these become relevant in interactive systems design, development and evaluation. Conversely, this is also why some parts of the tree have not (yet) been expanded beyond the generic level, i.e., the arbitrary and explicit structure modalities, or the gesture modality.

Property inheritance. The taxonomy shows that modalities can be analysed and described at different levels of abstraction. We do this all the time when working with modalities but may not be used to thinking about what we do in these terms. It follows that the taxonomy enables property inheritance. For instance, we can analyze the *linguistic* modality quite generally at super level, discovering the general properties of linguistic modalities we need for some purpose. Having done that, these properties get inherited by all *linguistic* modalities at lower levels. Thus, e.g., *spoken discourse* (atomic level) inherits all properties of the super-level *linguistic* modality as well as the more detailed properties of *acoustic linguistic* representation (generic level). Having analysed the parent properties, all we need to do to analyze the *spoken discourse* modality is to analyze, at atomic level, the new emergent properties of spoken discourse at this level. The origin of these emergent properties is clear from Table 2.1: it's the distinctions which enabled expansion of generic-level node 6 (*linguistic static/dynamic non-analogue acoustics*) into atomic-level node 6a (*spoken discourse*), i.e., between *discourse*, *text*, *labels/keywords* and *notation*.

Completeness at generic level. The taxonomy is claimed to be complete *at generic level* for the three media it addresses. No disproof of this claim has been found so far. However, the taxonomy is not complete at lower levels and might never be in any provable sense.

2.3.3.3 Some General Properties of Modalities

Modalities and levels of abstraction: it seems clear that a unimodal modality, i.e., a type of information representation in some physical medium, is always being thought of at some specific level of abstraction. Since the taxonomy makes this very clear, it might act as a safeguard against common errors of over-generalization and under-specification in modality analysis.

Enumerability: modalities can only be finitely enumerated at generic level, and to do that, one has to go back to their derivation prior to the pragmatic fusions done in the taxonomy shown in Table 2.1. This is of little interest to do, of course, but, otherwise, it would always seem possible, in principle, to create and validate new distinctions and hence generate new modalities at sub-generic levels. In practice, though, there is a far more important question of enumeration, and this one will always have a rough-and-ready answer determined by the state of the art, i. e.: *how many different unimodal modalities do we need to take into account in interactive systems design, development, and evaluation for the time being?*

Validation: modalities are generated through distinctions, and these distinctions are more or less scientifically validated at a given time. The primary method of validation is to apply a set of generated categories to phenomena in data corpora and carefully analyze the extent to which the categories are able to account for all observed phenomena both exhaustively and unambiguously.

2.3.4 Information Channels

The notion of an 'information channel' marks the most fine-grained level of modality theory and the level at which the theory, when suitably developed beyond its current state, links up with signal processing in potentially interesting ways.

Since a modality is a way of representing information in a physical medium, we can ask about the physical properties of that medium which *make it possible* to generate different modalities in it. These properties are called *information channels*. In the graphics medium, for instance, basic information channels include shape, size, position, spatial order, colour, texture, and time. From these basic properties, it is possible to construct higher-level information channels, such as a particular font type which is ultimately being used to represent the *typed text* modality.

Thus, information channels are the media-specific building blocks which define a modality in a particular medium. For instance, we could easily define a *static graphic black-and-white image* modality at atomic level (could be the new modality 9a1 in Table 2.1) and define its difference from modality *static graphic colour image* (9a2) by the absence of the information channel *colour*. In another example, the FACS, the Facial Action Coding System http://face-and-emotion.com/dataface/facs/description.jsp, starts from the fact that facial expression is being generated by some 50+ facial muscles used in isolation or in combination to form facial expressions of our mental and physical states, and specifies Action Units (AUs) for representing the muscular activity that produces momentary changes in facial appearance. The possible contraction patterns of the muscles are the information channels with which FACS operates.

2.3.5 Interaction Devices

Knowledge about modalities and multimodality is about how abstract information is, or can be, physically represented in different media and their information channels, and in different forms, called modalities. This knowledge has nothing to do with knowledge about *interaction devices*, and for good reason, because interaction devices come and go but modalities remain unchanged.

However, and this is the point to be made here, this simply means that designers and developers must go elsewhere to solve their problems about which physical devices to use for enabling interaction using particular modalities. And these problems remain interrelated with the issue of modality choice. It is often counter-productive to decide to use a particular modality for interaction, such as different 3D gesture modalities for input, if the enabling camera and image processing technologies currently cannot deliver reliable recognition of those gestures.

2.3.6 Practical Uses of the Theory

At this point, we have addressed three of the six desiderata listed at the start of the present *Modalities* section, i.e.: (i) identify all *unimodal* or elementary modalities which could be used to build multimodal interfaces and enable multimodal interaction; (ii) group modalities in one or several sensible ways, hierarchically or otherwise; and (iv) a practical toolbox of all possible unimodal input/output modalities to choose from. This, arguably, is quite useful in practice because it enables us to know which unimodal modalities there are in the three media scoped by the theory, how they are hierarchically interrelated, and how to decompose any multimodal representation into its constituent unimodal modalities. Issue (iii) on basic information on each modality goes beyond the scope of the present chapter but will soon be available at www.nislab.dk in the form of clickable taxonomy trees providing access to basic information on each modality.

Equally clearly, however, a practical toolbox of modalities to choose from would be far more useful if it came with information about *which modality to choose for which purpose* in interactive systems development. This takes us to desideratum (v): guidelines or something similar for which modality to use for a given development purpose, or what might be called the issue of *modality functionality*.

2.3.6.1 Modality Functionality

The issue about the *functionality* of a particular modality lies at the heart of the practical question about which modality or which set of modalities to use for a given development purpose. As it turns out, modality functionality, and hence modality choice, raises a whole family of issues, i. e.:

- is modality M(a) useful or not useful for development purpose P?
- is modality M(a) more or less useful for purpose P than an alternative modality M(b)?
- is modality M(a) *in combination with* modalities M(c, c+1, ... c+n) the best multimodal choice given purpose P?

Moreover, it must be taken into account if modality M(a) is considered to be used for *input* or *output* because this may strongly affect the answers to the above questions.

Far worse, even, as regards the complexity of the modality choice problem which we can now see emerging, is the fact that, in interactive systems design, development, and evaluation, development purpose P is *inherently highly complex*. In [5], we argue that P unfolds into *sets* of component parameters each of which has a *set* of possible values, of the following generic parameters:

- application type
- user
- user group (user population profile)
- user task or other activity
- application domain
- use environment
- interaction type
- interaction devices

Now, if we multiply such multiple sets of development purpose-specific values by the modality function questions above *and* by the sheer number of unimodal modalities in our toolbox, we get a quasi-intractable theoretical problem. Furthermore, intractability is not just due to the numbers involved but also to the fact that many of those sets of component parameters and their sets of values are likely to remain ill-defined forever.

Still, we need *practical answers* to the question about which modalities to use for a given development purpose. When, after many failed attempts to make modality theory directly applicable, we finally discovered the intractability problem just described, we identified (functional) *modality properties* as the primary practical contribution which modality theory can make.

2.3.6.2 Modality Properties

Modality properties are functional properties of modalities which characterize modalities in terms that are directly relevant to the *choice* of input/output modalities in interactive systems design, development, and evaluation.

To study the potential usefulness of modality properties, we made a study [14] of *all speech functionality claims* made in the 21 paper contributions on speech systems and multimodal systems involving spoken interaction in [15]. In this and a follow-up study which looked at a cross-section of the literature on speech and multimodality 1993-1998 [16, 17], we analysed a total of 273 claims made by researchers and developers on what particular modalities were good or bad for. An example of such a claim could be: "Spoken commands are usable in fighter cockpits because the pilot has hands and eyes occupied and speech can be used even in heads-up, hands-occupied situations". It turned out that more than 95% of those claims could be evaluated and either justified, supported, found to have problems, or rejected by reference to a relatively small number of 25 modality properties. These are exemplified in Table 2.2.

Regarding the modality properties listed in Table 2.2, it is important to bear two points in mind in what follows: (i) the listed modality properties are selected examples from the longer list of properties made to evaluate those 273 claims made in the literature; and, more importantly, (ii) that longer list was made *solely* in order to evaluate those 273 claims. So the list does not have any form of theoretical or practical closure but simply includes the modality properties which happened to be relevant for evaluating a particular set of claims about speech functionality. In other words, modality theory has more to say about speech functionality, and far more to say about modality functionality in general, than Table 2.2.

Now back to those studies. Moreover, even though (i) speech in multimodal combinations was the subject of only few claims in the first study which evaluated 120 claims, and (ii) there were many more claims about speech in multimodal combinations in the batch of 157 claims analysed in the second study, the number of modality properties used in claims evaluation only went up from 18 in the first study to 25 in the second study. The reason why these numbers are potentially significant is that *modality properties may remain tractable in number while still offering significant support* to interactive systems designers, developers and evaluators.

This likely tractability may be contrasted with several other familiar approaches. On is the *strict experimentalist approach*, in which a careful, expensive, and time-consuming study may conclude, e. g., that X % of Y children, aged between 6 and 8, the group being normal in its distribution of English speech and manual dexterity skills, in a laboratory setting, using a map application, were found to stumble considerably (and here we get tables, numbers and percentages) in their speech articulation when they used speech to indicate map locations, whereas they found it significantly easier to use pen-pointing to input the same information. – Incidentally, this short description includes values of all the generic parameters described in the previous section. – But what does this result tell practitioners who are specifying a different application, with different users and user profiles, for a different environment, for a different task, and using partially different input devices? The answer is that *we don't know*. Modality property MP1 in Table 2.2, on the other hand, directly says that speech is unsuited for specifying detailed spatial information. Furthermore, MP1 does so without even mentioning speech. Instead, if relies on property inheritance from linguistic representation of information in general.

2.3 Which Modalities are There?

Table 2.2 Modality properties.

No.	Modality	Modality Property
MP1	Linguistic input/output	Linguistic input/output modalities have interpretational scope. They are therefore unsuited for specifying detailed information on spatial manipulation.
MP3	Arbitrary input/output	Arbitrary input/output modalities impose a learning overhead which increases with the number of arbitrary items to be learned.
MP4	Acoustic input/output	Acoustic input/output modalities are omnidirectional.
MP5	Acoustic input/output	Acoustic input/output modalities do not require limb (including haptic) or visual activity.
MP6	Acoustic output	Acoustic output modalities can be used to achieve saliency in low-acoustic environments.
MP7	Static graphics	Static graphic modalities allow the simultaneous representation of large amounts of information for free visual inspection.
MP8	Dynamic output	Dynamic output modalities, being temporal (serial and transient), do not offer the cognitive advantages (with respect to attention and memory) of freedom of perceptual inspection.
MP11	Speech input/output	Speech input/output modalities in native or known languages have very high saliency.
MP15	Discourse output	Discourse output modalities have strong rhetorical potential.
MP16	Discourse input/output	Discourse input/output modalities are situation-dependent.
MP17	Spontaneous spoken labels/keywords and discourse input/output	Spontaneous spoken labels/keywords and discourse input/output modalities are natural for humans in the sense that they are learnt from early on (by most people). Note that spontaneous keywords must be distinguished from designer-designed keywords which are not necessarily natural to the actual users.
MP18	Notation input/output	Notation input/output modalities impose a learning overhead which increases with the number of items to be learned.

Another contrasting approach is *guidelines of the if-then type*. An if-then guideline might say, e.g., that if the environment is an office environment, if the person is alone in the office, and if the person is not a fast typist, then speech dictation might be considered as an alternative to typing. This guideline is nice and concrete, it might be true, and it might be helpful for someone who develops office applications. The problem is tractability, because how many guidelines would we need of this kind? Clearly, we would need a catastrophic number of guidelines.

On the other hand, modality properties come at a price to be paid in natural intelligence. It is that they focus on the modality itself and its properties, rather than mention values of the parameters listed in the previous section. That's why modality properties are comparatively economical in number: they leave it to the natural intelligence of developers to apply them to the parameter values which characterize the development project at hand. Arguably, however, there seems to be a rather trivial but important point here, i.e., if we want to know what a modality is or is not suited for, we need to understand, first of all, the *information representation properties* of that modality. So it would seem to be a likely prediction that the more useful a guideline is for guiding modality choice in practice, the more it resembles a statement of a modality property.

2.4 Multimodal Information Representation

Modality theory is fundamentally about the *unimodal* modalities that, as building blocks, go into the construction of multimodal information representation. Before we look into the process of construction, it is useful to briefly discuss the advantages offered by the growing number of input/output modalities that are becoming available.

2.4.1 Advantages of Multimodality

Since *no two modalities are equivalent*, all modalities differ amongst each other, as we have seen, in terms of their individual combination of expressive strengths and weaknesses *and* their relationship with the human perceptual, cognitive, emotional, etc. system. The central implications are that the more modalities we can choose from, (i) the *wider the range of information* it becomes possible to express as input or output; and (ii) the higher our chances become of identifying a modality combination which has a suitable, if not optimal, relationship with the human system for at given application purpose.

If we *combine* two or several modalities, we ideally get the sum of their expressive strengths and are able to overcome the expressive weaknesses of each of them taken individually. However, it still remains necessary to make sure as well that the combination is possible for, and acceptable to, the human users.

Also, the more modalities we have at our disposal as developers, the more we can develop applications for *all users*, including people with perceptual, cognitive and other disabilities, people with different degrees of computer literacy, the 1 billion people who are illiterate, as well as users with sometimes widely different preferences for which modalities to use. This is often done by *replacing* information expressed in one modality by, practically speaking, the same information expressed in a different modality, like when the blind get their daily newspaper contents read aloud through text-to-speech synthesis.

Given these limitless opportunities, it is no wonder that multimodality is greeted with excitement by all.

2.4.2 Constructing Multimodality from Unimodal Modalities

Since we have a theory of unimodal modalities, and, it would appear, *only because* we have something like that, it makes sense to view multimodal representation as something which can be *constructed* from unimodal representations, analogous to many other constructive approaches in science – from elements to chemistry, from words to sentences, from software techniques to integrated systems.

This view is sometimes countered by some who do not use modality theory, by an argument which goes something like this: the whole point about multimodality is to create something *entirely new*. When modalities are combined, we get *new emergent properties* of representations which cannot be accounted for by the individual modalities on their own. Now unless one is inclined towards mysticism, this argument begs the question about whether and to which extent multimodal combinations can be analysed, or even predicted, as resulting from an ultimately transparent process of combining the properties of unimodal modalities and taking the relationship with the human system into account. As repeatedly stressed above, this process is provably a very complex one in general, but so is the field of synthetic chemistry.

In the remainder of this chapter, we introduce some distinctions in order to approach the issue of multimodal construction and remove some of the mystery which still seems to surround it.

2.4.3 Linear Modality Addition and Replacement

Let us define a concept of modalities which can be *combined linearly* so that the combination *inherits* the expressive strengths of each modality and does not cause any significant negative *side-effects* for the human system. It is very much an open research question which modalities can be combined in this fashion and under which conditions. However, to the extent that modalities *can* be combined linearly, it is straightforward to use the modality properties of the constituent unimodal modalities to describe the properties of the resulting multimodal representation. The modalities simply add up their expressive strengths, and that's that. Let us look at some examples.

2.4.3.1 Modality Complementarity

In a first, non-interactive example, we might take a *static graphic* piece of *text* describing, say, a lawnmower, and add a *static graphic image* of the lawnmower to it, letting the text say what the lawnmower can do and how to use and maintain it, and the picture show how it looks. For good measure, we might throw in a *static graphic compositional diagram* of the lawnmower, showing its components and how they fit together, and cross-reference the diagram with the text.

In another, interactive, example, we might put up a large screen showing a *static graphic* Sudoku gameboard and have users play the game using *spoken numbers and other spoken input keyword commands* in combination with 3D camera-captured and image-processed *pointing gesture input*. A number is inserted into, or deleted from, the gameboard by pointing to the relevant gameboard square and uttering a command, such as "Number seven" or "Delete this". A recent usability test of the system showed rather unambiguously that this multimodal input/output combination works well both in terms of input/output modality expressiveness and in terms of fitting the human system [18]. Modality theory would predict that the spoken input keywords (other than the numbers 1 through 9 which are familiar to all Sudoku players), being designer-designed, might cause memory problems for the users, cf. Table 2.2, MP 17. However, since there were only a couple of them in the application, the problems caused would be predicted to be minimal, which, in fact, they were. In addition, it would be easy to put up the "legal" keywords as an external memory on the screen next to the Sudoku gameboard to solve the problem entirely.

These two examples are among the classical examples of *good multimodal compounds*: they work well because they use the *complementary* expressive strengths of different modalities to represent information which could not easily be represented in either modality on its own. In the first example, the complementarity i *a-temporal* because of the freedom of visual inspection afforded by the static graphics. In the second example, the complementarity is *temporal* because the speech is being used dynamically and therefore the pointing gestures have to occur at the appropriate time intervals lest the meaning of the message would be a different one – if, indeed, any contextually meaningful message would result at all. In the first example, since all component modalities are static, the multimodal representation causes neither working memory problems nor any perceptual or cognitive conflicts due to pressure to process too much information at the same time – indeed, the representation as a whole acts as an external memory. The second example makes use of a modality combination which is as natural as natural language and is, in fact, learned in the process of learning a natural language. These examples do seem to well illustrate the notion of linear addition of modalities, i. e., of *gaining novel expressiveness without significant side-effects*, cognitive or otherwise, through modality combination.

Good multimodal compounds need not be as simple and as classical as the examples just discussed. We recently tested the usability of a treasure hunting game system prototype in which a blind user and a deaf-mute user collaborate in finding some drawings essential to the survival of an ancient Greek town [19]. For the blind user alone, the input/output modalities are: *spoken keywords output* to help the blind navigate the 3D townscape and its surroundings to find what s/he needs; *non-speech sound* musical instrument output acting as arbitrary codes for the colours of objects important in the game; *haptic 3D force-feedback output* providing the blind user with data for navigating the environment, locating important objects, and building a mental map of the environment; *haptic 3D* navigation robot arm *input* through which the blind orientates in the environment; and *haptic click notation input* through which the blind acts upon objects important to the game.

The usability test of this system showed that this multimodal compound worked excellently except for the problem of remembering the arbitrary designer-designed associations between colours and musical instruments, something which, again, is predicted by modality theory [18]. For the purpose of the evaluated game, the colour problem would disappear if the objects simply described their colour through speech rather than using arbitrary non-speech sounds.

2.4.3.2 Modality Redundancy

Linear addition also often works well in another abstract kind of modality combination which is often termed *redundancy*. Sometimes it is useful to represent more or less the same information in two different modalities, for instance because the information is particularly important or because of particular values of the *user* or *environment* parameters (cf. above). So, for instance, we add an acoustic alarm (Table 2.1, 14) to a visual alarm Table 2.1, 13/16) for increased security in a process plant; or we add visual speech (Table 2.1, 8d-f) to speech output because the user is hard-of-hearing or because the environment is or can be noisy. The visual speech is actually *not* information-equivalent to the speech but comes close enough for the redundant multimodal representation to provide significant help to users when the speech is difficult to hear. Again, no significant side-effects would be predicted in these cases.

2.4.3.3 Modality Replacement

In linear *modality replacement*, one modality (or a combination of modalities) is replaced by another for the purpose of achieving practically the same representation of information as before. Again, many examples can be mentioned where this works sufficiently well, such as replacing *spoken discourse* for the hearing by *sign language discourse* for the deaf and hard-of-hearing; or replacing *static graphic written text* for the seeing by *static haptic Braille text* for the blind or hard-of-seeing. Yet again, no significant side-effects would be predicted.

It should be noted that several have distinguished other abstract types of modality combination in addition to complementarity, redundancy, and replacement, but space does not allow discussion of these, see, e. g., [20, 21].

The conclusion at this point is that in *a large fraction of cases* in which several modalities are combined into multimodal representations, the resulting multimodal representation (i) is largely a straightforward *addition* of modalities, or a straightforward replacement by functionally equivalent modalities, with no significant side-effects upon the human system; and (ii) that the knowledge of unimodal modalities can be used to good effect in predicting the functionality of the constructed multimodal compound. In other words, combining modalities can be a straightforward and predictable process of construction rather than the creation of a magical concoction with mysterious properties and totally unpredictable effects.

2.4.4 Non-linear Effects, Users, Design Detail, Purpose

However, we now need to return to the huge theoretical complexity of the general modality choice problem, this time in a multimodal context.

In fact, modality choice complexity is such that we do not recommend that *any* novel multimodal combination be launched without thorough usability testing. Modality theory can be very helpful in the analysis and specification phase, suggesting modality combinations to be used or left out, predicting their expressiveness and potential problems in relationship to the human system. But modality theory, however much further developed, cannot guarantee unimodal or multimodal application success. There are many, partly overlapping reasons for exercising caution.

The first reason to be mentioned is *non-linear effects*. For instance, nothing might appear more straightforward than to add a voice interface to an email system so that the user can access emails without having a static graphic text interface at hand. This kind of modality replacement is sometimes called *interface migration*. However, the user soon discovers things like that the overview of the emails received is gone and not replaced by any different mechanism, and that the date-time information read aloud by the system is (i) unnatural and (ii) takes an exceedingly long time to listen to. In other words, while the modality replacement no doubt preserves *information equivalence*, something has gone wrong with the relationship with the human system. In this case, successful modality replacement is not straightforward at all because the entire structure of the email information representation has to be revised to arrive at a satisfactory solution. Modality theory can tell the developer that, despite their information equivalence-in-practice in the present case, the (non-situated) *text* modality is fundamentally different from the (situated) *discourse* modality, the former having a tendency to being far more explicit and elaborate, as illustrated by the lengthy absolute date-time information provided in the email list; and (ii) that speech, being largely a *dynamic* modality, does not enable anything like the information overview provided by *static* graphics. So the theory can advise that developers should be on the alert for non-linear effects and test for them using early mock-ups, but the actual effects just described can hardly be predicted due to their detailed nature.

A second reason is the *users*. From a theoretical point of view, *both* in terms of practical information equivalence *and* in terms of the absence of side-effects on the human system, it may be a perfect example of modality replacement to replace *static graphic text* by *static haptic* Braille *text*. But then it turns out that, for instance, only 5% of the blind Danish users know Braille whereas some 80+% of Austrian blind users know Braille. Given this marked difference in generic parameter *user population profile*: component parameter *user background skills*, the modality replacement above might be a good idea in Austria whereas, in Denmark, one would definitely recommend replacing static graphic text by *text-to-speech* instead. Modality theory, of course, has no notion of Braille skill differences between Austrian and Danish populations of blind users. Or, to mention just

2.4 Multimodal Information Representation

one more example, the theory has little to say about the notoriously hard-to-predict *user preferences* which might render an otherwise theoretically well-justified modality addition or replacement useless.

A third reason is *design detail*. It may be true in the abstract that, for a large class of applications, the addition of a dynamic graphic animated human representation adds a sense of social interaction to interactive systems. But this advantage might easily be annulled by an animation that is perceived as being unpleasant of character, daft, weird, overly verbose, occupying far too much screen real-estate for what it contributes to the interaction, or equipped with a funny voice.

A fourth reason is the *purpose* of adding or replacing modalities. If the purpose is a more or less crucial one, such as providing blind users with text-to-speech or Braille access to the linguistic modality, this is likely to overshadow any non-linear effects, specific user preferences, or design oddities. But if the purpose is less essential or inessential – such as adding entertaining animations to web pages, small-talk to spoken dialogue applications, or arbitrary musical instrument coding of colours which could just as well be described through another output modality which is actually used in the application, such as spoken keywords – then users are likely to be far less tolerant to the many other factors which are at play in creating user satisfaction.

Chapter 3
Information-Theoretic Framework for Multimodal Signal Processing

Jean-Philippe Thiran[1], Torsten Butz[2], and A. Murat Tekalp[3]

[1] Ecole Polytechnique Fédérale de Lausanne (EPFL), Signal Processing Institute, 1015 Lausanne, Switzerland
JP.Thiran@epfl.ch
[2] Cisco Technology Center, 1180 Rolle, Switzerland
tbutz@cisco.com
[3] Koç University, Multimedia, Vision and Graphics Laboratory, Istanbul, Turkey
mtekalp@ku.edu.tr

3.1 Introduction

The signal processing community has shown to be increasingly reliant upon information theoretic concepts to develop algorithms for a wide variety of important problems ranging e. g. from audio-visual signal processing to medical imaging [22–31]. When comparing the proposed algorithms, two facts are particularly surprising. First, the range of practical problems which, are solved with the fundamentally very compact mathematical concepts of information theory seem to be very broad and unrelated. For example mutual information is having a big success in multimodal medical image registration, but has also been successfully used for information theoretic classification and feature extraction. The second striking fact is that the mathematical expressions governing the final algorithms seem not to be much related, even though the employed fundamental concepts were identical. For instance in [22] feature extraction by maximization of mutual information has been derived from the general concept of error probabilities for Markov chains. On the other hand information theoretic image registration algorithms with all their different optimization objectives [27 30] never made reference to error probabilities or Markov chains.

The main aim of this chapter, largely inspired from [32], consists of developing an information theoretic framework for multimodal signal processing, closely related to information theoretic feature extraction/selection. This important relationship will indicate how we can unify to a large extent multimodal medical image processing, e. g. multi-channel segmentation and image registration, and extend information theoretic registration to other features than image intensities. The framework is not at all restricted to medical images though and we will illustrate this by applying it to multimedia sequences as well.

The chapter is structured as follows: First we recall some information theoretic concepts which build our mathematical framework (section 2). In section 3 we present our framework for multimodal signal processing which allows to derive optimization objective functions used in multimodal medical image processing and to study their mathematical relationships. A short section on genetic optimization (section 4) will build the bridge to the final section where we show some interesting results in medical imaging (section 5.1) and multimedia signal processing (section 5.2). The discussion section 6 will wrap-up the results, before concluding in section 7.

3.2 Some Information Theoretic Concepts

We want to start by recalling some important information theoretic concepts which will be used extensively thereafter. All the presented notions are well known and widely used in several fields of information technology and computer science. We would also like to emphasize that the random variables throughout this work refer to discrete random variables, except when they are explicitly specified to be continuous.

3.2.1 Stochastic Process and Error Probability

A stochastic process is an indexed sequence of random variables (RV) with in general arbitrary mutual dependencies [33]. For the specific case of information theoretic signal processing, we construct the following stochastic process:

Let us define the discrete random variables X and X^{est} on the same set of possible outcomes Ω_X. Let us also define N discrete random variables Y_i on Ω_{Y_i} for $i = \{1,..,N\}$ resp. We consider the following stochastic process:

$$X \to Y_1 \to ... \to Y_N \to X^{est} \to E, \tag{3.1}$$

where in general the transition probabilities are conditioned on all the previous states. E is a binary RV defined on $\Omega_E = \{0,1\}$ and is 1 if the estimation X^{est} of X from Y is considered as an error. This stochastic process will be the fundamental model of our following developments and examples.

An important characteristic of equation 3.1 is the probability of error $P_e = P(E = 1)$, which equals the expectation μ_E of E:

$$P_e = \mu_E = 1 \cdot P(E=1) + 0 \cdot P(E=0). \tag{3.2}$$

We can use the conditional probabilities defining the transitions of equation 3.1 to write

$$P_e = \sum_{x \in \Omega_X} \sum_{y_1 \in \Omega_{Y_1}} ... \sum_{y_N \in \Omega_{Y_N}} \sum_{x^{est} \in \Omega_X} P(E=1 | x^{est}, y_N, ..., y_1, x) \\ \cdot P(x^{est} | y_N, ..., y_1, x) \cdot ... \cdot P(y_1 | x) \cdot P(x). \tag{3.3}$$

3.2 Some Information Theoretic Concepts

If any of the random variables is defined over a continuous interval and is therefore given by a continuous probability density function, the corresponding sum in equation 3.3 has to be replaced by an integral.

Notice that the error probability P_e has a close connection to the well known concept of signal distortion, as shown in [32].

It is important to note that so far no hypothesis about the specific transition probabilities has been set. This is in particular the case for the error variable E: the fact that for example $x^{est} \neq x$ does not necessarily imply that $E = 1$ with probability one. This generality might look quite artificial and impractical, but we can show that specific hypotheses about the different steps in the stochastic process of equation 3.1, including the last one, will result in well known mathematical formulas of quantization, classification and multimodal signal processing [32]. Nevertheless a complete study of equation 3.1 would go beyond the scope of this paper. Therefore we want to restrict ourselves to the case where all the transition probabilities of equation 3.1 besides the last one are Markovian [33]:

$$P(x^{est} | y_N, ..., y_1, x) = P(x^{est} | y_N)$$
$$P(y_N | y_{N-1}, ..., y_1, x) = P(y_N | y_{N-1}) \quad (3.4)$$
$$...$$
$$P(y_1 | x) = P(y_1 | x).$$

This implies that the stochastic process $X \rightarrow Y_1 \rightarrow ... \rightarrow Y_N \rightarrow X^{est}$ forms a Markov chain. The Markovian condition is obviously not fulfilled for the last transition probability, as the error probability has at least to depend on the input to the chain x and on its final output x^{est}. In what follows, we suppose that the Markovian conditions of equation 3.4 are fulfilled, so that the error probability P_e becomes

$$P_e = \sum_{x \in \Omega_X} \sum_{y_1 \in \Omega_{Y_1}} ... \sum_{y_N \in \Omega_{Y_N}} \sum_{x^{est} \in \Omega_X} P(E = 1 | x^{est}, y_N, ..., y_1, x)$$
$$\cdot P(x^{est} | y_N) \cdot P(x^{est} | y_{N-1}) \cdot ... \cdot P(y_1 | x) \cdot P(x). \quad (3.5)$$

Furthermore we will consider the special case that $P(E = 1 | x^{est}, y_N, ..., y_1, x)$ of equation 3.5 equals just $P(E = 1 | x^{est}, x)$, i.e. that the error probability of the chain only depends on the relationship between the input value and its estimated value after having gone though the chain.

3.2.2 Fano's Inequality and the Data Processing Inequality

The model proposed in the previous section and in particular the exact evaluation of the error probability (equation 3.5) requires knowledge of all the transition probabilities. Sometimes though we might not have access to the data to such a deep extend. Still we want at least to approximately estimate the error probability $P_e = P(E = 1)$. When the process $X \rightarrow Y_1 \rightarrow ... \rightarrow Y_N \rightarrow X^{est}$ fulfils the Markovian conditions of equation 3.4 and when the probability of error is given by $P_e = Pr(X^{est} \neq X)$, we can use an expression known as Fano's inequality [34] to compute a lower bound of P_e as a function of the input RV X and the last transmission RV Y_N only.

Let us state Fano's inequality, as we will use it extensively later in this chapter. Let $A \to B \to A^{est}$ be a Markov chain. A (and therefore A^{est}) has to be finitely (or countable-infinitely) valued. Then we have a lower bound on the error probability $P_e = Pr(A^{est} \neq A)$ that the output of the chain A^{est} is not the input A:

$$P_e \geq \frac{H(A|B) - H(P_e)}{\log|\Omega_A - 1|} \geq \frac{H(A|B) - 1}{\log|\Omega_A|}$$
$$= \frac{H(A) - I(A, B) - 1}{\log|\Omega_A|}, \tag{3.6}$$

where $|\Omega_A|$ is the number of elements in the range of A, $H(.)$ stands for the Shannon entropy of one RV and $I(.,.)$ for the Shannon mutual information between a pair of RVs. Therefore for the case of the Markov chain $X \to Y_1 \to ... \to Y_N \to X^{est}$ as introduced in chap. 2.1 and under the assumption that $P_e = Pr(X^{est} \neq X)$, this lower bound is written as

$$P_e \geq \frac{H(X|Y_N) - 1}{\log|\Omega_X|} = \frac{H(X) - I(X, Y_N) - 1}{\log|\Omega_X|}. \tag{3.7}$$

There is another very useful inequality which is applicable within the previously mentioned assumptions: the data processing inequality [33] which states that if $A \to B \to C$ is a Markov chain, we have

$$I(A, B) \geq I(A, C),$$
$$I(B, C) \geq I(A, C). \tag{3.8}$$

The combination of these expressions allows building a large number of resulting inequalities, such as

$$P_e \geq \frac{H(X) - I(X, Y_N) - 1}{\log|\Omega_X|} \geq \frac{H(X) - I(X, Y_{N-1}) - 1}{\log|\Omega_X|}$$
$$\geq \frac{H(X) - I(X, Y_{N-2}) - 1}{\log|\Omega_X|} \geq ...$$
$$P_e \geq \frac{H(X) - I(X, Y_N) - 1}{\log|\Omega_X|} \geq \frac{H(X) - I(Y_1, Y_N) - 1}{\log|\Omega_X|} \tag{3.9}$$
$$\geq \frac{H(X) - I(Y_2, Y_N) - 1}{\log|\Omega_X|} \geq ...$$

etc.

Under the specified assumptions, these expressions allow one to focus on a specific transition within the Markov chain $X \to Y_1 \to ... \to Y_N \to X^{est}$. This is of large interest if we have particular knowledge about one of the transitions in the chain or if some other transition is not sufficiently understood.

The concept of bounding the error probability P_e is by far not exploited with the presented inequalities. There are other entropies, especially Renyi entropy [35] that can be more appropriate for specific applications [36]. Furthermore the estimation of an upper bound could be very interesting [36], or specific assumptions on the RVs can result in very interesting specialised expressions. All this is almost a research domain on its own closely related to rate distortion theory. In this chapter we restrict ourselves to the theory described so far in this section.

3.2.3 Information Theoretic Feature Extraction.

We now want to shortly recall the basic concept of information theoretic feature extraction as presented in [22]. Let us assume we have a set S of n class prototypes, each labelled by one of the class symbols of $\Omega_C = \{1,..,n_c\}$, where n_c is again the number of classes. Furthermore we have a multi-dimensional feature vector $y_i \in \Omega_Y$ associated to every sample s_i within the set S. The RV modelling probabilistically the classes of the prototypes is called C and is defined over the set Ω_C. Its feature space representation, denoted Y, is defined over Ω_Y. Feature extraction aims to extract that subspace F_Y of the initial feature space Y that is most significant for the specific classification task. Formally we can represent this by the following stochastic process:

$$C \to Y \to F_Y \to C^{est} \to E. \qquad (3.10)$$

The initial transition $C \to Y$ can be interpreted as a feature selection step. The transition $Y \to F_Y$ corresponds to feature extraction, where we select a sub-space F_Y of Y. Thereafter we estimate the class C^{est} from the final feature representation F_Y and evaluate if the whole transmission process from C to C^{est} can be considered as erroneous.

One approach to select and extract the optimal features Y and F_Y for a particular classification task would be to minimize directly the error probability P_e of the selected learning prototypes. This would have the disadvantage though, that the optimization would get the most relevant features with respect to the chosen classification algorithm ($F_Y \to C^{est}$). We rather want to determine those features for which any "suitable" classification algorithm can obtain "good" results. Therefore we want to neglect the classification step $F_Y \to C^{est}$ during the optimization. To do this, we can take profit from Fano's inequality of chap. 2.2. In the context of the stochastic process of equation 3.10, this inequality is re-written as

$$P_e \geq \frac{H(C|F_Y) - 1}{\log n_c} = \frac{H(C) - I(C, F_Y) - 1}{\log n_c}. \qquad (3.11)$$

Note that $H(C)$, the entropy of the chosen prototypes, as well as $n_c (=|\Omega_C|)$ stay constant, as the number of classes as well as the initial set of samples stay fixed during the optimization. Therefore we have to maximize the mutual information $I(C, F_Y)$ in order to minimize the lower bound on the error probability P_e, which ensures that a particular classification algorithm can perform well.

It is important to note that we estimated a lower bound of the error probability. Upper bounds might appear more suited for the described problem of classification and, in fact, can be estimated in several cases for a fixed classifier. For example, if it is known from the beginning that a maximum likelihood classifier will be employed, the error probability P_e is upper-bounded by the conditional entropy $H(F_Y | C): P_e \leq H(F_Y | C)$. But information theoretic feature extraction attempts to extract those features which are most suited for the given classification task independently of a particular classifier.

Next, we would like to introduce information theoretic multimodal signal processing by multimodal feature extraction. Its close relationship to feature extraction for classification will be easily recognised, as we will also use lower error bounds as optimization objectives. Furthermore, the same arguments as for feature extraction will justify the use of a lower bound instead of an upper bound.

3.3 From Error Probability to Multimodal Signal Processing

There are several possibilities to apply the presented framework to multimodal signals. We want to explore one specific approach which we used extensively in several applications of multimodal signal processing. It is based on one very basic but intuitive hypothesis: *A pair of multimodal signals originates from the same physical source, even though the signals might have suffered from distortions, delays, noise and other artifact which hide there common origin.* As we will show, the approach seems to be particularly suited to derive and compare a large number of existing optimization objectives particularly well known in the multimodal medical imaging community.

3.3.1 Multimodal Stochastic Processes

First we will outline how we can build Markov chains from multimodal signals. The resulting chains should fulfil the conditions required to apply all the theory of chap. 2. In particular the conditional error probability $P(E=1|...)$ has to be 1, whenever the output from the chain differs from the corresponding input (Hamming distortion). We use the fact that multimodal signals originate from the same physical reality, even though the concrete representations of this reality might be quite different (Fig. 3.1). We can therefore expect that there exist features in a couple of multimodal signals which reflect this physical correspondence statistically.

3.3 From Error Probability to Multimodal Signal Processing

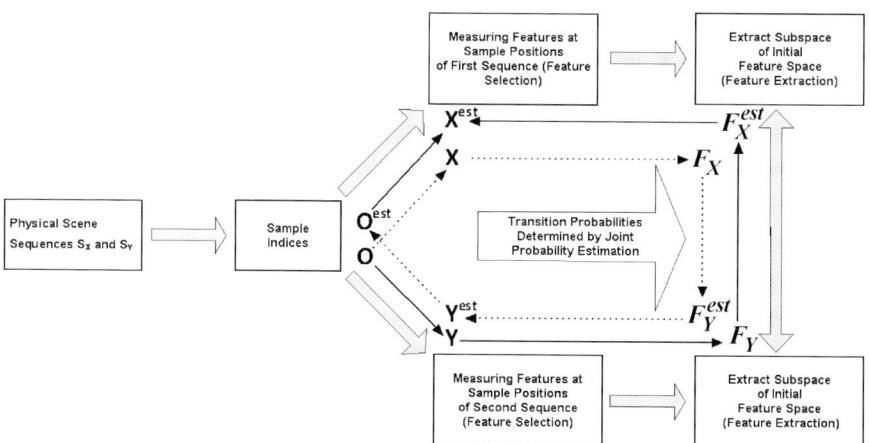

Fig. 3.1 Markov chains can be built from a pair of multimodal signals. The connecting block between a couple of multimodal signals (the transition probabilities for $F_X \to F_Y^{est}$ and $F_Y \to F_X^{est}$) is obtained by joint probability estimation

Let us consider the example of an audio-video sequence and assume we know a pair of features which show a statistical dependence. If now we take the feature value of one signal at a randomly chosen time in the sequence and the feature representation at an arbitrarily chosen second time in the other signal of different modality, we should be able to tell if the two measurements are likely to originate from the same physical reality, i. e. were acquired at the same physical time, or not.

In direct analogy, we also want to mention multimodal medical images, where the spatial coordinates of the images represent the physical correspondence and take over the role of the time coordinate in audio-video sequences.

A couple of multimodal signals is initially given by two signal sequences S_X and S_Y, both of length n, and taking on values in the sets Ω_X and Ω_Y respectively[1]. For instance a 3D image contains $n = n_x \times n_y \times n_z$ samples (or voxels) showing different intensities or colours. Let us define a uniform RV O on the set $\Omega_O = \{1,...,n\}$ labelling the samples in S_X and S_Y. This RV is used to model the fact that we will consider a random selection of pairs of samples in the sequences S_X and S_Y. Therefore for 3D images we have $P(o = (i, j, k)) = \frac{1}{n} = \frac{1}{n_x \cdot n_y \cdot n_z}, \forall o \in \Omega_O$ ("for all voxels in the image"). Starting from the RV O, we can build the following two Markov chains (see Fig. 3.1):

$$O \to X \to F_X \to F_Y^{est} \to Y^{est} \to O^{est} \to E, \qquad (3.12)$$

$$O \to Y \to F_Y \to F_X^{est} \to X^{est} \to O^{est} \to E, \qquad (3.13)$$

---

[1] Sometimes the sampling coordinates are not the same in both signals. For example two images of different modality might have different dimensions and therefore different numbers of samples. For such cases we just want to make reference to interpolators which can build the bridge between the two respective sequences [40, 59]

where X (resp. Y) model the specific feature values of the samples in the sequence S_X (resp. S_Y) as a RV conditioned on the outcome of the sample position o. Which sequence features are exactly considered, represents a feature selection step. E. g. in an image, for each sample position generated from the RV O we can consider the intensity at that position, but also the gradient, Gabor response, etc. Obviously X and Y can also model multi-dimensional feature spaces, which might ask for an additional feature extraction step. This means we project the measured features into lower dimensional sub-spaces of X and Y. Such sub-spaces are again RVs and we denoted them F_X and F_Y in Fig. 3.1 As the considered features of both sequences S_X and S_Y originate from the same sampling label o, we can link the two sequences through joint probability estimation [37, 38]. When on the average the chosen feature values f_X and f_Y of F_X and F_Y reflect maximally the fact that they originate from the same sampling position o, and then we minimize the error probability of the Markov chains. Therefore we want to select and extract those features F_X and F_Y from the initial sequences S_X and S_Y that show as much as possible of this physical correspondence.

Up to now we have constructed a couple of related Markov chains for general multimodal signals. Let us now see what we can say about the corresponding error probabilities $P_{e1} = Pr(O^{est} \neq O)$ (Markov chain equation 3.12) and $P_{e2} = Pr(O^{est} \neq O)$ (Markov chain equation 3.13) when we use Fano's inequality (equation 3.6) and the data processing inequality (equation 3.8) [39]:

$$P_{e1} = Pr(O^{est} \neq O)$$

$$\geq \frac{H(O|Y^{est}) - H(P_{e1})}{\log(n-1)}$$

$$\geq \frac{H(O|Y^{est}) - 1}{\log n}$$

$$= \frac{H(O) - I(O, Y^{est}) - 1}{\log n} \tag{3.14}$$

$$= \frac{\log n - I(O, Y^{est}) - 1}{\log n} \tag{3.15}$$

$$= 1 - \frac{I(O, Y^{est}) + 1}{\log n} \tag{3.16}$$

$$\geq 1 - \frac{I(F_X, F_Y^{est}) + 1}{\log n} \tag{3.17}$$

and

$$P_{e2} = Pr(O^{est} \neq O)$$

3.3 From Error Probability to Multimodal Signal Processing

$$\geq \frac{H(O|X^{est}) - H(P_{e2})}{\log(n-1)}$$

$$\geq \frac{H(O|X^{est}) - 1}{\log n}$$

$$= \frac{H(O) - I(O, X^{est}) - 1}{\log n} \qquad (3.18)$$

$$= \frac{\log n - I(O, X^{est}) - 1}{\log n} \qquad (3.19)$$

$$= 1 - \frac{I(O, X^{est}) + 1}{\log n} \qquad (3.20)$$

$$\geq 1 - \frac{I(F_Y, F_X^{est}) + 1}{\log n}. \qquad (3.21)$$

To get equation 3.15 from equation 3.14 (resp. equation 3.19 from equation 3.18) we used the fact that O is a uniform random variable over the set $\Omega_O = \{1,...,n\}$ of n possible sampling positions in the sequences S_X and S_Y, and has therefore entropy of $\log n$. The last inequality follows directly from the data-processing inequality for Markov chains [33].

The probability densities of F_X and F_X^{est}, resp. F_Y and F_Y^{est}, are both estimated from the same data sequences S_X, resp. S_Y. Therefore the estimations of the mutual information $I(F_Y, F_X^{est})$ and $I(F_X, F_Y^{est})$ are equal, and we can write $I(F_Y, F_X^{est}) \approx I(F_X, F_Y^{est}) \approx I(F_X, F_Y)$. The value of this mutual information $I(F_X, F_Y)$ is determined from the joint probability density which is estimated by non-parametric probability estimation [37, 38] from the sequences S_X and S_Y (for example joint histogramming). From the symmetry of mutual information it follows that both lower bounds are equal, so that minimizing them simultaneously equals maximizing the mutual information between the feature representations F_X and F_Y of the multimodal signals.

3.3.2 *Objective Functions for Multimodal Signal Processing*

Using the example of multimodal medical image registration, we will show that it is possible to derive a large class of objective functions of image registration and to theoretically determine their relationships and differences. In this respect, we will show how to derive normalised entropy, correlation ratio and likelihood directly from equations 3.17 and 3.21. We will also generalize normalised entropy to the general concept of *feature efficiency* for multimodal signal processing [39, 40].

Feature Efficiency. Taking a closer look at equations 3.17 and 3.21 reveals an important danger when simply maximizing the mutual information $I(F_X, F_Y)$ in order to minimize the lower error bounds of P_{e_1} and P_{e_2}. In order to visualize this danger, let us re-write the lower bounds in a different way and use the fact that for any pair of discrete random variables A and B it can be shown [33] that $H(A,B) \geq I(A,B)$ and $\frac{H(A)+H(B)}{2} \geq I(A,B)$ to weaken them:

$$P_{\{e1,e2\}} \geq 1 - \frac{I(F_X, F_Y) + 1}{\log n}$$

$$\geq 1 - \frac{H(F_X, F_Y) + 1}{\log n} \quad (3.22)$$

and

$$P_{\{e1,e2\}} \geq 1 - \frac{I(F_X, F_Y) + 1}{\log n}$$

$$\geq 1 - \frac{H(F_X) + H(F_Y) + 2}{2 \cdot \log n}. \quad (3.23)$$

Equations 3.22 and 3.23 both indicate that the error bounds can be decreased by increasing the marginal entropies $H(F_X)$ and $H(F_Y)$ without considering their mutual relationship (this is equivalent to maximizing the joint entropy $H(F_X, F_Y)$, as we also have $H(F_X, F_Y) \geq H(F_X)$ and $H(F_X, F_Y) \geq H(F_Y)$). This would result in adding superfluous information to the feature space RVs F_X and F_Y. What we really want though is adding selectively the information that determines the mutual relationship between the signals while discarding superfluous information. Mathematically we want to find a suitable trade-off between maximizing the bounds of equations 3.22 and 3.23 and minimizing the bounds of equations 3.17 and equation 3.21. Feature pairs which carry information that is present in both signals (large mutual information), but *only* information that is present in *both* signals (low joint entropy) are the most adapted features for several multimodal signal processing tasks such as for multimodal image registration. The described feature efficiency coefficient is a functional that extracts these features from multimodal signal pairs.

For this let us define a *feature efficiency coefficient* which measures if a specific pair of features is efficient in the sense of explaining the mutual relationship between the two multimodal signals while not carrying much superfluous information. The problem of efficient features in multimodal signals is closely related to determining efficient features for classification. Our proposed coefficient $e(A,B)$ for any pair of RVs A and B (in particular also for the feature space RVs F_X and F_Y) is defined as follows:

$$e(A,B) = \frac{I(A,B)}{H(A,B)} \in [0,1]. \quad (3.24)$$

Maximizing $e(A,B)$ signifies a trade-off between minimizing the lower bound of the error probability by maximizing the mutual information $I(A,B)$, but also minimizing the joint entropy $H(A,B)$ (resulting in maximizing the weakened bounds of equations 3.22 and 3.23). Looking for features that maximize the efficiency coefficient of equation 3.24 will therefore look for features which are highly related (large mutual information) but haven't necessarily much information (marginal entropy)[2].

Interestingly there is a functional closely related to $e(A,B)$ that has already been widely used in multimodal medical image processing, even though its derivation was completely different. It was called normalised entropy $NE(A,B)$ [41] and was derived as an overlap invariant optimization objective for rigid registration:

$$NE(A,B) = \frac{H(A)+H(B)}{H(A,B)} = e(A,B)+1 \in [1,2]. \quad (3.25)$$

The derivation was specific for image registration and arose from the problem that mutual information might increase when images are moved away from optimal registration when the marginal entropies increase more than the joint entropy decreases. This is equivalent to our mathematically derived problem above, but for the special case of image registration. Obviously maximizing $NE(A,B)$ of equation 3.25 is equivalent to maximizing the efficiency coefficient of equation 3.24.

It is very interesting to note that in the early years of information theoretic multimodal signal processing, joint entropy $H(A,B)$ was also an optimization objective of choice. Interestingly this statistic had to be minimised in order to get for example good registration. Looking at the deduced error bounds of equations 3.17, 3.21 and particularly 3.22, one realizes that minimizing joint entropy does *not* minimize these error bounds. On the contrary, it actually maximizes the weakened bound of equation 3.22 and therefore contradicts error bound minimization. The result were very "efficient" features, but with relatively large error bounds (e.g. mapping a black on a white image). This results for example in disconnecting the images during the registration process. We employed the same property in the previous section but only in combination with error bound minimization to separate the superfluous information in the signals from the predictive information.

These arguments are very general. Nevertheless they could have resulted in other definitions for feature efficiency than equation 3.24, such as

$$e(A,B) = \frac{I(A,B)}{H(A)+H(B)}, \quad (3.26)$$

or

$$e(A,B) = \frac{I(A,B)^{\frac{2}{3}}}{H(A,B)^{\frac{1}{3}}}. \quad (3.27)$$

---

[2] Because of the range [0,1] of $e(A, B)$, this functional is sometimes called "normalised measure of dependence" [61].

While the first example is a variant equivalent to equation 3.24, as it simply uses the weakened inequality of equation 3.23 instead of equation 3.22, the second is an extension of $e(A,B)$, which can be generalised as follows:

$$e_k(A,B) = \frac{I(A,B)^k}{H(A,B)^{1-k}}, k \in [0,1]. \quad (3.28)$$

We call an element of this class of functions the *feature efficiency coefficient of order k*. The three cases of $k=0$, $k=1$ and $k=\frac{1}{2}$ represent:

- $k=0$: We put emphasis entirely on the feature efficiency without caring about the resulting lower bound of the error probabilities (minimizing joint entropy). The algorithm will always converge towards signal sequence representations where the same single feature value is assigned to all the samples.
- $k=1$: We put emphasis on minimizing the lower error bound without caring about the efficiency of the features (maximizing mutual information). The algorithm would converge towards signal representations where all samples get assigned a different feature value.
- $k=\frac{1}{2}$: We put equal emphasis on minimizing the lower error bound and on feature efficiency (maximizing normalised entropy).

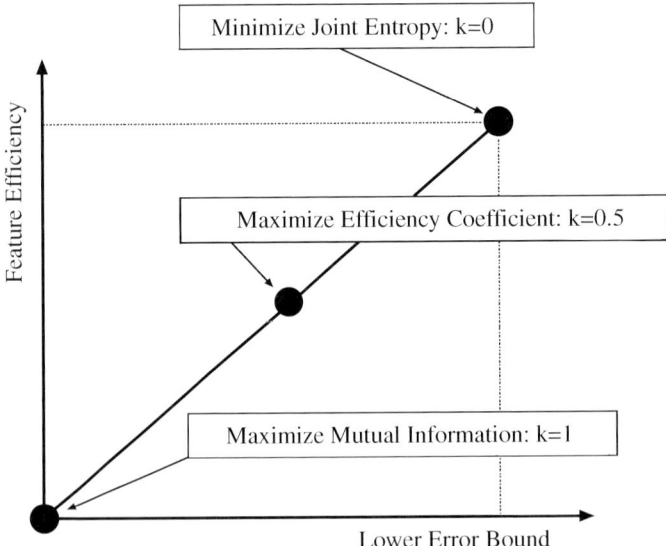

Fig. 3.2 The sketch puts the efficiency coefficients for different orders k into a quantitative relationship. The contradictory optimization objectives of minimizing the lower error bound, but maximizing the feature efficiency has to be combined in a suitable way for a given problem. In the case of medical images, $k=\frac{1}{2}$ has shown to work fine, as it results into an optimization functional equivalent to normalised entropy

3.3 From Error Probability to Multimodal Signal Processing

The two objectives of on the one hand minimizing the lower error bounds and on the other hand maximizing feature efficiency are therefore contradictory. The user has to choose an appropriate order k of equation 3.28 for a given problem. For example order $\frac{1}{2}$ has shown to be very interesting for medical image registration [41, 42]. In Fig. 3.2 we show a quantitative sketch of feature efficiency for different orders of k. In fact this trade-off between feature efficiency and error probability has an interesting analogy in rate-distortion theory, where on the one hand we want to transmit as little information as possible, but on the other hand keep the transmission error as small as possible.

Let us add a synthetic example that illustrates nicely how the feature efficiency of equation 3.28 varies with different orders k. For this we take the initial Magnetic Resonance (MR) images of Fig. 3.3 a) and b) and plot the feature efficiency coefficients of their image intensities for different orders k versus the number of uniform image quantization levels (Fig. 3.3 c). Image content dependent optimal quantization can be looked at as image segmentation. Our approach of quantizing both images with the same number of uniform bins is very crude, but very illustrative in the context of feature efficiencies. In chap. 5.1, we show results with a more practical quantization scheme (Fig. 3.5).

In Fig. 3.3 c) we see that the maximum feature efficiency varies significantly with its order k. For $k = 0$ the optimum suggests using one single quantization bin (very efficient image representation) while for $k = 1$ we would keep all the initial data, including the un-related image noise. For intermediate levels, in particular for $k = \frac{1}{2}$, we get an intermediate optimal number of bins (7 bins) which conserves the anatomical information of the initial scans, while discarding the un-related noise of the images (Fig. 3.4). This is exactly the behaviour outlined in Fig. 3.2 in the special context of image quantization.

Correlation Ratio: We will show now that optimization objectives other than mutual information and normalised entropy can be derived from the proposed framework, starting with correlation ratio [30]. Let us start by recalling Fano's inequality for the Markov chain of equation 3.12 for multimodal signals (equation 3.17) and then lower the bound under the condition that F_Y is characterised by a continuous Gaussian probability density:

$$P_{e1} = P(O^{est} \neq O)$$

$$\geq 1 - \frac{I(F_X, F_Y) + 1}{\log n}$$

$$= 1 - \frac{H(F_Y) - H(F_Y | F_X) + 1}{\log n} \tag{3.29}$$

$$\geq 1 - \frac{\log(\sqrt{2\pi e Var(F_Y)}) - H(F_Y | F_X) + 1}{\log n}, \tag{3.30}$$

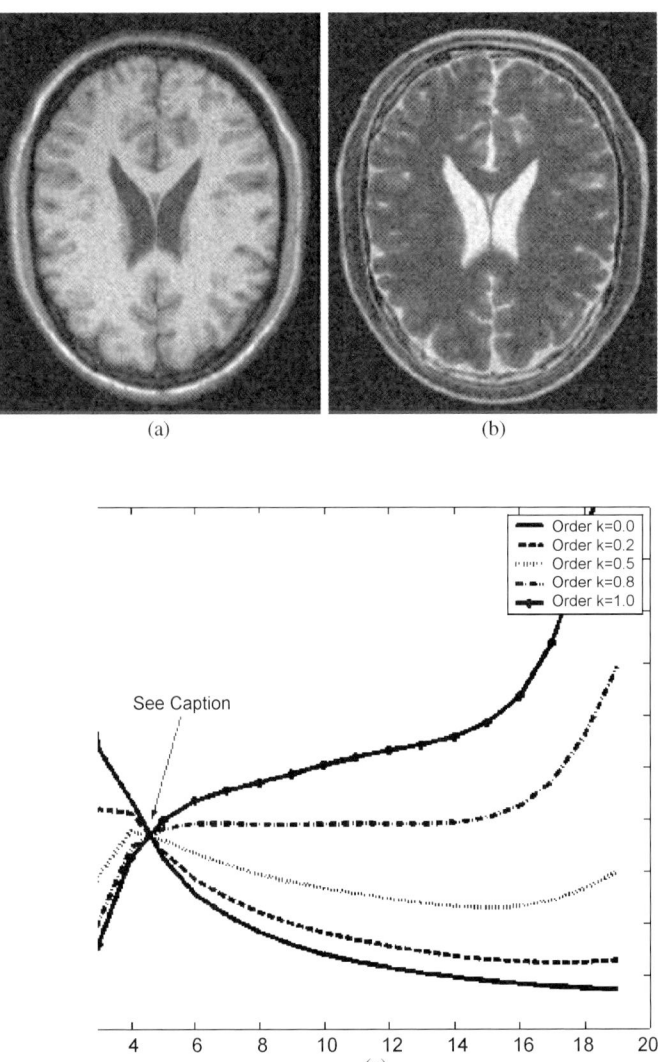

Fig. 3.3 In a) and b) we show a corresponding T1 and T2 dataset. In c) we show the feature efficiency coefficients of equation 3.28 for different orders k as a function of uniform quantization levels. We see that the maximum varies heavily with the order k. For $k = 0.2$ the optimum lies at 2 quantization intervals (Fig. 3.4 a) and b)), while for $k = 0.5$ we get 7 levels which conserves anatomical information while discarding unrelated noise (Fig. 3.4 c) and d)). All the different graphs cross where mutual information time's joint entropy equals one (see arrow)

3.3 From Error Probability to Multimodal Signal Processing

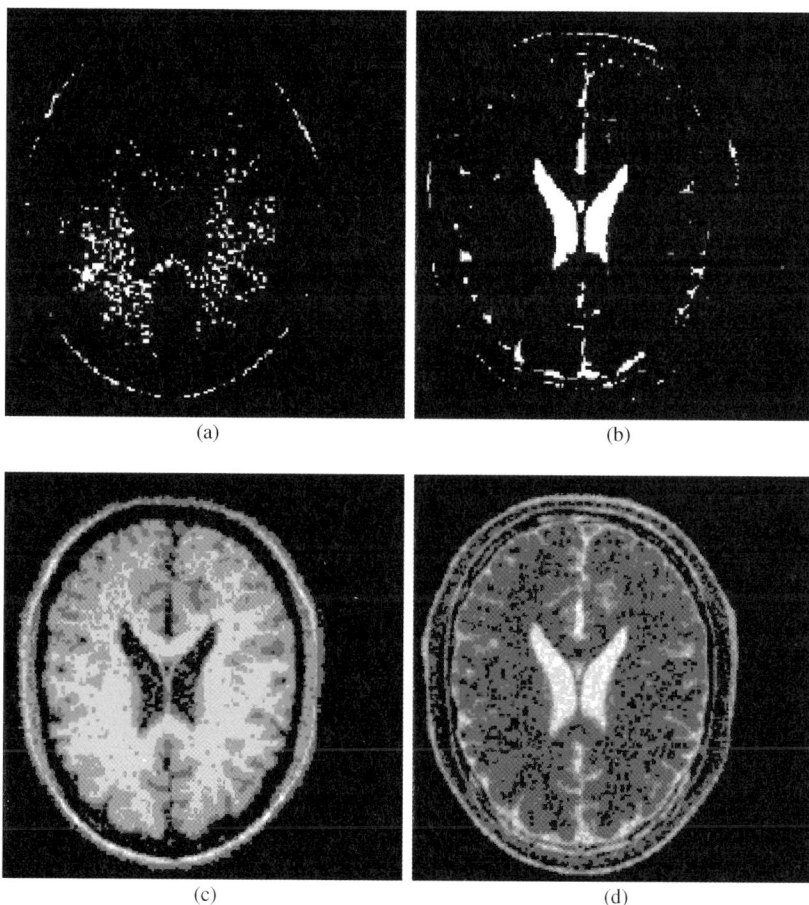

Fig. 3.4 In the images a) and b), respectively c) and d), we show the uniform quantization results at optimal feature efficiency (Fig. 3.3c)) for the feature efficiency coefficients of order $k = 0.2$ and $k = 0.5$ respectively. The former conserves very little but efficient information, while the latter keeps most of the anatomically relevant structures

where F_X is a discrete random variable so that Fano's inequality still holds. It's important to note that in contrast to equations 3.17, 3.21, 3.24 and 3.28, the last lower bound is not symmetric anymore with respect to F_X and F_Y.

Instead of minimizing the lower bound of equation 3.29, we can minimize the weakened lower bound of equation 3.30 by maximizing $\log(\sqrt{2\pi e Var(F_Y)}) - H(F_Y | F_X)$. Let us now assume that the probability density function $P(f_Y | f_X)$ of the transition $F_X \to F_Y$ is given by

$$F_Y = E(F_Y | F_X) + N(0, E(Var(F_Y | F_X))), \tag{3.31}$$

where $N(\mu,\sigma^2)$ is an additive Gaussian noise of mean μ and variance σ^2 and $E(F_Y | F_X)$ is the conditional expectation of F_Y knowing F_X. Then the conditional probability $P(F_Y = f_Y | F_X = f_X)$ is given by

$$P(F_Y = f_Y | F_X = f_X) = \frac{1}{\sqrt{2\pi}\sigma} e^{-\frac{(f_Y - E(F_Y|F_X))^2}{2\sigma^2}}, \qquad (3.32)$$

with $\sigma^2 = E(Var(F_Y | F_X))$. Therefore we can easily calculate the conditional entropy $H(F_Y | F_X)$:

$$H(F_Y | F_X) = -\sum_{f_X} \int_{f_Y} P(f_X, f_Y)$$

$$\cdot \log(\frac{1}{\sqrt{2\pi} \cdot \sigma} \cdot e^{-\frac{(f_Y - E(F_Y|F_X))^2}{2\sigma^2}}) df_Y$$

$$= \log(\sqrt{2\pi e E(Var(F_Y | F_X))}). \qquad (3.33)$$

This means that we can minimize the lower bound of equation 3.30 by maximizing

$$\log(\sqrt{2\pi e Var(F_Y)}) - H(F_Y | F_X) =$$
$$= \log(\sqrt{2\pi e Var(F_Y)}) \qquad (3.34)$$
$$- \log(\sqrt{2\pi e E(Var(F_Y | F_X))}),$$

which is equivalent to maximizing its squared exponential

$$\eta_1(F_Y | F_X) = \frac{Var(F_Y)}{E(Var(F_Y | F_X))}, \qquad (3.35)$$

or maximizing

$$\eta_2(F_Y | F_X) = 1 - \frac{E(Var(F_Y | F_X))}{Var(F_Y)}. \qquad (3.36)$$

It's important to note that $\eta_2(F_Y | F_X)$ is just the correlation ratio as proposed in [30] for multimodal medical image registration, when the employed features F_X and F_Y are the image intensities.

Maximum Likelihood. In the previous paragraph on correlation ratio, only assumptions about the underlying transition probabilities were taken. On the other hand we didn't use any prior on the specific feature representations to be used. Let

3.3 From Error Probability to Multimodal Signal Processing

us now relax the prior on the transitions, but assume that we can fix the feature representation F_X. In direct analogy to equation 3.29, we have also

$$P_{e1} = P(O^{est} \neq O)$$

$$\geq 1 - \frac{I(F_X, F_Y) + 1}{\log n}$$

$$= 1 - \frac{H(F_X) - H(F_X | F_Y) + 1}{\log n}, \quad (3.37)$$

where $H(F_X)$ remains constant during the minimization as F_X is fixed. Therefore we want to find the feature representation F_Y so that the conditional entropy

$$H(F_X | F_Y) = -\sum_{f_Y, f_X} P(f_Y, f_X) \cdot \log P(f_X | f_Y) \quad (3.38)$$

is minimal. Let us now use histograming to estimate the joint probabilities $P(f_Y, f_X)$:

$$P(f_Y, f_X) = \frac{|(f_X, f_Y)|}{n}, \quad (3.39)$$

where $|(f_X, f_Y)|$ is the number of samples that have feature value f_X in one modality and feature value f_Y in the other modality. Therefore we can rewrite equation 3.38 as follows [43]:

$$H(F_X | F_Y) = -\sum_{f_Y, f_X} \frac{|(f_X, f_Y)|}{n} \cdot \log P(f_X | f_Y)$$

$$= -\frac{1}{n} \sum_{o \in \Omega_O} \log P(f_X(o) | f_Y(o))$$

$$= -\frac{1}{n} \log(\prod_{o \in \Omega_O} P(f_X(o) | f_Y(o))), \quad (3.40)$$

where Ω_O indexes the n data samples (section 3). Equation 3.40 is, up to the negative constant $-\frac{1}{n}$, exactly the log-likelihood of obtaining a signal F_X from a signal feature representation F_Y for a given transition probability distribution $P(F_X = f_X | F_Y = f_Y)$ and a fixed feature space representation F_X.

Image Registration as Feature Selection. In the previous paragraphs, we derived several optimization objectives within the framework of error probability. This allows us to analyze their mutual relationship on a theoretical basis and facilitates the choice of the optimization objective for a particular problem. For example we derived the differences between mutual information and normalised entropy, also for applications outside medical image registration. Even more general, these developments show a very general concept of multimodal signal processing based on feature selection and extraction which determines those feature space representations of the initial signal sequences that confirm most of the basic and natural hypothesis that multimodal signals have the same physical origin. On the other hand though, it might not seem very clear yet how the presented framework is actually applied to a particular problem, such as image registration. Let us therefore have a closer look at multimodal medical image registration.

So far we have just been discussing feature selection and extraction but not image transformations, such as rigid or affine deformations. The step from the proposed framework of multimodal feature selection and extraction to image registration is straight forward though. In fact, it's possible to identify image registration as a special case of feature selection. We want to select that image transformation that best reflects the fact that that the images are acquired from the same physical scene. This transformation will minimize the error probabilities $P_{\{e_1,e_2\}}$.

If S_Y is the sequence of the floating image and S_X of the reference image, the corresponding Markov chains can be re-written:

$$\begin{aligned} O \to X = S_X(o) \to F_X &= X = S_X(o) \to F_Y^{est} = S_Y(T(o)) \\ &\to Y^{est} = S_Y(o) \to O^{est} \to E, \\ O \to Y = S_Y(o) \to F_Y &= S_Y(T(o)) \to F_X^{est} = S_X(o) \\ &\to X^{est} = F_X^{est} = S_X(o) \to O^{est} \to E. \end{aligned} \quad (3.41)$$

We see that most of the transitions are deterministic and that several of them are parameterised by the transformation parameters T of the floating image. In fact T can be looked at as the optimization parameters of a feature selection step of the floating image (sequence) S_Y. The optimal T should confirm the most of the basic multimodal hypothesis, that the two multimodal images (sequences) S_X and S_Y have the same physical origin. It's important to note that "the same physical origin" is a flexible hypothesis. This means that sometimes two brain images, even though of different patients, are considered to come from the same physical scene, which is not a particular patient's anatomy but the brain anatomy in general.

3.4 Optimization

The presented framework of feature selection and extraction for multimodal signal processing leads quite frequently to very demanding optimization objectives (chap. 5). Resulting objective functions can have very distinct shapes de-

3.4 Optimization

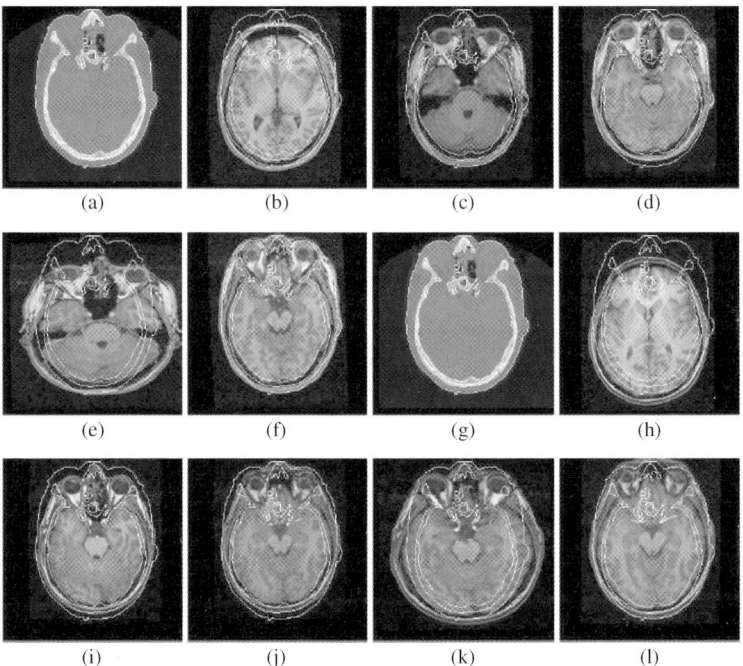

Fig. 3.5 a) is the CT-target image. In b), the contours of the target image are superposed on the floating MR-scan. In c) and d) we see the results after a rigid optimization, when using respectively the intensity based mutual information and the edginess mutual information. In e) and f) we show the corresponding results for affine registration. Figures g) through l) show the results for a second MR-scan. In e) and f) (respectively k) and l)) we recognize a significant improvement with the edginess based mutual information; respectively that the global maximum of intensity based mutual information doesn't correspond to good registration

pending on the chosen feature space representations from which the optimal representation should be selected and extracted. As a result it's mostly not clear that local optimization schemes would be sufficient to lead to robust results. That's why we use a globally convergent genetic optimization algorithm [44] to be able to study the global behaviour of the optimization functions and to ensure that we avoid local optima to get "good" results for a specific application, such as image registration. Thereafter we refine the results locally using a steepest gradient algorithm.

The problem with this approach is that genetic optimization is very time consuming. Therefore we parallelised an existing genetic optimization library [45] for distributed memory architectures, using the MPICH implementation of MPICH implementation of MPI ("Message Passing Interface") [46, 47].

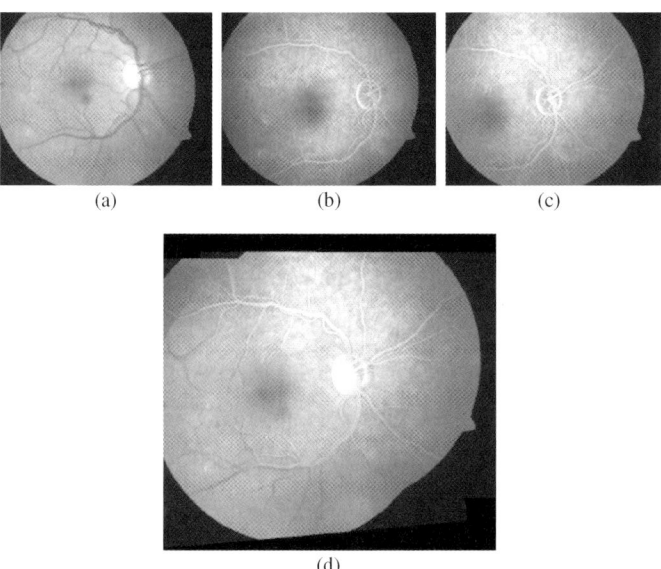

Fig. 3.6 The figures a), b) and c) had first to be registered in order to reconstruct the extended view shown in d)

3.5 Results

This chapter is mainly dedicated to multimodal signal processing. Therefore we only want to show results in this field and discard examples of quantization or classification. In particular we want to illustrate the field of multimodal medical image registration and show how the feature based framework enlarges the vision of image registration and results in a unifying framework for multimodal medical image processing in general.

The second example is based on speech-video sequences, where we show the importance of adequate features to interpret both signals of a multimodal sequences simultaneously.

3.5.1 Multimodal Medical Images

We will show two examples for medical image registration [48] which show the importance of the feature based framework for multimodal signal processing. First of all, we want to show that the quality of registration results can depend heavily on the employed features. For example to rather use edge instead of intensity information for the registration can be very beneficial for some cases.

3.5 Results

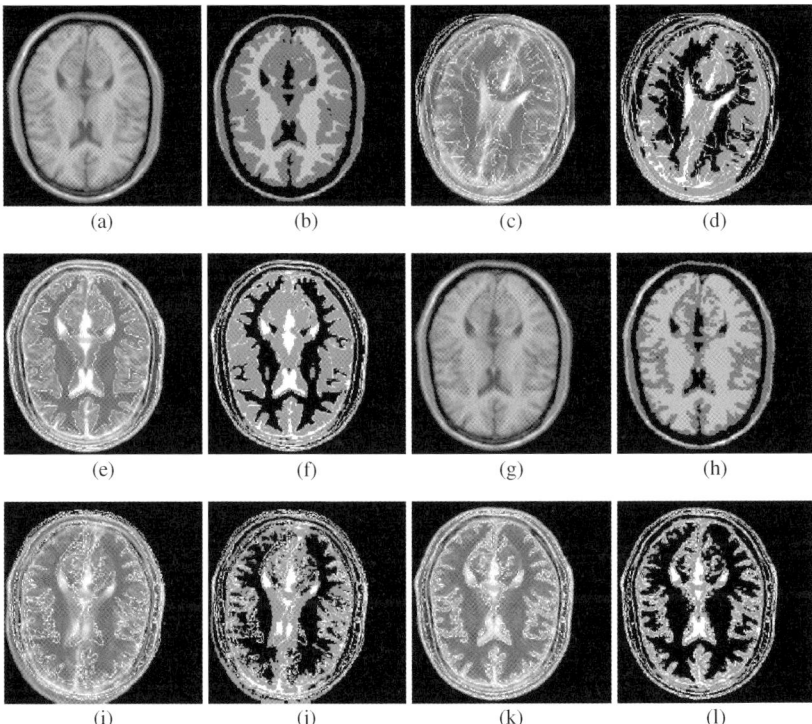

Fig. 3.7 Image a) shows the reference image and c) the initial floating image. In e) we show the rigidly registered result. Images b), d) and f) show the quantised outputs of a), c) and e) with the optimal number of bins. Images g) to l) show an experiment equivalent to a) to f), but with noisier datasets. The contours of b), resp. h), are outlined in c) to f), resp. i) to l)

The second point will show that the proposed framework enlarges the view of image registration and leads to an integrated approach on multimodal medical image processing in general. We will show examples where image registration, segmentation and artifact correction of medical images can be incorporated into one generalised algorithm. In the paragraphs on registration with quantization and registration with bias correction, we used synthetic MR-scans from the *BrainWeb* database [49, 50].

Feature Space Image Registration. Conventionally multimodal medical image registration determined optimal transformation parameters by maximizing image intensity based information theoretic quantities, particularly those re-derived in chap. 3.2 [29, 30, 51]. In the sense of error probability this seams to be the right thing to do. Nevertheless it's important to note that maximizing mutual information minimizes the error probability on the average. This means that the statistical matching of large structures is much more emphasised than other small but anatomically important regions of the patients anatomy. Therefore we propose to use rather the edginess information in the images [52], so that the probability estima-

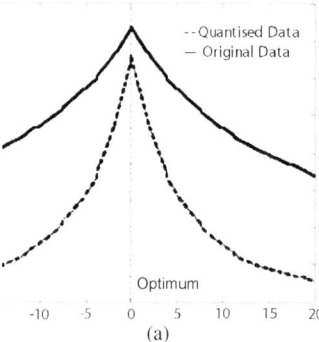 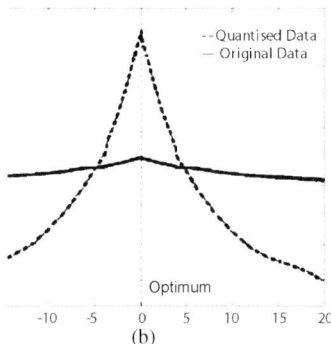

Fig. 3.8 Figure a) compares mutual information of the original noisy images of Fig. 3.7 g) and k) with the mutual information of their quantization results for translations away from optimal registration. For better comparison, the plot of the mutual information between the original datasets in a) was moved down by one unity (-1.). In b) we see feature efficiency of order ½ for the same images as for a). Plot a) shows that maximization of mutual information is unable to perform simultaneous registration and segmentation, contrary to the feature efficiency of b). Also we see that simultaneous segmentation improves the general behaviour of the optimization objectives, in particular for feature efficiency of order ½. Let us recall that feature efficiency of order ½ is equivalent to the well known and widely used normalised entropy

tion does not use the volumetric information anymore, but rather information on the volume boundaries.

We considered affine registration to maximize mutual information between the two feature space representations of the initial images (their gradients) and compared the results to the conventional intensity based mutual information registration. The results for CT-MR inter-subject registration are shown in Fig. 3.4. As we can see, while the rigid registration of the MR image with the CT image is achieved correctly when considering either the voxel intensities (Fig. 3.4(c) and (i)) and the edginess (Fig. 3.4 (d) and (j)) as feature space, the maximization of the intensity-based mutual information yields aberrant results in case of non rigid (affine) registration (Fig. 3.4 (e) and (k)) while using edginess as a feature space for this optimization leads to correct and stable results (Fig. 3.4 (f) and (l)). The interpretation of those results is presented in section 6.

Besides improving the robustness for MR-CT registration, we will show that edginess information combined with globally convergent genetic optimization makes mutual information based image registration applicable to the registration of blood vessel images in the human retina. In Fig. 3.4 we show how three partial views of the retinal vascular system can be combined to provide a virtually extended view. The different intensity distributions in the images are caused by an injected contrast agent which enables the study of the retinal blood flow for diabetic retinopathy.

Image Registration and Quantization. Medical images are more or less noisy representations of the patient's anatomy. The noise has a negative impact on sta-

3.5 Results

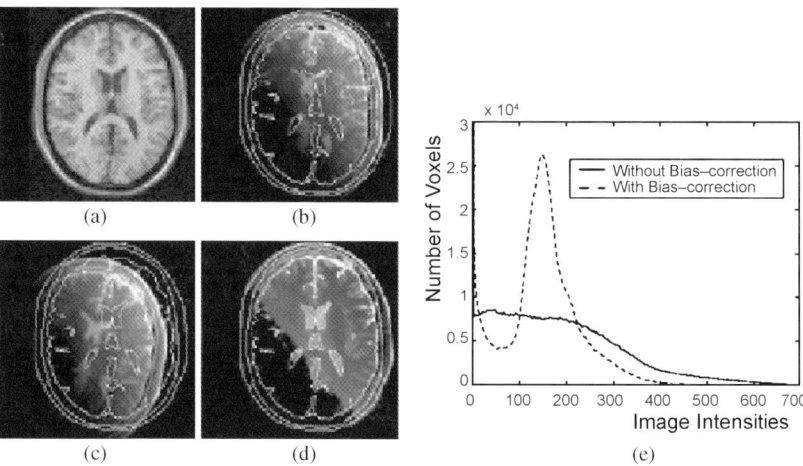

Fig. 3.9 We rigidly registered the image of b) onto the reference image shown in a). c) shows the bad result without simultaneous bias-correction and d) shows good registration with simultaneous bias-correction (contours of the reference image a) are shown in white). In e) we show the histograms of c) and d) respectively, showing the effect of the bias correction on the grey level distribution

tistical image registration. Some approaches to minimize the influence of noise are based on initial filtering of the datasets (e.g. anisotropic filtering [53]) or even on anatomical segmentation to extract the information of the images that is really relevant for registration. In this example we therefore try a very naive way of extracting the representative anatomical information in the medical images while discarding the dispensable noise. We use simple image intensity quantization which varies the number of bins for both axes of the joint probability distribution independently. But decreasing the number of bins obviously decreases the marginal entropies of the image representations, therefore simply maximizing the mutual information of equations 3.17 and 3.21 is dangerous. We rather use the feature efficiency coefficient of order $\frac{1}{2}$ (equation 3.28) of the quantised images to find the optimal number of quantization intervals as well as the geometrical registration parameters. Let us recall again that the feature efficiency of order $\frac{1}{2}$ is equivalent to the widely used normalised entropy. Mathematically we can write the optimization objective as follows:

$$[\vec{t}^{opt}, q_X^{opt}, q_Y^{opt}] = $$
$$= \arg\max_{\vec{t} \in \mathbf{R}^d, q_X \in \mathbf{Z}^+, q_Y \in \mathbf{Z}^+} e_{\frac{1}{2}}(Q_{q_X}(X), T_{\vec{t}}(Q_{q_Y}(Y))), \quad (3.42)$$

where X and Y are the RVs associated to the image intensities of the reference and floating image respectively. q_X and q_Y are the number of bins used for the density estimation of X and Y and $\vec{t}$ are the parameters of the geometric transformation T of the floating image. d is the dimension of $\vec{t}$ and is determined by the particular transformation model, e. g. for rigid-body we have 6 and for affine 12 parameters. Results for rigid registration are shown in Fig. 3.5.

In order to indicate more the benefits and importance of combining segmentation and registration into one single optimization scheme, let us sketch the mutual information and feature efficiency of order $\frac{1}{2}$ for the initial noisy images of Fig. 3.5 g) and i) as well as for their quantization results with respect to translations away from their optimal registration. The plots are shown in Fig. 3.5. In particular two important facts of the theoretical expectations outlined in chap. 3.2 can be reconfirmed. First of all mutual information is as expected unable to segment the initial noisy images during the registration task: mutual information of the initial data is larger than mutual information of the optimally quantised data (Fig. 3.5 a)). And second, feature efficiency of order $\frac{1}{2}$ has a much more pronounced maximum at optimal registration for the quantised images than for the original data (Fig. 3.5 b)). The second point is in fact to a smaller degree also true for mutual information.

Image Registration with Bias correction. Interventional imaging modalities suffer frequently from a large bias field. Bias field is a standard term used in Magnetic Resonance imaging to define a smooth variation of the grey level values along the acquired image. This may be due to different causes linked to the MR scanner, such as poor RF coil uniformity, static field inhomogeneity, RF perturbation, etc. [54] This makes image registration particularly difficult. Nevertheless it would be of particular interest to register pre-operatively acquired scans of different modalities onto the interventional datasets. In this section we want to show that the presented framework easily allows registering images with large bias fields. The approach simply combines minimum entropy bias-correction [55, 56] with mutual information based image registration. From the developed theory, one can recognize immediately that mutual information is not appropriate for this task as minimizing entropy contradicts obviously the maximum mutual information principle of equations 3.17 and 3.21. Therefore maximizing directly mutual information would not correct the bias field even though the error bounds of equations 3.17 and 3.21 would be minimised. Just as in the previous paragraph, this is a typical example of inefficient features. Rather than maximizing mutual information, we want to maximize the efficiency coefficient of the bias-corrected image intensities/features (equation 3.24).

The resulting mathematical formalism can thereafter be written as follows: Let $\vec{p} \in \mathbf{R}^{d_1}$ parameterize the polynomial bias-correction of [54, 55], where d_1 is determined by the degree of the polynomials. Furthermore we have to determine the parameters $\vec{t} \in \mathbf{R}^{d_2}$ of the geometric transformation T, where d_2 is the number of parameters that determines the transformation. In our specific application, we optimised over rigid-body transformations (d_1 equalled 6). Mathematically we have:

$$[\vec{t}^{opt}, \vec{p}^{opt}] = \arg \max_{[\vec{t},\vec{p}] \in \mathbf{R}^{d_1+d_2}} e_{\frac{1}{2}}(X, T_{\vec{t}}(P_{\vec{p}}(Y))), \qquad (3.43)$$

3.5 Results   55

where the parameters $\vec{p}^{\,opt}$ specify the optimal bias-correction and $\vec{t}^{\,opt}$ determines the optimal rigid transformation. Here X and Y refer to the RVs associated to the image intensities. Figure 3.5 presents the results.

3.5.2 Speech-Video Sequences

In this application we want to determine the region in a video scene that contains the speaker's mouth, i.e. where the motion seen in the image sequence corresponds to the audio signal [56].

With the framework presented before we will find the features in the audio and video signal that minimize the lower bounds on the error probabilities of equations 3.17 and 3.21 in the region of the speaker's mouth [57]. The sampling RV O of the Markov chains of equations 3.12 and 3.13 now refer to the time index in the sequence and not to spatial coordinates as for medical image registration, as the signal acquisition is now performed along a time interval.

From equations 3.17 and 3.21 it follows that small lower error bounds in the region of the speaker's mouth is equivalent to a large feature space mutual information $I(F_X, F_Y)$ in this region. F_X and F_Y stand for the audio and video features respectively. On the other hand a large bound should result in the regions where the movements are not caused by the speaker's lips and are therefore unrelated to the speech signal. So that's where $I(F_X, F_Y)$ should be small.

To represent the information of the audio signal, we first converted it into a power-spectrum (Fig. 3.5a).

In order to deal with this multi-dimensional audio signal, we included a linear feature extraction step in the algorithm. As for any couple of RVs A and B, we have $H(A) \geq I(A,B)$ and from equations 3.17 and 3.21 we get a weakened lower bound for the error probabilities $P_{\{e_1,e_2\}}$:

$$P_{\{e1,e2\}} \geq 1 - \frac{I(F_X, F_Y) + 1}{\log n}$$

$$\geq 1 - \frac{H(F_X) + 1}{\log n}. \quad (3.44)$$

Therefore we looked for the linear combination of the power spectrum coefficients $W(f_i, t)$ (Fig. 3.5a) that carries most entropy. The finally obtained audio-feature is therefore defined by

$$F_X(t) = \sum_i \alpha_i^{opt} \cdot W(f_i, t). \quad (3.45)$$

$$\vec{\alpha}^{\,opt} = \arg \max_{\vec{\alpha}:|\vec{\alpha}|=1, \alpha_i \geq 0} H(\sum_i \alpha_i \cdot W(f_i, t)). \quad (3.46)$$

In Fig. 3.5 b), we show for one sequence the weights α_i^{opt} that maximize the entropy of equation 3.45 and therefore define the audio-features F_X of the audio signal.

We want to show two important points about the presented theory: first of all that there exist features that relate the mouth movements of a speaker directly to the corresponding speech signal. On the other hand we want to show that the choice of a particular feature representation is very crucial for the performance of the algorithm. There are features that contain lots of information (have lots of entropy), but are unrelated to the other signal. Other features represent this dependency much better and yield very good results.

The straight forward approach to quantify the dependency (in the sense of equations 3.17 and 3.21) between an audio and video signal of a speaker would consist of calculating the mutual information between the intensities of each pixel in the frame and the audio-feature of equation 3.45. In Fig. 3.6 we show the corresponding results.

We can see that this straight forward approach doesn't lead to the result we could have expected. It seems that the pixel intensities of the speaker's mouth don't carry much information about the audio signal. Instead we propose a local feature that is more related to intensity changes (and therefore also to motion in the scene) than to the intensities themselves:

$$F_Y(i,j,t) = \sum_{l,m=-1}^{1} g_{t+1}(i+l,j+m) - g_{t-1}(i+l,j+m), \quad (3.47)$$

where $g_t(i,j)$ stands for the intensity of a pixel at coordinates (i,j) in the frame at time t.

Thereafter we calculated for each pixel in the scene the mutual information between the resulting audio- and video-feature $I(F_X, F_Y)$. As shown in Fig. 3.6 a clear relationship between the speech and the speaker's mouth is obtained.

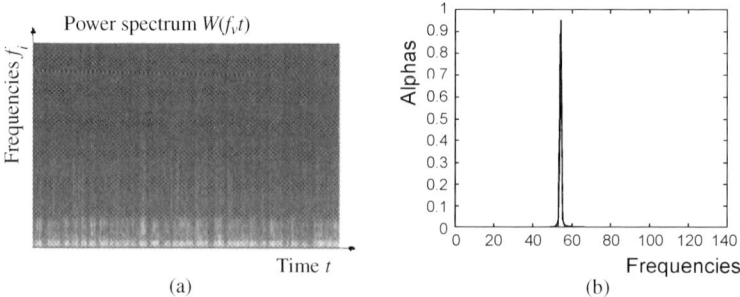

Fig. 3.10 a) The power spectrum of the video sequence. At each time point we have the power coefficients of several frequencies. b) The alphas for which the weighted sum of equation 3.45 has maximum entropy

3.5 Results

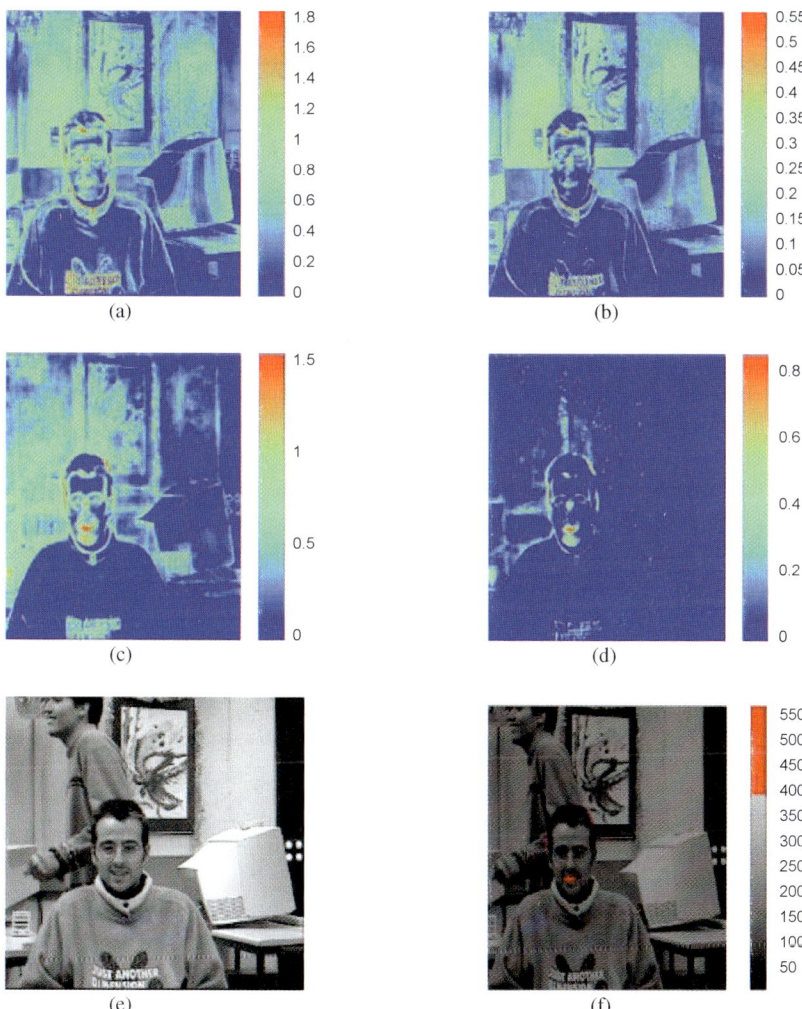

Fig. 3.11 a) We show the intensity entropies for each pixel of the sequence. It shows that our sequence contained lots of motion in the background of the scene (people passing, waving arms, etc.). b) The mutual information between the pixel intensities and the audio-feature has been calculated. We see that there is not a particularly high mutual information in the region of the speaker's mouth. c) shows the entropy of the video-features F_Y for each pixel in the video scene. d) relates this video information to the extracted audio-features F_X of equation 3.45 by calculating the feature space mutual information $I(F_X, F_Y)$ for each pixel. e) shows a typical frame of the sequence. f) is simply the thresholded image of d) super-posed on the frame of e). It shows that the mutual information maxima lie clearly at the speaker's mouth

3.6 Discussion

We have shown that for medical image registration the choice of a good set of features to guide the registration is crucial. The results can vary significantly depending on what information the selected features carry. The comparison of Fig. 3.4 between image intensities and edginess information shows that for some images the matching of boundaries can be more appropriate than the statistical matching of image intensities. Boundaries are small but very significant matching criteria, while image intensities might over-emphasize the large and therefore statistically important regions in the data sets. In terms of maximization of the mutual information between feature spaces, it appears that if we consider non-rigid registration using the voxel intensity as feature space, the maximum of MI is achieved not because of a good correspondence between the images but because of an increase in the marginal entropy of the MR image. This problem does not appear if we consider edginess as feature (figure 12(f)), whose entropy is not increased by the geometric transformation.

Thereafter we used the developed framework to guide multimodal signal processing in a unified framework. This means e.g. that image segmentation or bias correction can be incorporated into a generalised registration algorithm. We show that our framework is able to extract the information of the medical images that is most important for the registration task and to get rid of bias artifacts or background noise, information that can just corrupt the registration results.

The last experiment was performed on speech-video sequences, where we show two main points. First of all we demonstrate that there exists a direct relationship between the speech signal and the video frames which can be explored for e.g. multimodal speaker localization. It is very important that our proposed approach doesn't make any hypothesis about their underlying relationship. The information theoretic framework rather quantifies non-parametrically their mutual dependency. As a second important point we show that just as in the case of medical images, the choice of the right features is very crucial. Our framework gives a very flexible approach to construct feature selection algorithms such as proposed for medical image registration. Using this framework we were able to re-confirm (Fig. 3.6) that motion estimation gives features much more related to the corresponding speech signal than pure image intensities (compare for example with [24]).

3.7 Conclusions

This chapter presented some important points of information theoretic signal processing. The first one consisted of unifying a large class of algorithms in one single mathematical framework, using the information theoretical concepts of stochastic processes and their error probabilities. Combined with Fano's inequality and the data processing inequality, this mathematically compact framework allows the

3.7 Conclusions

derivation and interpretation of optimization objectives which govern a wide range of information theoretic signal processing tasks.

The second main subject consisted of applying the introduced framework to the large field of multimodal signal processing. We applied the theory successfully to several important and revealing problems of medical image and speech-video sequence processing. In the next chapter, we will illustrate in more details how this framework can be used in some important applications including audio-visual speech recognition.

Chapter 4
Multimodality for Plastic User Interfaces: Models, Methods, and Principles

Jean Vanderdonckt[1], Gaëlle Calvary[2], Joëlle Coutaz[2], Adrian Stanciulescu[1]

[1] Université catholique de Louvain, Louvain School of Management(LSM), Belgian Lab. of Computer-Human Interaction (BCHI), Place des Doyens, 1 – B-1348 Louvain-la-Neuve, Belgium
{jean. vanderdonckt, Adrian. stanciulescu}@uclouvain.be
[2] Université Joseph Fourier, Grenoble Informatics Laboratory (LIG), 385 rue de la Bibliothèque, BP 53 – F-38041 Grenoble cedex 9, France
{gaelle.calvary, joelle.coutaz}@imag.fr

4.1 Introduction

In the early nineties, research in multimodal user interface design was primarily concerned with the goal of mimicking human interpersonal capabilities [62], reminiscent of the "Put That There" paradigm [63]. More recently, multimodality has been viewed as a means to improve interaction flexibility, adaptation, and robustness [64]:

- *Interaction adaptation* denotes the ability of the system to adapt itself or to be adapted according to the constraints imposed by the context of use, where the context of use in understood as the combination of a user using a given computing platform in a particular physical environment [65]. For example, when the user switches from a desktop platform to a mobile platform, the user interface may be adapted by relying on other interaction techniques than those previously available on the initial platform.
- *Interaction flexibility* denotes the multiplicity of ways the user and the system exchange information [66]. For example, commercial navigation systems use verbal outputs as well as graphics maps and icons to provide drivers with understandable instructions. The system, as well as the user, can choose the most convenient and effective way to acquire information and convey meaning.
- *Interaction robustness* covers features of the interaction that support the successful achievement and assessment of goals, whether these goals are human goals or system goals) [66]. For example, in a noisy environment, the system may draw upon visual facial features to improve its understanding of spoken utterances, making the interaction more robust.

Today, context-aware processing capabilities are being developed to enhance system adaptation, flexibility, and robustness. As a result, from user interfaces whose spectrum of modalities is pre-defined by-design, research is now concerned with multimodal user interfaces (UIs) where the set of modalities and their combination may evolve dynamically to accommodate and take advantage of the changes in the real world. By extension, we call *plasticity*, the capacity of interactive systems to adapt to the context of use while preserving, or even enhancing, the usability of the system [67, 68]. Usability expresses the *use worthiness* of the system that is the value that this system has in the real world [69]. Therefore, the major difference between adaptation and plasticity is that there is no guarantee that the adaptation will lead to a value in the real world in the first case as opposed to an explicit guarantee in the second case.

The *context of use* is a dynamic structured information space whose finality is to inform the adaptation process. It includes a model of the *user* who is intended to use (or is actually using) the system, the social and physical *environments* where the interaction is supposed to take place (or is actually taking place), and the hardware-software *platform* to be used (or is actually being used). The latter covers the set of computing, sensing, communication, and interaction resources that bind together the physical environment with the digital world.

UI adaptation has been studied in Artificial Intelligence (AI) for many years but the context of use has often been restricted to a user model [70]. Today, Model Driven Engineering (MDE) is being rediscovered and explored as a generic framework to address the problem of plastic UIs [65]. Significant results have been demonstrated for plastic distributed graphics UIs [71, 72]. On the other hand, multimodality has been covered in a rather limited way; rather as a by-product than as a driving force for adaptation [72–74]. Similarly, work in multimodal UI has focused primarily on the problem of synergistic blending at multiple levels of abstraction [75], rarely on adaptation issues or viewing multimodality as an explicit means to adapt UI according to their context of use [76, 77]. In this article, we bring together the main results from the developments in plastic UIs and in multimodal UIs using a theoretic and conceptual perspective as a unifying approach. It is aimed at defining models useful to support UI plasticity by relying on multimodality, at introducing and discussing basic principles that can drive the development of such UIs, and at describing some techniques as proof-of-concept of the aforementioned models and principles. In the following section, we introduce running examples that will serve as illustration throughout our discussion of the use of multimodality to support plasticity.

4.2 Running Examples

This section includes a brief description of two examples we have implemented using multimodality as a means for UI plasticity: a neuro-surgery system and a web site.

4.2.1 The Assisted Neuro-surgery System

Our neuro-surgery system supports a craniotomy task. This task consists of drawing a surgery path over the patient's head. Three technical stages are necessary to provide the surgeon with the appropriate guidance:

4.2 Running Examples

1. In the pre-operative stage, the surgeon selects the structures of interest that must be segmented from pre-operative images. He may also define additional information, such as annotations that are relevant to the task execution as well as commands for auditory output.
2. At the pre-operative calibration stage, the segmented structures are registered along with the physical scene as well as any movement of the patient or microscope compensated for. The optics of the microscope is calibrated.
3. Finally, during the surgical intervention, the virtual structures augmented with the annotations are visualised and/or rendered via the audio channel.

In order to constantly display useful information, the patient head is decomposed into three zones (Fig. 4.1): the region labelled as 1 has the lowest priority, the region labelled as 2 has medium priority and the region labelled as 3 has the highest priority. The algorithm first tries to place annotations only in the level 1 area to avoid intersection with the level 2 zone. If the available area is not sufficient, the algorithm tries to place annotations by spanning over the level 2 zone while avoiding overlying annotations onto the level 3 zone. If this is still not possible, annotation is automatically rendered through the audio output. Annotation localization is preserved by visualizing the corresponding anchor point onto region 3.

Switching between the different renderings shown in Fig. 4.2 depends on the distance between the patient's head and the surgeon's head as well as on the angle of the surgeon's body with respect to the vertical axis.

1. When the surgeon is in the background context (which means a maximum degree of 5° with respect to vertical axis and more than 60 cm from the patient's head), the rendering is performed as in Fig. 4.2a.
2. When the surgeon is in the background/Real object context (which means between 5° and 10° with respect to vertical axis and between 30 and 60 cm from the patient's head), Fig. 4.2b is rendered.

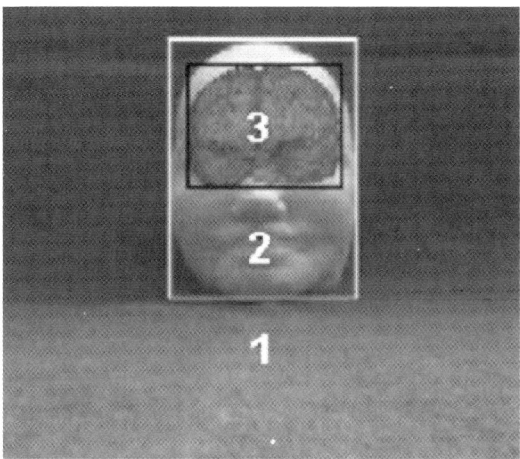

Fig. 4.1 Priority regions of the neuro-surgery system

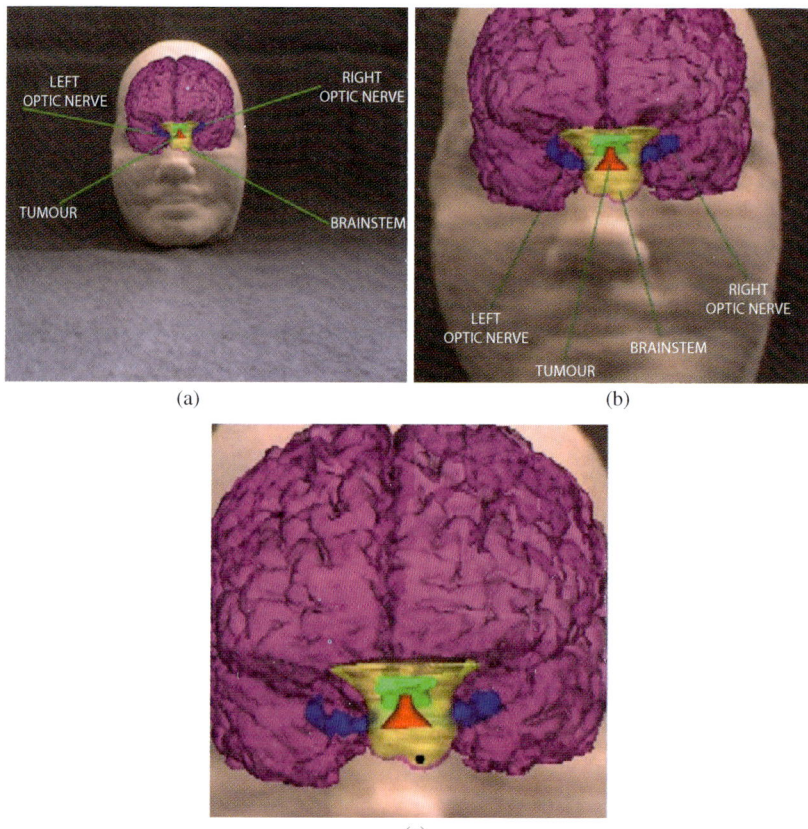

Fig. 4.2 Context-aware annotation rendering: (a) Background Context, (b) Background/Real Object Context, (c) Virtual Object Context

3. When the surgeon is in the virtual Object context (which means more than 20° with respect to vertical axis and less than 30 cm from the patient's head), graphics text is replaced with vocal rendering.

4.2.2 The Sedan-Bouillon Web Site

The Sedan-Bouillon Web site (www.sedan-bouillon.com, a web site developed by UJF within the European IST Cameleon project – http://giove.cnuce.cnr.it/-cameleon.html) provides tourists with information for visiting and sojourning in the regions of Sedan and Bouillon. Figure 4.3a shows a simplified version of the actual Web site when a user is logged in from a PC workstation. By dynamically logging to the same web site with a PDA, users are informed on the PDA that they can allocate UI components across the interaction resources currently available.

4.2 Running Examples

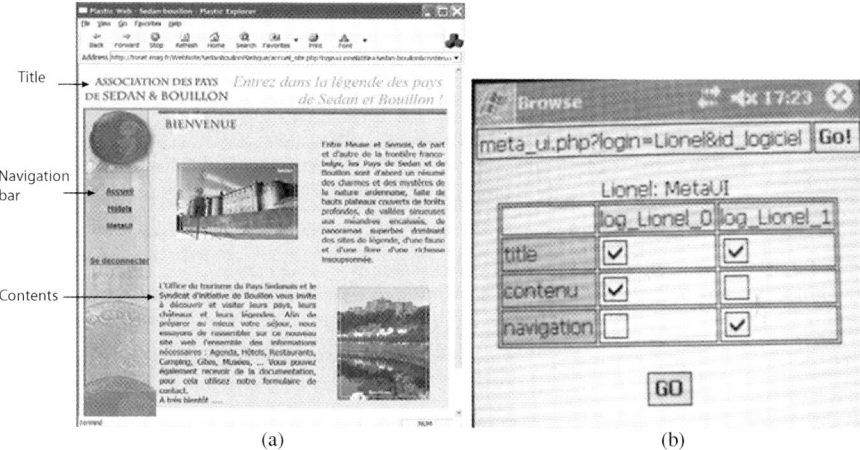

Fig. 4.3 (a) The Sedan-Bouillon Web site when centralised on a PC. (b) The control panel to distribute UI components across the interaction resources of the platform (PC+PDA)

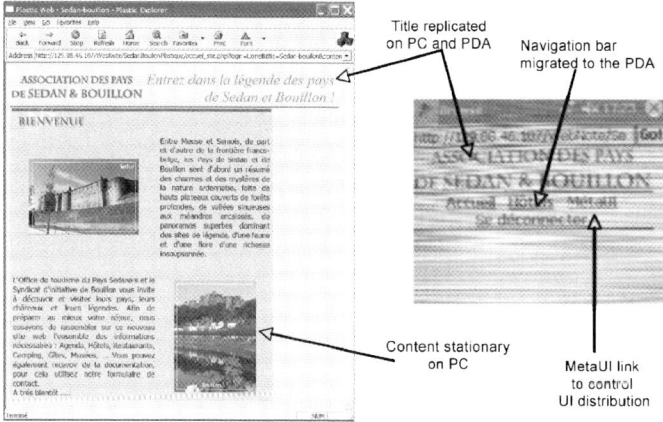

Fig. 4.4 The UI of the Sedan-Bouillon Web site when distributed across the resources of the PC and the PDA

In the example of Fig. 4.3b, the user asks for the following distribution: the title of the page must appear on the PDA as well as on the PC (the title slots are ticked for the two browsers), the main content should stay on the PC, and the navigation bar should be rendered on the PDA. Figure 4.3 shows the resulting UI. In this configuration, the PDA is used like a remote controller to browse the information space displayed on a shared public surface. At any time, the user can ask for a reconfiguration of the UI by selecting the "meta-UI" link in the navigation bar of the PDA.

How do these systems support multimodality and plasticity? To answer these questions, we need first to define these notions more carefully.

4.3 Modality and Multimodality

The scientific community has now debated definitions and uses of multimodality for more than twenty years without reaching clear consensus. The concepts of "modality" and "multimodality" mean different things to different stakeholders. In cognitive psychology, a *modality* denotes a human sense (e. g., vision, audition, taste, etc.) whereas in Human-Computer Interaction (HCI), multimodality corresponds more or less to interaction techniques that involve multiple human senses simultaneously. Much depends on the perspective, e. g., from a user or a system point of view, or on the degree of precision needed to solve or discuss a particular problem. In this section, we present our choices using a system perspective.

4.3.1 Definitions

Reusing Nigay and Coutaz's definition [78], we define a modality M as the coupling of an interaction language L with a physical device d:

$$M ::= < L, d > \qquad (4.1)$$

An *interaction language* defines a set of well-formed expressions (i. e. a conventional assembly of symbols) that convey meaning. The generation of a symbol, or a set of symbols, results from actions on physical devices. A *physical device* is an element of the host platform of the interactive system that plays the role of an interaction resource: it acquires (input device) or delivers (output device) information. For instance:

- A speech input modality is described as the couple *<pseudo natural language NL, microphone>*, where *NL* is defined by a specific grammar.
- A written (typed) natural language corresponds to *<pseudo natural language NL, keyboard>*.
- A speech output modality is modelled as *<pseudo natural language NL, loudspeaker>*.
- A graphics output modality as *<graphics language, screen>*.
- A direct manipulation modality as the couple *<manipulation language, mouse>*.

A more detailed expression of the interaction modality as an interaction language coupled to a device is expressed in terms of a UML class diagram in [79]. From a system perspective, the distinction between "interaction language" and "device" is relevant for two reasons:

1. It is consistent with the logical and physical separation of concerns advocated by the Arch software architecture model for structuring the UI rendering [80] as explained in Chapter 9. The *Logical Presentation* (LP) level is interaction language-dependent but device-independent, whereas the *Physical Presentation* (PP) level is both language and device dependent. Thus, coupling an interaction language

with a different device would result in the modification of the PP level. Modifying the interaction language would have an impact on both the LP and PP levels.
2. It is consistent with human experience. From the system perspective, coupling an interaction language with a different device would result in a new modality. This is true as well for the human. For example, manipulating a graphical object with a mouse, or performing the same task with a brick [81] would provide users with a very distinctive experience.

In addition, a modality can play the role of an input device to produce sentences of an interaction language. For example, the words of a sentence of a pseudo-natural language may be produced by selecting the appropriate words from pull-down menus using a mouse. In this case, the keyboard device is replaced by the modality *<select within pull-down menus, mouse>*. The general definition of a modality is thus expressed in the following way:

$$M::= \text{<interaction language, d>} \mid \text{<interaction language, M>} \quad (4.2)$$

Multimodality can be loosely defined by the many ways of combining modalities:

- An *input multimodal system* is a system that combines two or more user input modes in a coordinated manner with multimedia system output [62, 82–85]. For instance, MultimodaliXML [74, 86] enables the user to rely on graphical interaction, vocal interaction, and/or tactile interaction to provide information to an information system. The multimodal UI is specified in UsiXML [87], a XML-compliant language for describing multimodal UIs based on semantics of multimodal UIs [88].
- An *output multimodal system* is a system that presents information in an "intelligent" way by combining two or more communication modalities [76, 77]. This is the case for our neuro-surgery system where graphics and audio outputs are used.

In the following section, we analyze the combination of modalities in a more precise way using the CARE properties. Other taxonomic spaces such as TYCOON [89] could equally be used. The motivation for using the CARE properties is that they have served to structure the development of software tools for multimodal interfaces (See Chapter 9).

4.3.2 The CARE Properties

- The CARE properties (**C**omplementarity, **A**ssignment, **R**edundancy, and **E**quivalence) offer a formal way of characterizing and assessing aspects of multimodal UIs [90]. Before providing the formal definition of the CARE properties, we need to introduce the following notions:
- A *goal* is a state that an agent (i.e. a user or the system) intends to reach by producing an *expression*. An *input expression* is produced by a user who needs

to reach a particular goal. It is then processed by the system, which in turn produces an *output expression*.
- A *temporal window* is a time interval used as a temporal constraint on the use of modalities for producing input and output expressions. A *temporal relationship* characterizes the use over time of a set of modalities. The use of modalities may occur simultaneously or in sequence within a temporal window.
- The function *Reach(s, m, s')* models the expressive power of a modality *m*, that is, its capacity to allow an agent to reach state *s'* from state *s* in one step. A sequence of successive steps (or states) is called an *interaction trajectory*.
- Modalities of a set *M* are *used simultaneously* (or in parallel) if, within a temporal window, they happen to be active at the same time. Let *Active (m,t)* be a predicate to express that modality *m* is being used at some instant *t*. The simultaneous use of modalities of a set *M* over a finite temporal window *tw* can be formally defined as:

$$\text{Parallel } (M, tw) \Leftrightarrow (\text{Card } (M) > 1) \land (\text{Duration } (tw) \neq \infty) \land \\ (\exists\, t \in tw, \forall m \in M, \text{Active } (m, t)) \quad (4.3)$$

where *Card (M)* is the number of modalities in set *M*, and *Duration (tw)* is the duration of the time interval *tw*.

- Modalities *M* are *used sequentially* within temporal window *tw* if there is at most one modality active at a time, and if all of the modalities in the set are used within *tw*:

$$\text{Sequential } (M, tw) \Leftrightarrow (\text{Card } (M) > 1) \land (\text{Duration } (tw) \neq \infty) \\ \land (\forall t \in tw, (\forall m, m' \in M, \text{Active } (m, t) \Rightarrow \neg \text{Active } (m', t)) \quad (4.4) \\ \land (\forall m \in M, \exists\, t \in tw, \text{Active } (m, t)$$

Temporal windows for parallelism and sequentiality need not have identical durations. The important point is that they both express a constraint on the pace of the interaction. The absence of temporal constraints is treated by considering the duration of the temporal window as infinite. We are now able to define the meaning for equivalent modalities, for modality assignment, as well as for redundancy and complementary between multiple modalities.

Modality equivalence: modalities of set *M* are equivalent for reaching state *s'* from state *s*, if it is necessary and sufficient to use any one of the modalities of *M*. *M* is assumed to contain at least two modalities. More formally:

$$\text{Equivalence } (s, M, s') \Leftrightarrow (\text{Card } (M) > 1) \land (\forall m \in M, \text{Reach } (s, m, s')) \quad (4.5)$$

Equivalence expresses the availability of choice between multiple modalities but does not impose any form of temporal constraint on them.

Modality assignment: modality *m* is assigned in state *s* to reach *s'*, if no other modality can be used to reach *s'* from *s*. In contrast to equivalence, assignment expresses the absence of choice: either there is no choice at all to get from one state to another, or there is a choice but the agent always opts for the same modality to get between these two states. Thus, there exist two types of assignment:

4.3 Modality and Multimodality

$$\text{StrictAssignment}(s, m, s') \Leftrightarrow \text{Reach}(s, m, s') \wedge (\forall m' \in M, \text{Reach}(s, m', s') \Rightarrow m' = m)$$
$$\text{AgentAssignment}(s, m, M, s') \Leftrightarrow (\text{Card}(M) > 1) \wedge (\forall m' \in M, (\text{Reach}(s, m', s')$$
$$\wedge (\text{Pick}(s, m', s')) \Rightarrow m' = m)) \qquad (4.6)$$

where *Pick(s, m, s')* is a predicate that expresses the selection of *m* among a set of modalities to reach *s'* from *s*.

Modality redundancy: modalities of a set *M* are used redundantly to reach state *s'* from state *s*, if they have the same expressive power (they are equivalent) and if all of them are used within the same temporal window *tw*. In other words, the agent shows repetitive behaviour:

$$\text{Redundancy}(s, M, s', tw) \Leftrightarrow \text{Equivalence}(s, M, s') \wedge$$
$$(\text{Sequential}(M, tw) \wedge \text{Parallel}(M, tw)) \qquad (4.7)$$

Redundancy can comprise two distinct temporal relationships – sequentiality or parallelism – which may have different implications for usability and software implementation. In particular, parallelism puts restrictions on the types of modalities that can be used simultaneously: modalities that compete for the same system or human resources cannot be activated in parallel. The agent can then act sequentially only provided that it can comply with the temporal constraints (i.e., it must act quickly in order for its inputs to be treated as if they were parallel). When parallelism is possible, we have *Concurrent-Redundancy* and *Exclusive-Redundancy* for sequential behaviour:

$$\text{Concurrent-Redundancy}(s, M, s', tw) \Leftrightarrow \text{Equivalence}(s, M, s') \wedge \text{Parallel}(M, tw)$$
$$\text{Exclusive-Redundancy}(s, M, s', tw \Leftrightarrow \text{Equivalence}(s, M, s') \wedge \text{Sequential}(M, tw)$$
$$(4.8)$$

Modality complementarity: Modalities of a set *M* are used in a complementary way to reach state *s'* from state *s* within the same temporal window *tw*, if all of them are used to reach *s'* from *s*, i.e., none of them taken individually can cover the target state. Deictic expressions such as "Put that there" [63] are characterised by cross-modality references, are examples of complementarity.) To express this adequately, we need to extend the notion of reachability to cover sets of modalities: *Reach(s, M, s')* means that state *s'* can be reached from state *s* using the modalities in set *M*.

$$\text{Complementarity}(s, M, s', tw) \Leftrightarrow (\text{Card}(M) > 1) \wedge (\text{Duration}(tw) \neq \infty)$$
$$\wedge ((\forall M' \in PM, (M' \neq M \Rightarrow \neg \text{Reach}(s, M', s'))) \qquad (4.9)$$
$$\wedge \text{Reach}(s, M, s') \wedge (\text{Sequential}(M, tw) \vee \text{Parallel}(M, tw))$$

where *PM* denotes the parts of set *M*.

As for redundancy, complementarity may occur in parallel or sequentially within a temporal window corresponding to *Synergistic-Complementarity* and *Alternate-Complementarity* respectively. For example, with Alternate-Complementarity, a "put that there" expression would be understood by the system provided that the user first utters "put that", then show the object, then utter "there" followed by a gesture pointing at the new location, and provided that all these actions are performed within the same temporal window!

$$\text{Synergistic-Complementarity}(s, M, s', tw) \Leftrightarrow (\text{Card}(M) > 1) \wedge (\text{Duration}(tw) \neq \infty)$$
$$\wedge \; (\forall M' \in PM, (M' \neq M \Rightarrow \neg\text{Reach}(s, M', s'))) \wedge \text{Reach}(s, M, s') \wedge \text{Parallel}(M, tw)$$
(4.10)

$$\text{Alternate-Complementarity}(s, M, s', tw) \Leftrightarrow (\text{Card}(M) > 1) \wedge (\text{Duration}(tw) \neq \infty)$$
$$\wedge (\forall M' \in PM, (M' \neq M \Rightarrow \neg\text{Reach}(s, M', s'))) \wedge \text{Reach}(s, M, s') \wedge \text{Sequential}(M, tw)$$
(4.11)

Whether it is synergistic or alternate, input complementarity requires the system to perform *data fusion* whereas output complementarity requires *data fission* (see Chapter 9 for more details). Having introduced a formal definition for the notions of modality and multimodal UI, we now go once step further with the notion of plastic multimodal UI.

4.4 The Problem Space of Plastic Multimodal UIs

Figure 4.5 synthesizes the problem space for plastic multimodal user inter-faces. As shown in the figure, the problem space is characterised (but is not limited to) the following dimensions:

- The means used for adaptation (i. e., re-molding and re-distribution).
- The UI component granularity representing the smallest UI units that can be adapted by the way of these means (from the whole UI considered as a single piece of code to the finest grain: the interactor).

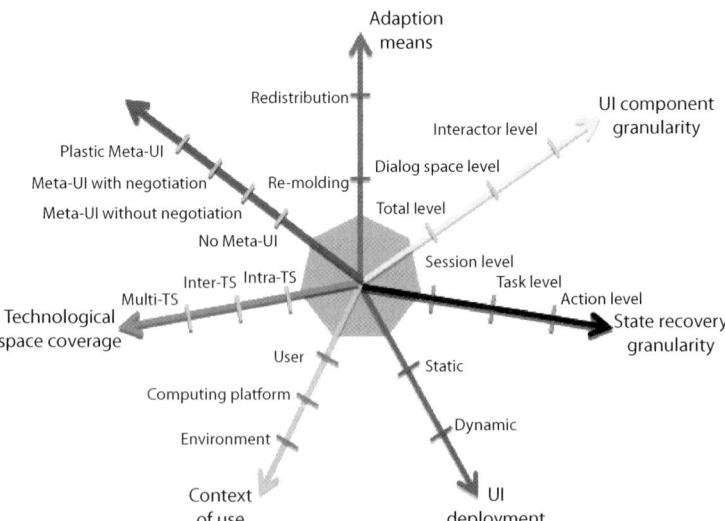

Fig. 4.5 The problem space of plastic multimodal user interfaces

4.4 The Problem Space of Plastic Multimodal UIs

- The state recovery granularity after adaptation has occurred (from the session level to the user's last action).
- The UI deployment (static or dynamic) as a way to characterize how much adaptation has been pre-defined at design-time vs. computed at runtime.
- The context coverage to denote the causes for adaptation with which the system is able to cope.
- The coverage of the technological spaces as a way to characterize the degree of technical heterogeneity supported by the system.
- The existence of a meta-UI to allow users to control and evaluate the adaptation process.

These dimensions are now developed in details in the next subsections.

4.4.1 Two Adaptation Means: UI Re-molding and UI Re-distribution

UI re-molding denotes any UI reconfiguration that is perceivable to the user and that results from the application of transformations on the UI. It may result in a *graceful degradation* or in a *graceful upgradation*. Transformations include:

- *Insertion* of new UI components to provide access to new services relevant in the new context of use. For instance, if more screen space becomes available, more information could be displayed.
- *Deletion* of the UI components that become irrelevant in the new context of use. For instance, removing unnecessary UI elements to accommodate screen constraints of a PDA is a frequent technique.

Reorganization of UI components by revisiting their spatial layout and/or their temporal dependency. Reorganization may result from the insertion and/or deletion of UI components. For instance, substituting a UI by another one is a particular case of reorganization: a full list box may be reduced to a drop-down list box if the user switches from a PC to a PDA. On the other hand, switching from a portrait to a landscape view requires spatial reorganization only. Reorganization may affect different UI portions depending on the UI type. For instance, reorganizing the UI of a web site may affect the style, the layout, the contents, the structure, the navigation as well as user interaction capabilities, all of these in a possible independent manner.

Graceful degradation consists of a UI reorganization following a change of computing platform, from a less constrained platform to a more constrained one, e.g., from a PC to a PDA, from a Screen Phone to a mobile phone. *Graceful upgradation* is the inverse process: it consists of improving the UI when the user switches from a more constrained platform to a less constrained one.

Re-molding may result in using different modalities, or in exploiting multimodality differently. UI adaptation is often assimilated to UI re-molding. This is true as long as we live in a closed world where the interaction resources are limited to that of a single computer at a time. In ubiquitous computing, the platform may be a dynamic cluster composed of multiple interconnected computing devices whose interaction resources, all together, form a habitat for UI components. In this kind of situation, instead of being centralised, the user interface may be distributed across the interaction resources of the cluster. Sedan-Bouillon (Fig. 4.3 and Fig. 4.4) illustrates this situation. We thus need to consider an additional means to UI adaptation: UI re-distribution.

UI re-distribution denotes the re-allocation of the UI components of the system to different interaction resources. For example, the Sedan-Bouillon Web site shown in Fig. 4.3 whose UI is centralised on a single PC screen, is re-distributed in Fig. 4.4 across the interaction resources of the PC and the PDA. Note that, as a consequence of re-distribution, all or parts of the UI may require re-molding. In the Sedan-Bouillon example, the PC version of the navigation bar has been re-molded into a new navigation bar resulting from its re-allocation to the PDA.

UI re-distribution as well as UI re-molding may be applied at different levels of UI components granularity.

4.4.2 UI Components Granularity

UI components granularity denotes the smallest software UI unit that can be affected by re-molding and/or re-distribution:

- Re-molding and/or re-distribution may be total (the whole UI is affected) or partial.
- When partial, re-molding and/or re-distribution may work at the workspace level, or at the interactor level.

A workspace is a logical space that supports the execution of a set of logically connected tasks. *In fine*, workspaces are populated with interactors (e. g., graphics widgets, speech widgets). For example, a UI expressed in SVG supports total re-molding only: it is not possible to resize subparts of the UI to maintain text legibility as the display area is shrunk. In Sedan-Bouillon, re-molding, as well as re-distribution, is performed at the workspace level: the title, content, and the navigation bar are the smallest units for re-molding and re-distribution. Whatever the means used for adaptation (re-molding and/or re-distribution) and the UI components granularity, another important dimension is the granularity of the state recovery.

4.4.3 State Recovery Granularity

State recovery granularity characterizes the effort users must apply to carry on their activity after adaptation has occurred:

- When the system state is saved at the *session level*, users have to restart their activity from the system initial state, that is, the state that the system enters when it is launched.
- At the *task level*, the user can pursue the job from the beginning of the current interrupted task (provided that the task is attainable in the retargeted UI).
- At the *physical action level*, the user is able to carry on the current task at the exact point within the current task (provided that the task is attainable in the new version of the user interface).

Sedan-Bouillon supports a task level state recovery: if re-distribution occurs while filling a form, users have to refill the form.

4.4.4 UI Deployment

UI deployment denotes the installation of the UI components in its habitat. The installation may be static or dynamic.

A *static deployment* means that UI adaptation is performed at the time the system is launched and from then, no re-molding and no re-distribution can be performed.

At the opposite, a *dynamic deployment* means that re-molding and re-distribution can be performed on the fly.

Note that a dynamic deployment that supports a session level state recovery is different from a static deployment. In the first case, re-molding and/or re-distribution can occur on the fly while bringing the user back to the entry state. On the other hand, static deployment forces the user to quit the session and to launch the appropriate version of the system.

Typical examples of static deployments are web sites that recognize the use of a PDA through the "user agent" field of the http header, and then select the appropriate version of the web site. This version has been pre designed and pre-computed for that particular platform. Rekimoto's pre-distributed pick and drop is an example of a static UI deployment [92]. Sedan-Bouillon, on the other hand, support dynamic UI deployment.

4.4.5 Coverage of the Context of Use

We have defined the context of use as the triple: <user, physical and social environment, computing platform>. Each one of these dimensions is a rich information space. For example, in Sedan-Bouillon, a platform is a cluster of connected computing devices. In turn, this cluster may be static or dynamic (computing devices may appear and disappear on the fly), and the cardinality of this cluster is an important factor to consider in relation to scalability and usability. Computing devices are also characterised by the interaction resources they support. The platform dimension of the context of use is probably the most frequently addressed portion of the context to accommodate the UI with. We will discuss coverage of context of use in more detail in Section 5 with the notion of domain of plasticity.

4.4.6 Coverage of Technological Spaces

"A technological space is a working context with a set of associated concepts, body of knowledge, tools, required skills, and possibilities." [93]. Examples of technological spaces include documentware concerned with digital documents expressed in XML, dataware related to data base systems, ontologyware, and so on. Most UIs are implemented within a single Technological Space (TS), such as Tcl/Tk, Swing, or HTML. This homogeneity does not hold any more for plastic multimodal UIs since re-distribution to different computing devices may require crossing technological spaces. For example, a Java-based UI must be transformed into WML 2.0 when migrating from a PDA to a WAP-enabled mobile phone.

TS coverage denotes the capacity of the underlying infrastructure to support UI plasticity across technological spaces:

- *Intra-TS* correspond to UIs that are implemented and adapted within single TS. Sedan-Bouillon (HTML, PHP) is an example of intra-TS coverage.
- *Inter-TS* corresponds to the situation where the source UI, which is expressed in a single TS, is transformed into a single distinct target TS.
- *Multi-TS* is the flexible situation where the source and/or the target user interfaces are composed of components expressed in distinct technological spaces.

4.4.7 Existence of a Meta-UI

A *meta-UI* is to ambient computing what the desktop metaphor is to conventional workstations [71]. The concept of meta-UI is defined here as a simplification of an end-user development environment. A full fledge meta-UI should allow end-users to program (configure, and control) their interactive spaces, to debug (evaluate) them, and to maintain and re-use programs. It binds together the activities that can be performed within an interactive space. In particular, it provides users with the means to configure, control, and evaluate the adaptation process. It may, or may not, negotiate the alternatives for adaptation with the user. It may or may not be plastic.

A *meta-UI without negotiation* makes observable the state of the adaptation process, but does not allow the user to intervene. The system is autonomous.

A *meta-UI incorporates negotiation* when, for example, it cannot make sound decisions between multiple forms of adaptation, or when the user must fully control the outcome of the process. This is the case for Sedan-Bouillon where the user can decide on the allocation of the UI components.

The balance between system autonomy and too many negotiation steps is an open question. Another issue is the *plasticity of the meta-UI* itself since it lives within an evolving habitat. Thus, the recursive dimension of the meta-UI calls for the definition of a *native bootstrap meta-UI* capable of instantiating the appropriate meta-UI as the system is launched. This is yet another research issue.

So far, we have defined the primary dimensions of the problem space for UI plasticity. In the next two sections, we refine re-molding with regard to the use of modalities and to its impact on the levels of abstractions of an interactive system.

4.4.8 UI Re-molding and Modalities

Re-molding a UI from a source to a target UI may imply changes in the set of the supported modalities and/or changes in the way the CARE properties are supported. For example, because of the lack of computing power, the synergistic-complementarity of the source multimodal UI may be transformed into an alternate-complementarity, or complementarity itself may disappear.

UI re-molding is *intra-modal* when the source UI components that need to be changed are retargeted within the same modality. Note that if the source user interface is multimodal, then, the target UI is multimodal as well: intra-modal remolding does not provoke any loss in the modalities set.

Re-molding is *inter-modal* when the source UI components that need to be changed are retargeted into a different modality. Inter-modal retargeting may engender a modality loss or a modality gain. Thus, a source multimodal UI may be retargeted into a mono-modal UI and conversely, a mono-modal UI may be transformed into a multimodal UI. The reconfiguration of the Sedan-Bouillon web site between Fig. 4.3 and Fig. 4.4 is an example of an inter-modal remolding: the navigation bar uses the same interaction language but the device has changed from a PC to a PDA. In addition, if the user decides to replicate the navigation bar on the PC and on the PDA, we then have redundancy (i. e. two modalities to navigate through the web site). The neuro-surgery system supports inter-modal re-molding (i. e. from <text, screen> to <text, loudspeaker>) when the surgeon leans over the patient's head (see Fig. 4.2c).

Re-molding is *multi-modal* when it uses a combination of intra- and inter-modal transformations. For example, TERESA supports multi-modal re-molding between graphics and vocal modalities [72]. As for inter-modal re-molding, multi-modal re-molding may result in a modality loss or a modality gain.

Given that a modality is defined as $M::=$ <L, d> | <L, M>, a modality change may result:

- From a device substitution (provided that the new device supports the original interaction language),
- From an interaction language substitution (provided that the original device supports the new interaction language),
- Or from a radical change with both a device and interaction language substitution.

4.4.9 UI Re-molding and Levels of Abstraction

UI re-molding may range from cosmetic changes to deep software reorganization. We suggest using the levels of abstraction of the Arch architectural model as a classification scheme (See Chapter 9 for more details).

At the *Physical Presentation (PP)* level, physical interactors (widgets) used for representing functions and concepts are kept unchanged but their rendering and behaviour may change. For example, if a concept is rendered as a button class, this concept is still represented as a button in the target UI. However, the look and feel of the button or its location in the workspace may vary.

At the *Logical Presentation (LP)* level, adaptation consists in changing the representation of functions and concepts. Changes at the LP level imply changes at the PP level.

At the *Dialogue Component (DC)* level, the tasks that can be executed with the system are kept unchanged but their organization is modified. As a result, the structure of the dialogue structure is changed. AVANTI's polymorphic tasks [94] are an example of a DC level adaptation. Changes at the DC level imply changes at the LP and PP levels.

At the *Functional Core Adaptor (FCA)* level, the entities as well as the functions exported by (or adapted from) the functional core (the services) are changed.

The *Functional Core (FC)* level corresponds to the dynamic composition of services such as Web services. This is a very difficult problem that we are not in a position to solve in the short term.

4.4.10 Summary

The problem space expresses a set of requirements for the design and implementation of plastic multimodal user interfaces. To summarize, the most demanding case corresponds to the following situation:

- All aspects of the CARE properties are supported, from synergistic-complementary multimodality to mono-modal user interfaces.
- The set of the supported modalities includes a mix of "conversation-oriented" modalities such as speech and gesture, and of "world model" modalities such as GUI-WIMP and post-WIMP modalities.
- Re-molding and re-distribution are both supported.
- Re-molding and re-distribution both operate at the interactor level while guaranteeing state recovery at the user's action level.
- Re-molding and re-distribution cover all three aspects of the context of use (i. e. user, environment, platform). In particular, the platform can be composed of a dynamic set of heterogeneous resources whose cardinality is possibly "high".
- Re-molding and re-distribution are able to cross over technological spaces, and they include a plastic meta-UI.

- The meta-UI is a complete end-user development environment.
- Re-molding draws upon multiple modalities that may impact all of the levels of abstraction of an interactive system from the Functional Core to the Physical Presentation level.
- And all this can be deployed dynamically.

Figure 4.5 provides a simple way to characterize and compare interactive systems with regard to plasticity and multimodality. The coverage of context of use by a particular system may be refined, if needed, with the notion of domain of plasticity.

4.5 Domain of Plasticity of a User Interface

UI plasticity can be characterised by the sets of contexts it is able to accommodate. Contexts at the boundaries of a set define the *plasticity threshold* of the user interface for this set. The sum of the surfaces covered by each set, or the sum of the cardinality of each set, defines an overall objective quantitative metrics for plasticity. In other word, this sum can be used to compare solutions to plasticity: A user interface U1 is more plastic than a user interface U2 if the cardinality of the set of contexts covered by U1 is greater than that of U2.

We suggest additional metrics to refine the overall measure of plasticity in relation to discontinuity [66]. These include:

- *The size of the largest surface*: large surfaces denote a wide spectrum of adaptation without technical rupture.
- *The number of distinct sets*: a large number of sets reveal multiple sources for technical discontinuities. Are these discontinuities compatible with user's expectation? Typically, GSM does not work everywhere.
- *Surface shapes*: a convex surface denotes a comfortable continuous space (Fig. 4.6a). Conversely, concave curvatures may raise important design issues (Fig. 4.6b). Typically, ring shape surfaces indicate that the interior of the ring is not covered by the user interface. It expresses a technical discontinuity for contexts that are contiguous in the ranking scheme. Is this inconsistency, a problem from the user's perspective? A hole within a surface depicts the case where the user interface is nearly plastic over both sets of contexts, but not quite. Is this "tiny" rupture in context coverage expected by the target users.

There are different ways of characterizing the frontiers of a plasticity domain, depending on the types of variable coming from the context of use. The simplest form to define a plasticity domain, consists of identifying variables from the context of use and, for each variable, define values or intervals that correspond to particular interests. In 1D, respectively 2D, 3D, nD (where n >3), each plasticity domain represents an interval, respectively, a rectangle, a parallelepiped, or an hypercube. Plasticity domains are delineated by domain frontiers whose intersections are located at plasticity inflexion points.

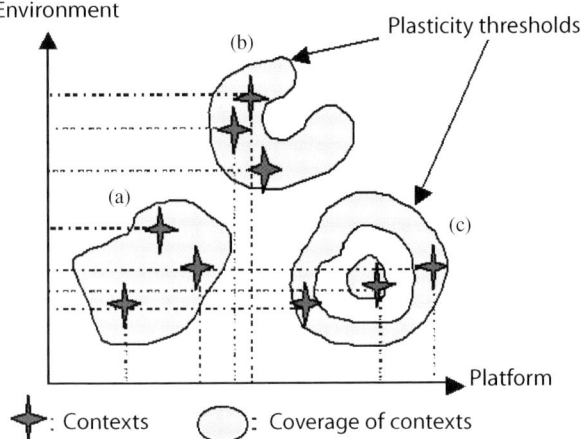

Fig. 4.6 Measuring plasticity from the system's perspective. Greyed areas represent the sets of contexts that a particular technical solution covers. Environments and platforms are ranked against the level of constraints they impose on the user interface

Figure 4.6 illustrates a 2D space where the platform and the environment are of interest. In Fig. 4.7a, the plasticity domain is decomposed into intervals according to one dimension. If three values, say low, medium and high, represent significant value changes, then this axis is decomposed into four plasticity domains such as C1 = {x < low}, C2 = {low <= x < medium}, C3 = {medium <= x < high}, and C4 = {x > high}.

In Fig. 4.7b, the *plasticity space* is characterised along two dimensions (i.e. two physical properties taken from the context of use). If, in turn, each property has three values that are representative of a significant change, then the plasticity space

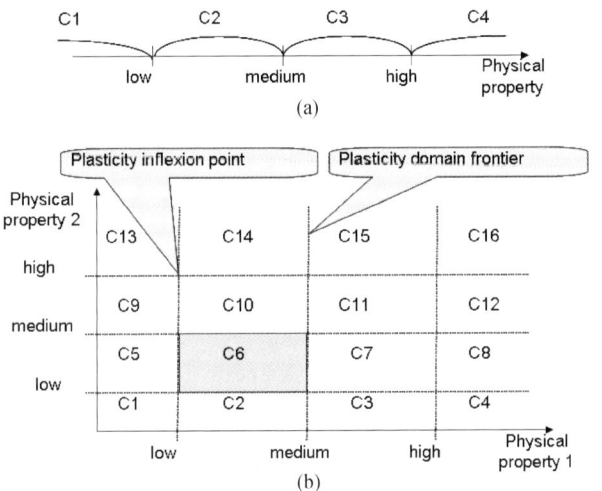

Fig. 4.7 Plasticity domains in 1D (a), 2D (b)

4.5 Domain of Plasticity of a User Interface

is partitioned into sixteen plasticity domains. When the value of such a property changes to cross the frontier between two plasticity domains, then it goes through a *plasticity inflexion point.*

For example, in the neuro-surgery system, two physical properties from the context of use have been selected to denote changes in the context of use:

- The distance between the patients' head and the surgeon's head,
- The angle of the surgeon's body with respect to the vertical axis.

According to Fig. 4.7b, there should be sixteen different plasticity domains available. Since not all of them are significant, only three of them are finally kept as represented in Fig. 4.9. Fig. 4.10 shows which modality is used depending on which plasticity domain is active:

1. When the surgeon leans over the patient's head at a maximum of 5° with respect to the vertical axis and is located more than 60 cm from the patient's head, Fig. 4.9a is activated.
2. When the surgeon is leaning between 5° and 10° with respect to vertical axis and between 30 and 60 cm from the patient's head, Fig. 4.9b is activated.
3. When the surgeon leans more than 20° with respect to vertical axis and less than 30 cm from the patient's head, Fig. 4.9c is activated.

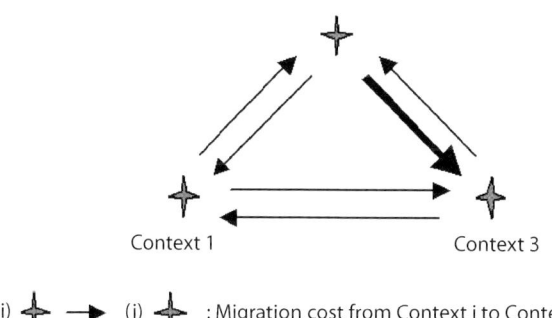

Fig. 4.8 Measuring plasticity from the human perspective. An arrow expresses the capacity of migrating between two contexts. Thickness denotes human cost

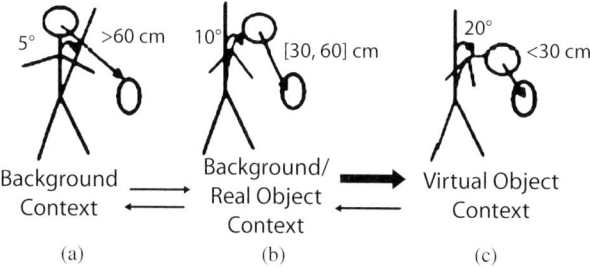

Fig. 4.9 Inter-context continuity in the neuro-surgery system

Fig. 4.10 Domains of plasticity of the neurosurgery system

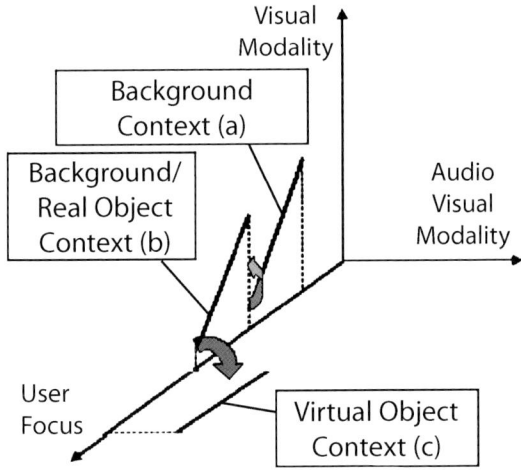

- For each view, different modalities are used, thus implying that a particular combination of interaction modalities is used for each domain of plasticity. For instance, a switch from the <text, screen> modality to the <text, loudspeaker> modality for rendering annotations.

Intuitively, from a technical point of view, a large unique convex surface characterizes a "good" plastic user interface whereas a large number of small concave surfaces denote a large number of technical discontinuities. Although size, shape, cardinality, and topology of surfaces, are useful indicators for reasoning about the plasticity of a particular technical solution, we need to consider a complementary perspective: that of users. To this end, we suggest two indicators: context frequency and migration cost between contexts.

Context frequency expresses how often users will perform their tasks in a given context. Clearly, if the largest surfaces correspond to the less frequent contexts and/or if a multitude of small surfaces is related to frequent contexts, then designers should revise their technical solution space: the solution offers too much potential for interactional ruptures in the interactional process.

Migration cost measures the physical, cognitive and conative efforts [95] users have to pay when migrating between contexts, whether these contexts belong to the same or different surfaces (cf. Fig. 4.8). Although this metrics is difficult to grasp precisely, the notion is important to consider even in a rough way as informal questions. For example, do users need (or expect) to move between contexts that belong to different surfaces? If so, discontinuity in system usage will be perceived. Designers may revise the solution space or, if they stick to their solution for well-motivated reasons, the observability of the technical boundaries should be the focus of special attention in order to alleviate transitions costs.

4.6 Three Principles for the Development of Plastic Multimodal UIs

As *plasticity threshold* characterizes the system capacity of continuous adaptation to multiple contexts, so *migration cost threshold* characterizes the user's tolerance to context switching. The analysis of the relationships between the technical and the human thresholds may provide a useful additional perspective to the evaluation of plastic user interfaces.

To address the challenges raised by the development of plastic multimodal UIs, we propose a set of principles that are presented in the next section.

4.6 Three Principles for the Development of Plastic Multimodal UIs

We have identified three principles that we have applied to the development of plastic multimodal UIs: (1) Blur the distinction between design-time and run-time, (2) Mix close and open adaptiveness and (3) Keep humans in the loop.

4.6.1 Blurring the Distinction between Design-time and Run-time

An interactive system is a graph of models that expresses and maintains multiple perspectives on the system. As opposed to previous work, an interactive system is not limited to a set of linked pieces of code. The models developed at design-time, which convey high-level design decision, are still available at runtime. A UI may include a task model, a concept model, an Abstract UI model (expressed in terms

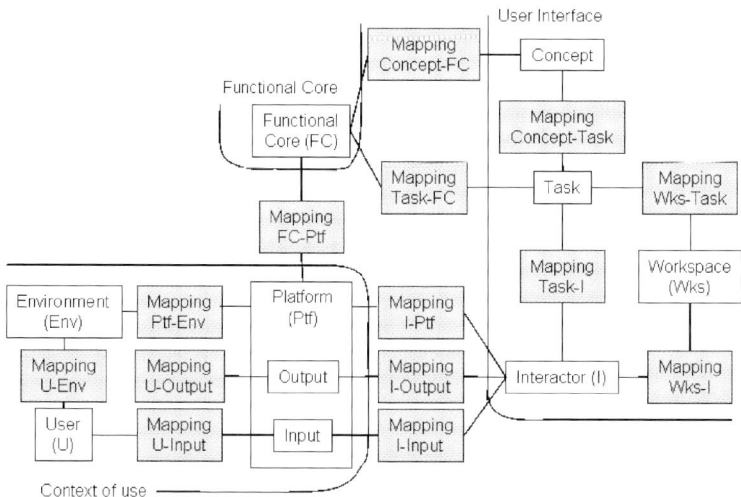

Fig. 4.11 An interactive system is a graph of models related by mappings

of workspaces), and a Concrete UI model (expressed in terms of interactors) all of them linked by mappings. Tasks and Concepts are mapped to entities of the Functional Core of the interactive system, whereas the Concrete UI interactors are mapped to input and output (I/O) devices of the platform. Mappings between interactors and I/O devices support the explicit expression of centralised versus distributed UIs.

Transformations and Mappings are models as well. In the conventional model-driven approach to UI generation, transformation rules are diluted within the tool. Consequently, "the connection between specifications and final results can be quite difficult to control and to understand" [96]. In our approach, transformations are promoted as models expressed in ATL [67]. QVT and XSLT are interesting options as well.

4.6.2 Mixing Close and Open Adaptiveness

A system is close-adaptive when adaptation is self-contained. The system is open-adaptive "if new adaptation plans can be introduced during runtime" [97]. By design, an interactive system has an innate domain of plasticity: it is close-adaptive for the set of contexts of use for which this system can adapt on its own (see Section 6 for a detailed discussion on domain of plasticity). If the system is able to learn adjustments for additional contexts of use, then the domain of plasticity extends dynamically, but this extension relies only on the internal computing capabilities of the system.

In ubiquitous computing, unplanned contexts of use are unavoidable, forcing the system to go beyond its domain of plasticity. If continuity must be guaranteed, then the interactive system must call upon a tier infrastructure that takes over the adaptation process [98]. The role of this infrastructure is to sense, perceive and identify the current context of use (including the interactive system state), to detect changes in the context of use that the interactive system cannot cope with, compute the best possible adaptation by applying new transformations and mappings on the system models and/or by looking for the appropriate UI executable components sitting somewhere in the global computing fabric. The infrastructure then suppresses the "defective" UI components of the interactive system, and inserts the new computed ones.

4.6.3 Keeping Humans in the Loop

On the one hand, the criteria of user explicit control states that end-users should always be in control of the UI, even if the UI is executing tasks for them. This means in theory that no UI operation should escape from being controlled by end-users. This is also valid when UI adaptation is executed. The user should stay in control of the adaptation process as well.

On the other hand, HCI design methods produce a large body of contemplative models (i. e. models that cannot be processed by a machine) such as scenarios, drawings, storyboards, and mock-ups. These models are useful reference material during the design process. On the other hand, because they are contemplative, they can only be transformed manually into productive models. Manual transformation supports creative inspiration, but is prone to wrong interpretation and to loss of key information. On the other hand, experience shows that automatic generation is limited to very conventional user interfaces.

To address this problem, we advocate a mix of *automated*, *semi-automated*, and *manual* transformations. Semi-automated and manual transformations may be performed by designers and/or end-users. For example, given our current level of knowledge, the transformation of a "value-centred model" into a "usability model" can only be performed manually by designers. Semi-automation allows designers (or end-users) to adjust the target models that result from transformations. For example, a designer may decide to map a workspace with UI services developed with the latest post-WIMP toolkit. The only constraint is that the hand-coded executable piece be encapsulated as a service. This service can then be dynamically retrieved by a tier infrastructure and linked to the models of the interactive space by the way of mappings. Thus, the UI services of a particular system at runtime can be a mix of generated and hand-coded highly tuned pieces of UI. By the way of a meta-UI, end-users can dynamically inform the adaptation process of their preferences.

4.7 Conclusion

In an ever-evolving world, multimodality deserves the best attention since it delivers much more capabilities to adapt a user interface to human needs than traditional Graphical User Interfaces (GUIs). Identifying the appropriate interaction modalities and combining them in a way that improves the usability of the whole system remain an open problem for which no general solution exists today. Many individual solutions demonstrate the superiority of bi- or tri-modal user interfaces over mono-modal user interfaces. But these combinations should be carefully operated to take the best benefit from multimodality. Otherwise, there is a risk of over-complexifying a user interface by adding more and more modalities which, at the end, do not improve the whole usability.

Apart from the problem of combining interaction modalities, the problem of finding the appropriate interaction modalities as well as their combination for UI plasticity raises many new questions that have not been explored so far. The use of multiple modalities, especially at run-time, opens new doors for more usable user interfaces than ever. Let us hope that this odyssey in a world of multiple modalities will bring us with clear decisions for the ultimate benefit of the end-user, not only for designers or developers.

4.8 Acknowledgments

We gratefully acknowledge the support of the Request research project under the umbrella of the WIST (Wallonie Information Société Technologies) program under convention n°031/5592 RW REQUEST) and the support of the SIMILAR network of excellence (http://www. similar.cc), the European research task force creating human-machine interfaces similar to human-human communication of the European Sixth Framework Program (FP6-2002-IST1-507609) as well as the E-MODE ITEA-1 project.

Chapter 5
Face and Speech Interaction

Mihai Gurban[1], Veronica Vilaplana[2], Jean-Philippe Thiran[1] and Ferran Marques[2]

[1] Ecole Polytechnique Fédérale de Lausanne (EPFL)
[2] Universitat Politècnica de Catalunya (UPC)

Two of the communication channels conveying more information in human-to-human interaction are face and speech. Whereas the usefulness for speech as communication modality is clear, it is necessary to further define the concept of face and facial expression as communication modalities. The absolute and relative positions as well as the orientation of human faces provide relevant information to interpret people's behaviours and their relationships. Moreover, people's expressions (and, in turn, feelings) can be estimated by analyzing the short-term temporal evolution of their facial features.

Nevertheless, a more robust interpretation of the information being expressed by people can be obtained by the combined analysis of both sources of information: short-term facial feature evolution (face) and speech information. Face and speech combined analysis is the basis of a large number of human computer interfaces and services. Regardless of the final application of such interfaces, there are two aspects that are commonly required: detection of human faces (and, if necessary, of facial features) and combination of both sources of information. In this chapter, we review both aspects, propose new alternatives and illustrate their behaviour in the context of two specific applications: audio-visual speaker localization and audio-visual speech recognition.

5.1 Face and Facial Feature Detection

In this section, we present the state of the art in face and facial feature detection from the perspective of the type of image model being assumed. We start by presenting face detection techniques since most common facial feature detection approaches are applied once faces have been detected. Nevertheless, it should be commented that there exist approaches that address directly a given face and speech interaction problem (e.g.: speaker localization) obtaining the face and facial feature detection as a byproduct of the approach [101].

5.1.1 Face Detection

The problem of face detection was relatively unexplored for its own sake until the mid 1990s. An excellent review which focuses on face recognition but also describes detection approaches developed prior to 1995 is presented in [102]. Research activities on face detection have intensified over the last decade and the problem has been approached with a diversity of strategies. This fact and the huge amount of publications related to the task make difficult to organize the different approaches. In [103] they are classified in four categories: *knowledge-based*, *feature invariant*, *template matching* and *appearance-based* methods, though categories overlap, since several methods fall in more than one category. In [104], methods are organised in two broad classes distinguished by how they use face knowledge: *feature-based* methods, which rely on attributes that are defined using knowledge about how faces look like, and *image-based* methods, that incorporate knowledge implicitly through learning.

However, we understand that the main differences between strategies originate from the different models that are used to represent images and patterns. Therefore, we will present the methods in four categories according to the following four image models:

- *Pixel based*: the image is understood as a set of independent pixels,
- *Block based*: the image is seen as a set of rectangular arrays of pixels. The rectangular areas are defined by scanning the image with a sliding window of fixed size,
- *Compressed domain based*: the image is seen as the set of coefficients in a particular transform domain, such as the DCT,
- *Region based*: the image is represented as a set of homogeneous connected components. The regions are obtained by segmenting the original image following a certain criterion of homogeneity.

The image model defines a set of potential candidates (e.g. pixels, blocks or regions). For instance, in a block-based image model, this set is formed by all the possible rectangular subimages (at all positions, of any size). This set is called the *search space*. Over this search space, the *selection of candidates* performs several tasks with the final goal of defining a set of candidate patterns that will be evaluated by a final *classifier*:

1. Simplification of the search by reducing the initial set of patterns. This step is not always applied: in some face detectors all the patterns proposed by the image model are evaluated.
2. Precise definition of the region of support for the remaining patterns. In some cases, regions of support are defined by masking the patterns with an elliptical mask.
3. Preprocessing or normalization of patterns. Typically, applying histogram equalization, illumination correction, etc.

5.1 Face and Facial Feature Detection

4. Measurement of features on the normalised patterns. Attributes are measured on the image by their associated *descriptors* to obtain a set of features that represents the pattern.

Figure 5.1 presents the analysis of an image using the four different image models and a set of likely candidates obtained for each model. In the sequel, for each of the image models, we will give a general description of the main parts of

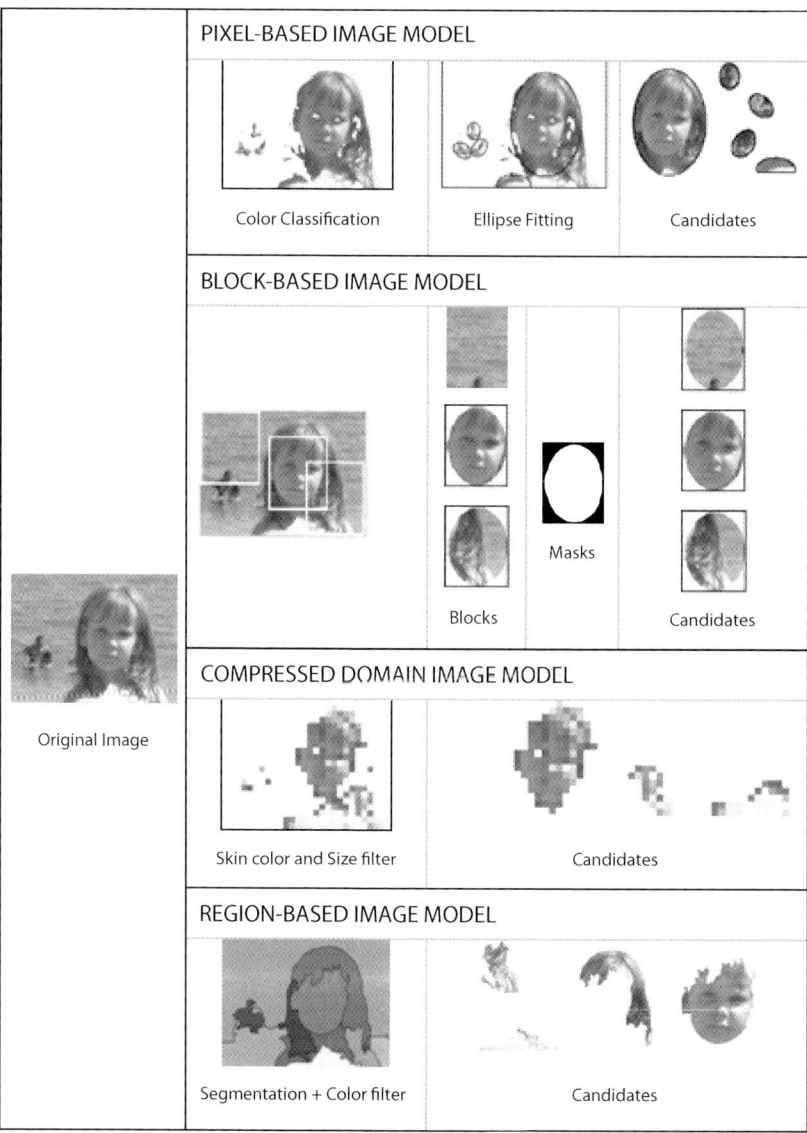

Fig. 5.1 Example of the four image models and selection of candidates

the systems: search strategy, selection of candidates, attributes used in the face models and type of classifiers. Then, details on the most representative techniques will be provided.

Note that this analysis allows comparing the techniques from the point of view of their design. However, the comparison of the different techniques in terms of performance (usually measured in terms of detection and false alarm rates) is not possible for all the methods, due to the lack of standardised tests. The interested reader is referred to each publication for details about training or test data sets and performance details.

5.1.1.1 Pixel-based Image Models

The heaviest part of the systems within this category is the selection of candidates. Explicit attributes of faces are employed to simplify the search by defining a very reduced set of candidate patterns, which are then classified by relatively simple classifiers.

Most of these techniques segment faces in colour images. They use the colour attribute of faces as a first cue and classify each pixel in the image into two classes, skin and no-skin, using a skin colour model estimated with a training data set. An example of this initial step is shown in Fig. 5.2. Then, skin-like pixels are grouped using connected component analysis or clustering algorithms. Next, a simple shape descriptor is commonly used: if the shape of a connected region has an elliptic or oval shape, the pattern becomes a face candidate and is evaluated by a final classifier. Some methods use also a size descriptor to introduce additional constrains. The following are some representative works that have adopted this approach.

Chai and Ngan [105, 106] use colour information for face localization and segmentation in 'head and shoulders' type images. Skin colour values are modelled with two ranges in Cr and Cb components. Image pixels are classified as

Fig. 5.2 Classification of pixels into skin and no-skin classes in pixel-based methods. Original image (a) and skin-coloured pixels (b)

skin colour pixels if both Cr and Cb values fall inside these ranges. The spatial distribution of detected skin-colour pixels is analysed, and the largest clusters are kept, defining the candidate patterns. For the classification, the variance of the luminance values is analysed and candidate regions that are too uniform to be part of a face are discarded.

Garcia and Tziritas [107] also use a generic colour descriptor as first cue. Colour clustering is performed to extract a set of dominant colours and to quantize the image according to this reduced set. Quantised pixels are classified as skin or no-skin coloured. A merging stage is iteratively applied on the set of homogeneous skin colour regions in the colour quantised image, in order to provide a set of candidate areas. Shape (aspect ratio of bounding box) and size descriptors are used to reduce the number of candidates. For the classification, a set of simple statistical data is extracted from a wavelet packet decomposition on each face candidate, and a probabilistic metric derived from the Bhattacharya distance is used to classify the feature vectors.

Terrillon et al. [108, 109] model skin colour with a Gaussian distribution, classify pixels and group them in clusters. The smallest clusters are discarded and for the remaining clusters invariant Orthogonal Fourier-Mellin moments are extracted. These features are used with two classifiers: a multilayer perception neural network [108] and a support vector machine (SVM) with different kernels [109].

Other techniques based on colour pixel classification are [110–114]. Li et al. [115] perform face detection in sequences, and combine colour with motion information to perform a first classification of pixels, assuming that faces are moving objects in a static background. Motion information helps to find a robust definition of patterns of interest.

5.1.1.2 Block-based Image Models

These techniques usually work with greyscale images, and represent patterns as squared windows. The search strategy consists in scanning the input image with a sliding window of fixed size (typically 19x19). In order to detect faces of different sizes, the input image is repeatedly down-scaled, and the analysis is performed, with a window of the same size, at each scale.

Some methods apply an exhaustive search, and analyze patterns over an exhaustive range of locations and scales in the image. In others, the search space is sampled and a scaling factor and steps for horizontal and vertical displacements of the scanning window are defined. This way, the analysis is faster but the localization is less accurate.

The selection of candidates is quite simple. All patterns defined by the image model are final candidates after being preprocessed. A feature vector is extracted from the preprocessed patterns and is fed into the classifier. In turn, to create the face model, a dataset of face images is used. Faces are scaled to the window size, aligned and preprocessed in the same way as the patterns.

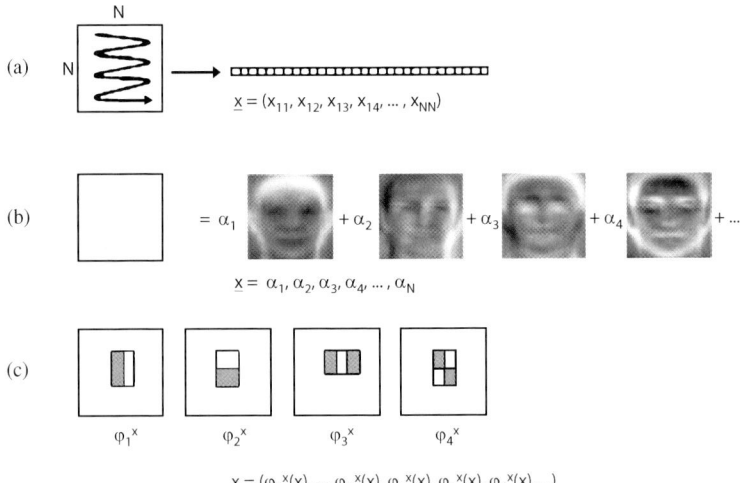

Fig. 5.3 Features for classification in block-based methods. In (a), pixel features obtained by a raw by raw reading of the pattern. In (b), a global approach where the feature vector is formed by the coefficients of the projection of the pattern on a PCA computed from training data. In (c), local features computed within two, three and four rectangular areas; at each position k of the rectangles, the sum of the pixels in the white rectangles are subtracted from the pixels in the dark rectangles

For this group of techniques, the most important part of the system is the classifier, since it has to analyze a very large number of candidates. We organize the techniques by the locality of the features used for the classification (see Fig. 5.3). Initially, pixel-based or global features were used. More recent strategies exploit the idea that each pixel has significative statistical dependence only with a reduced set of pixels and compute a set of local features. Finally, a fourth category is presented, describing techniques that tackle the computational burden inherent in block-based strategies with a coarse to fine cascade of classifiers, independently of the type of features.

Using pixel features: In these methods, all the pixels in the (preprocessed) pattern are used for the classification and the feature vectors are formed with the intensity values obtained by a raster scan of the preprocessed patterns.

Rowley, Baluja and Kanade report in [116] one of the earliest methods to detect upright frontal faces in greyscale images with a neural network based system. They approach the task as a two class classification problem, where the classes are implicitly modelled through learning. The method is extended in [117] to detect faces at any degree of rotation in the image plane. Osuna, Freund, Girosi [118] propose the first approach to face detection using Support Vector Machines (SVM). They detect frontal, un-occluded faces in greyscale images by estimating the decision boundary between the face and no face classes. Furthermore, Feraud et al. present in [119] a method based on a constrained generative model, Lin et al.

[120] propose a system based on probabilistic decision-based neural network, Roth et al. [121] apply an architecture called SNOW (sparse network of winnow).

Using global features: These methods imply a holistic approach and work with global features. They assume that images of human faces (of the same size) lie in a subspace of the overall image space. Principal component analysis (PCA), linear discriminate analysis or factor analysis are used to represent this subspace. The feature vector is extracted from the representation of the pattern in the subspace.

Moghaddam and Pentland propose in [122] a one-class approach, a probabilistic learning method for face localization based on density estimation to model the face class and locate faces in greyscale images. They assume a Gaussian distribution for frontal face patterns in the image space. Then, PCA is applied to find an estimate of the face likelihood of an input pattern using only the projections of the pattern into the subspace spanned by the first eigenvectors. The estimated probability density is used to formulate the face localization problem in a maximum likelihood framework.

Sung and Poggio [123] follow a two class classification approach to detect multiple frontal or slightly rotated faces in greyscale images. The distributions of a face class and a 'near-face' class are modelled using a training set of face and face-like patterns. In both cases they apply a modified k-means algorithm to find six Gaussian clusters. PCA is applied to each cluster to reduce the dimensionality of the representation; only the first 75 eigenvectors are considered. For each candidate, a feature vector of 12 distances between the normalised pattern and the 12 cluster centroid is computed. Finally, the feature vectors are input to a multilayer perception network trained with back propagation that performs the final classification.

A technique based on mixtures of factor analyzers and another one based on Fisher's linear discriminant are presented by Yang, Ahuja and Kriegman [124]. In turn, Popovicy and Thiran propose training an SVM with PCA features in [125]. Other linear subspace methods are presented in [126, 127].

Using local features: Techniques within this group are based on the idea that (aligned) face images have a sparse structuring of statistical dependence. Each pixel has a strong statistical dependency with a small number of neighbour pixels and a negligible dependency with the remaining ones [128].

Papageorgiou and Poggio [129] propose a new representation describing patterns in terms of a large set of local oriented intensity differences between adjacent areas in the image. The features are computed with an over-complete Haar wavelet transform. An SVM classifier is trained with a large set of positive and negative samples.

Schneiderman and Kanade [130] present a system based on a naive Bayes classifier. They choose a functional form of the posterior probability function that captures the joint statistics of local appearance and position. Using a discrete representation of local appearance the overall statistical model is estimated by counting the frequency of occurrence of the patterns over training images.

Colmenarez and Huang [131] also use local statistics in a system that models classes with a family of discrete Markov processes and selects the process that optimizes the information-based discrimination between the two classes.

Coarse to fine processing or Rejection based methods

A drawback of traditional block-based approaches is that the application of complex classifiers on an exhaustive search can be very time consuming. One way to reduce the computational cost is to build the classifier as a cascade of classifiers. At each stage, a classifier makes a decision to either reject the pattern or to continue the evaluation using the next classifier. Usually the complexity of the classifiers increases from the first to the last classifier. This strategy is designed to remove most of the candidates with a minimum amount of computation. Most of the image is quickly rejected as background, and more complex classifiers concentrate on the most difficult face-like patterns.

The most recent state-of-the-art classifiers combine this coarse to fine strategy with the use of local features to reach high performance, both in speed and detection rates. We will briefly describe the most relevant works that follow this approach.

Fleuret and Geman [132] propose a face detector based on a nested family of classifiers. The detection is based on edge configurations: the global face form is decomposed into a set of correlated edge fragments corresponding to the presence of some facial feature. Each classifier is a succession of tests dedicated to certain locations and orientations in the input pattern. When going through the different levels in the hierarchy, the classifiers are more dedicated to particular locations and orientations. The approach is coarse to fine both in terms of pose and complexity of the classifiers.

Viola and Jones [133] combine the idea of a cascade of classifiers of [132] with efficient computation of local features and a learning algorithm based on Adaboost [134], leading to one of the current state-of-the-art methods. They work with an intermediate representation of the image, called 'integral image' that allows a fast computation of a set of local, Haar-like features. The search is done by scanning this integral image and applying the classifiers at every location. A very large number of simple features is computed from the integral image. Each feature is associated to a 'weak' (very simple) classifier that applies a threshold on that feature. Three types of rectangle features are used, that compute differences between the sum of the pixels within rectangles. This produces over 180.000 features for patterns of size 24×24, a many times over-complete set of features.

The approach is based on combining a set of weak classifiers in order to build a stronger one. At each step of the training, and among all possible weak classifiers, one is selected such that its error rate on the training set is minimal. The parameters of this classifier are estimated and the global (strong) classifier is given as a linear combination of the selected (weak) classifiers.

The approach is coarse-to-fine in terms of complexity of the classifiers. Face processing is achieved by combining successively stronger classifiers in a cascade.

5.1 Face and Facial Feature Detection

Each classifier in the cascade is trained on the negative samples of the previous classifiers. This combination allows background zones in the image to be quickly discarded while spending more computational resources on 'face-like' candidates. To detect faces of different sizes, one classifier is trained and used at each scale.

The work is extended in [135] to handle profile views and rotated faces. Different detectors are built for different views and poses, and a decision tree is trained to determine the point of view for each candidate, so that only the detector for the selected view is run, an approach similar to Rowley's in [117].

The technique is very fast due to computation of features through the integral image and the use of the cascade to rapidly reject most of the candidates, and has a high performance thanks to the training with Adaboost. However, it has some weak points: it is very difficult and time-consuming in training, and requires a very large amount of no-face samples, since the classifiers in the cascade are trained on false detection samples from previous classifiers.

H. Sahbi, D. Geman and N. Boujemaa [136] present a face detector system for greyscale images based on a hierarchy of SVM classifiers. For each pattern, a feature vector with 8x8 low frequency coefficients of the Daubechies wavelet transform is computed. A hierarchy of SVMs, structured as a tree, provides a coarse to fine search. The complexity and discrimination of the classifiers increase in proceeding from the root to the leaves. To build each classifier, a simplified SVM decision function is learned using a subset of support vectors. The number of support vectors, which defines the complexity of each classifier, depends on its level in the hierarchy. A candidate is a face if there is a complete chain of positive answers – through a complete path from the root to one leave – and it is a no-face if there is a collection of negative responses for a set of classifiers that covers all poses.

Zhang et al. propose in [137] a modification to Viola and Jones's technique, using global features derived from PCA in the later stages of boosting (each weak classifier is based on a single feature, one of the PCA coefficients) to increase the detection rate of the last classifiers in the cascade. Romdhani et al. [138] merge the SVM approach of Osuna [118] with the idea of a cascade of evaluations of Viola and Jones and the Antifaces of [139]. The goal is to use the cascade to build a system faster than those based only on SVMs. A reduced set of vectors is used to obtain a set of SVM classifiers of increasing complexity. The classifiers are combined in a cascade. They use Gaussian kernels in the SVMs, and bootstrapping for the learning.

Finally, a semi-naive Bayes classifier that decomposes the input pattern into subsets and represent statistical dependency within each subset is described by Schneiderman in [128]. The technique is extended in [140] to work with a Bayesian network. The same author proposes in [141] a cascade method analogous to [133] that uses simple fast feature evaluation in early stages of the cascade and more complex discriminative features at later stages. A 'feature-centric' computational strategy is proposed to re-use feature evaluations across multiple candidate windows.

5.1.1.3 Compressed Domain Image Models

Motivated by the need of fast and efficient search and index algorithms for visual information stored in a compressed form in large databases, some researchers proposed to work directly in the compressed domain. Generic features (mainly for colour and shape attributes) are extracted from a downsampled version of the image, like in pixel-based methods. Then, specific features (texture) are extracted and evaluated at full resolution.

Wang and Chang [142] work with MPEG compressed video, describing a fast algorithm to detect faces in I frames. The technique works at the macroblock level, working directly with the DCT coefficients. Therefore, a minimal decoding of the video is required. The selection of candidates is performed by analyzing colour and shape information. The distribution of skin colour in the CbCr chrominance plane as well as face aspect-ratio constraints are learned from training data. The DC values of Cb and Cr blocks are used to represent the average chrominance of each macroblock. In a first stage, macroblocks are classified as skin or no-skin coloured. Then, morphological operations are performed to eliminate noise and fill holes in the resulting skin coloured (block-based) regions, and shape information

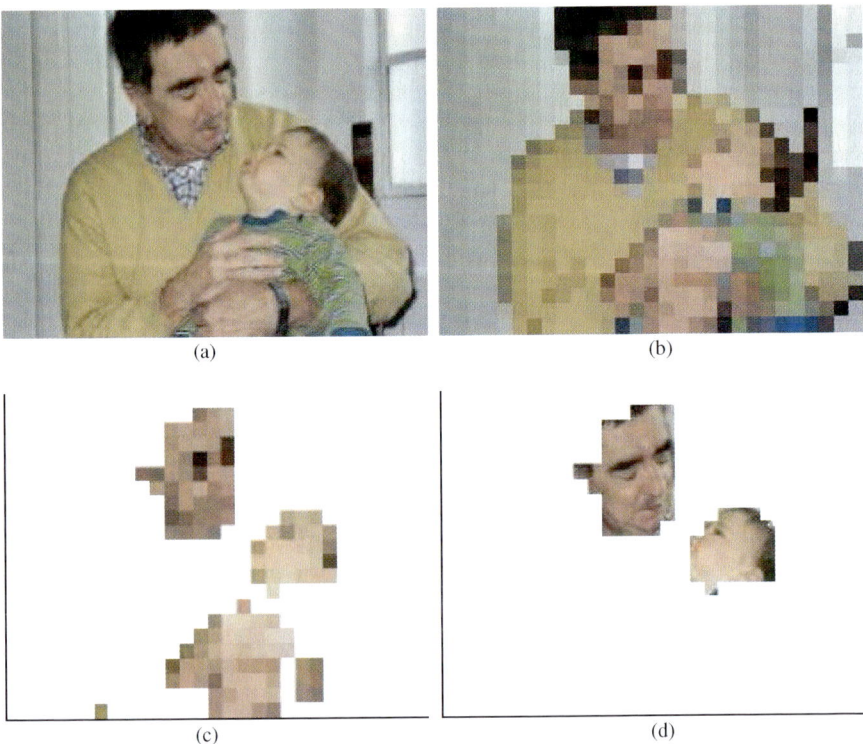

Fig. 5.4 Compressed-based method. Original image (a), macroblocks represented with the average colour (b), skin-coloured macroblocks (c) and final faces (d)

is used to select the final set of candidates (constraints on aspect ratio, minimum and maximum face size). The final classification is performed with two simple classifiers based on the energy distribution of the luminance DCT coefficients over different frequency bands.

The authors report relatively high detection rates, but a large number of false positives (mainly other skin coloured objects from the background). The system is restricted to work with colour images, since candidates are selected using chrominance information. The smallest detectable faces are about 48x48 pixels, since the minimum size for candidates is 3x3 macroblocks.

An important limitation of the algorithm is that, as it works at the macroblock resolution, the positions of detected faces (rectangles at borders of 16x16 macroblocks) are not perfectly aligned with the real faces (see Fig. 5.4(d)).

Lu and Eleftheriadis [143] combine the previous system [142] with the texture classifier of Sung and Pogio [123] mapped from the pixel domain to the DCT domain. They do not scale the image to detect faces of different size, as commonly performed in block-based approaches. To avoid the hard problem of resolution transforms in the DCT domain they train different classifiers for several image scales. Other techniques that work on the compressed domain are [144, 145].

5.1.1.4 Region Based Image Models

Some applications, like 3D face modelling or colour enhancement in the facial region, require the precise segmentation of faces present in the scene. In other applications, like person recognition, segmentation may improve the system performance.

Regions are useful for several reasons. First, they allow a simplification of the search space in terms of position and scale since the number of regions is typically much lower than the number of blocks. One possible approach is to create a hierarchy of regions representing the image at different resolution levels [146]. Second, regions help the selection of candidates by (i) providing a precise definition of the region of support and (ii) enabling a robust feature estimation on image supports that are homogeneous. Finally, by using regions, face detection and segmentation are jointly performed; that is, faces are separated from background and the real face contours are directly obtained.

Yang and Ahuja [147] perform a multiscale segmentation [148] to generate homogeneous regions at multiple scales. Using a Gaussian skin colour model, regions of skin tone are extracted by merging regions from the coarse to the finest scale and grouped into elliptical shapes. A candidate is classified as a face if facial features (eyes and mouth) exist within the elliptic region.

Vilaplana and Marques [149, 150] present a technique that uses a binary partition tree [151] as hierarchical region-based image representation. The approach starts building an initial partition of the image into regions that are homogeneous in colour. Next, from this partition, a binary partition tree is created. The system is built as a combination of one-class classifiers based on visual attributes which are

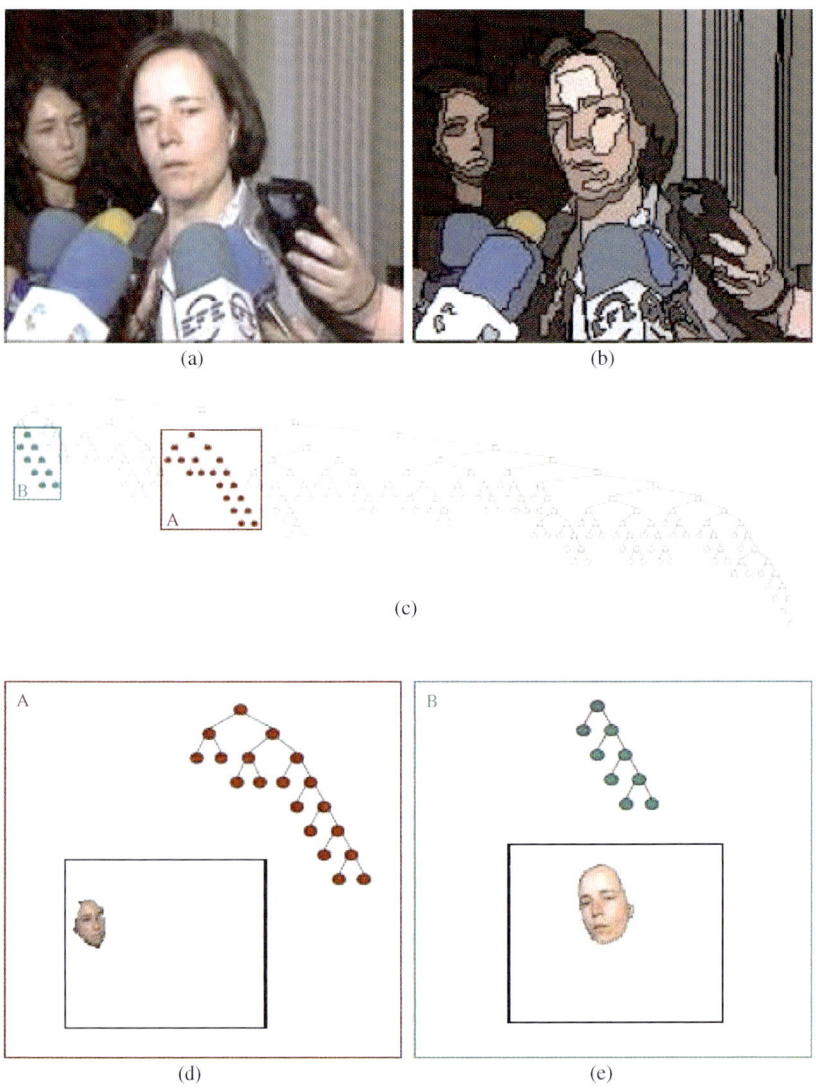

Fig. 5.5 Region-based face detection using binary partition trees. Original image (a), Initial partition (b), Binary partition tree (c) and final faces (d) and (e)

measured on the regions (nodes in the tree) by a set of descriptors. The selection of candidates analyzes the regions proposed by the tree, and performs a first simplification by discarding some of the regions and redefining the area of support of the remaining ones, if necessary. Then, a set of specific descriptors are computed and the features are input to the classifier for the final decision. The system has a very flexible structure since it is a combination of classifiers, with one classifier for each descriptor. Therefore, it is very easy to select, for each application, the most useful set of descriptors and even the best combination for the classifiers.

5.1.2 Facial Feature Detection

Facial feature detection techniques were, in some way, the earliest methods for face detection [104, 152–154]. They start by detecting facial features and then infer the presence of a face according to the configuration of the features. Facial features such as eyebrows, eyes, nose, mouse, hair-line or face contours are commonly extracted using edge detectors. Anthropometric measures or statistical models that describe the relationship between features are used to verify the candidate configurations. The main problem of these techniques is that the features may be corrupted due to noise, illumination and occlusion.

As previously commented, current approaches locate first the face position and, afterwards, perform a facial feature detection within the estimated face area. Facial feature detection algorithms could also be classified with respect to the image model they use. Nevertheless, although there exist techniques using the block-based or the region-based models, the most successful ones are based on the pixel-based model. Note that compressed domain based approaches are seldom used due to inherent accuracy problems.

Given a window containing a face, integral projection functions are commonly used to detect the facial features boundaries. The vertical (horizontal) integral projection of a given image window accumulates the pixel information in a 1D function by integrating row by row (column by column) the 2D data. Therefore, transitions in integral projection functions correspond to boundaries of objects in the window. Feng and Yuen [155] extend the previous concept to the case of variance projection functions in the context of eye detection, to improve the sensitiveness to the variation in the image. Zhou and Geng [156] define the generalised projection function by linearly combining the integral and variance projection functions. Comparative results of these techniques are presented in Fig. 5.6.

Cootes et al. [157] introduce the concept of active shape models (ASM) in the biomedical context. ASM are statistical models of the shapes of objects which iteratively deform to fit to an example of the object in a new image. Typically, model points are assumed to lie on strong image edges. The iterative process updates the pose and shape parameters to best fit the model instance to the found points.

Cootes et al. [158] generalize ASM introducing the concept of active appearance models (AAM) and apply them to the problem of face analysis [159]. AAM use all the information in the image area covered by the object, rather than only near edges. Therefore, AAM matches a statistical model of both object shape and appearance to the image under study. The matching process implies finding those model parameters which minimize the difference between the image and a synthesised model instance. Stegmann et al. [160] propose a new face model and provide

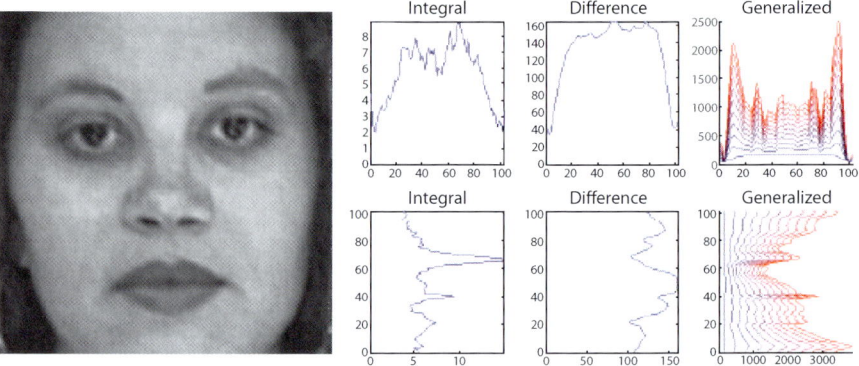

Fig. 5.6 Projection functions. Plots in the first (second) row correspond to the horizontal (vertical) projection. Plots in the first column correspond to the integral projection, in the second column to the difference projection and in the third column to different results of the generalised projection when varying the weights of the linear combination

Fig. 5.7 Example of final convergence of the active appearance model proposed by Stegmann et al. [160]

a complete public domain implementation, namely the Flexible Appearance Modelling Environment (FAME).

Note that both the active shape and active appearance models are local methods and may get trapped in local minima if initialised too far from the actual face position. Nevertheless, due to their excellent performance when initialised close enough to the correct position, AAM are widely used as face tracking approaches.

5.2 Interaction

In this section, we will present two applications of multimodal (joined face and speech) signal processing. The first application is speaker localization; that is, finding the active speaker in a video sequence containing several persons. The second application is audio-visual speech recognition; that is, determining exactly what the person is saying, based not only on the audio, but also on the movement of lips. The second application is more complex, as it also models the temporal aspect of speech. As it will be seen, the proposed approaches for these two applications have different requirements in terms of face and facial feature detection. Whereas in the first case, only a rough estimation of the mouth position is necessary for training, in the second case a more precise detection should be obtained. Finally, in the second part, we will also present a method of integrating modalities in such a way that the system is tolerant to noise or even the complete loss of one modality.

5.2.1 Multimodal Speaker Localization

Our first application is a method to find a speaker's mouth in a video sequence based on the correlation between the audio and the movement of his/her lips. This is an example of a simple multimodal application, showing how useful information can be extracted from the dependency between modalities. No attempt to determine what is being said is made. The temporal aspect of speech is not taken into account; that is, we only use the correlation between the movement of the lips and the sound at each time instant to determine the location of the speaker's mouth.

5.2.1.1 Prior Work on Speaker Localization

Hershey and Movellan [161] use an estimate of the mutual information between the average acoustic energy and the pixel value, whose joint probability density function (pdf) they assume to be Gaussian. Slaney and Covell [162] use Canonical Correlation Analysis to find a linear mapping which maximizes the audio-visual correlation on training data.

Audio-visual synchrony is also analysed by Nock et al. [163, 164]. The mutual information between the audio and the video is computed using two methods: one based on histograms to estimate the pdf, the other based on multivariate Gaussians.

Fisher et al. [165, 166] use a nonparametric statistical approach to learn maximally informative joint subspaces for multimodal signals. Their method uses no prior model and no training data.

Butz and Thiran [167, 168] propose an information theoretic framework for the analysis of multimodal signals. They extract an optimised audio feature as the maximum entropy linear combination of power spectrum coefficients. They show that the image region where the intensity change has the highest mutual information with the audio feature is the speaker's mouth. Besson et al. [169], use the same framework to detect the active speaker among several candidates. The measure that they maximize is the efficiency coefficient, i.e. the ratio between the audio-visual mutual information and the joint entropy. They use optical flow components as visual features, extracting them from candidate regions identified using a face tracker.

The disadvantage of methods that attempt to maximize an information theoretic measure *at test time* is that they need to use some time-consuming optimization procedure, such as gradient descent or a genetic algorithm. This means that, although these methods do not require a training procedure, the amount of computation that is needed during testing is important, making a real-time implementation unfeasible.

By contrast, the proposed multimodal approach does use a training procedure. The joint pdf of the audio energy and a visual feature based on optical flow is learned. This ensures that the number of operations performed while testing is reduced, and thus a real-time implementation would be possible.

Another advantage of the proposed approach is that, in contrast to methods that consider the audio and video of speech to have a Gaussian joint pdf, we can model any kind of probability density. The Gaussian mixture model that we use is a universal approximator of densities, even when using only diagonal covariance matrices, provided that enough Gaussians are considered.

5.2.1.2 The Proposed Speaker Localization Method

Feature Extraction

As we want to model the dependency between the audio and the video signals in the case of speech, we need to extract temporally synchronised features from both streams. The audio feature that we use is the logarithm of the energy (log-energy) of the audio signal. From the video, in the training phase, we only use the rectangular region of the mouth. This region is assumed to be extracted with any of the techniques presented in Section 1. We extract visual features as follows. We compute the optical flow from the luminance component of the images. A single vertical column of points is selected at the center of the mouth region, and only the vertical components of the motion field are retained, as shown in Fig. 5.8. The visual feature being used is the difference between the average optical flow on the top and bottom halves of this column.

What we observe is that the optical flow difference is closely related to the movement of the mouth. When the mouth is opening, the result is a large positive number, while when it is closing, the result is negative. However, when both vectors point in the same direction, they cancel each other out, thus neutralizing small

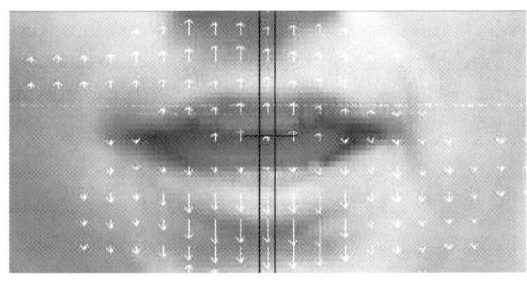

Fig. 5.8 A frame from the training sequences, with the corresponding optical flow

movements of the head. This tolerance to both vertical and horizontal displacement means that the extraction of the mouth region, required for training, does not need to be very accurate.

The Probability Distribution

In order to estimate the joint pdf of features extracted from the training sequences, we need an appropriate model. If $F_v^{train}(t)$ is the visual feature for the training frame t, and $F_a^{train}(t)$ the corresponding audio feature, we want to estimate the probability density function $p(F_a^{train}, F_v^{train})$. Assuming that $p(F_a, F_v)$ is Gaussian is too restrictive. Instead, we use a Gaussian mixture model (GMM), trained with an Expectation-Maximization (EM) procedure [170]. As mentioned before, the GMM can be used to represent any type of pdf, provided that enough components are included. Our trained model consists of four Gaussians with diagonal covariance matrices, which proved to be a good representation for our data without overfitting it.

The distribution of the audio-visual samples taken from the training sequence, as shown in Fig. 5.9, has a high concentration of points around zero audio energy. This is caused by pauses between words. As can be seen, the estimated pdf has a high peak in the same area, while the distribution of the remaining points is poorly modelled.

When searching the correspondence between the sound and the movement of the mouth, the silent samples (low audio energy) do not convey any useful information. Therefore, we removed these samples through thresholding.

In general, image points with low relative movement (low value of the video feature) are characteristic for a static background, even when associated with a high audio energy. As these samples cannot help determine the location of the speaker, we removed them as well.

Figure 5.10 shows the distribution of the remaining samples. Their pdf has an interesting property, that is, high audio energy is more often associated to positive values of the visual feature, while lower audio energy is associated to negative

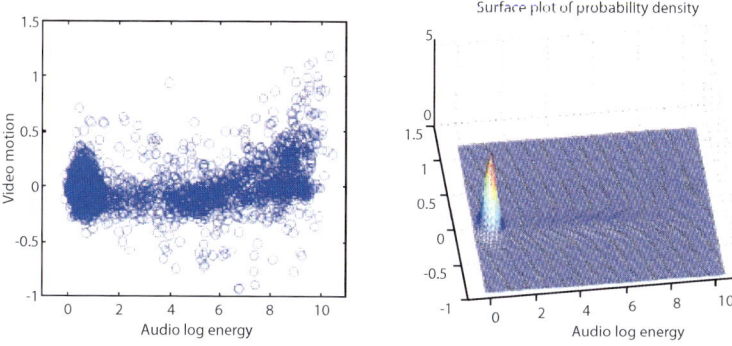

Fig. 5.9 The distribution of audio-visual samples and their estimated pdf

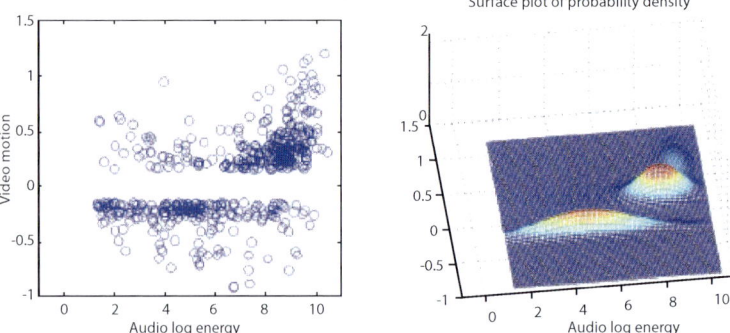

Fig. 5.10 The distribution of audio-visual samples and their estimated pdf, after removing the parts where there is either silence, or very little motion

values. Since our visual feature is the difference of vertical optical flow vectors, a positive value in the training samples represents the action of opening the mouth, while a negative one represents closing it. This confirms the intuition that opening the mouth should lead to louder sounds than closing it.

We can infer from the discrimination, based on the audio, between positive and negative values of the visual feature, that the audio-visual approach can offer more information than only video. This clearly shows the advantages of multimodal analysis. Our method finds the regions in the image where the motion corresponds to the audio value at that time instant; that is, the part of the image that most likely represents a speaking mouth.

Finding the Active Speaker

Our method of speaker localization is based on a maximum likelihood approach. We find the region of a test image where samples have the highest likelihood to have originated from our learned pdf. Our tests show that this region corresponds quite accurately to the active speaker's mouth.

The testing sequences consist of two speakers side by side, taking turns at speaking. They pronounce series of connected digits. Since we do not model the words themselves, it is not a requirement for testing to have the same vocabulary as the training set, but generally the same set of phonemes.

The testing procedure is as follows. From the optical flow in the video, only the vertical components are retained. We compute the value of the visual feature in all points on a grid (with a 10-pixel spacing), using the same method as in training. After selecting columns having the same height as the mouth regions from training, we compute the difference of average vertical optical flow between their top and bottom halves. The reason for using a grid is that the value of the visual feature does not differ much between neighbouring points, and we considered the 10-pixel accuracy as sufficient for speaker localization.

5.2 Interaction

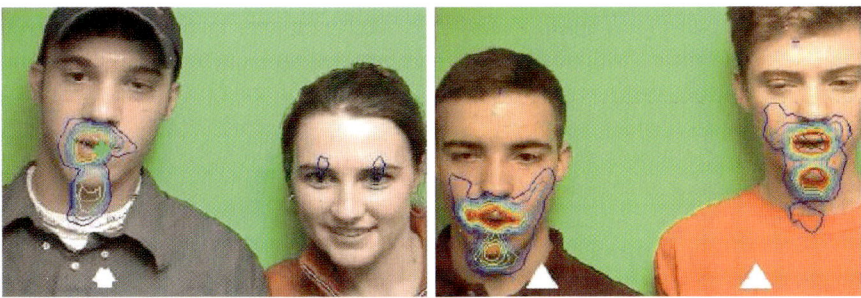

Fig. 5.11 Isocontours of likelihood maps, superimposed on frames from the corresponding temporal windows

For each video frame, the corresponding audio energy, together with the visual feature values on the points of the grid, are used to compute log-likelihoods from the learned joint pdf. If $F_a^{test}(t)$ is the audio feature for the test frame t, and $F_v^{test}(t,x,y)$ is the visual feature value at coordinates (x,y) in the same frame, then the obtained log-likelihood is:

$$l(t,x,y) = \log\left[p\left(F_a^{test}(t), F_v^{test}(t,x,y)\right)\right] \quad (5.1)$$

where p is the pdf obtained from training.

We sum the log-likelihoods resulting from several consecutive frames at each image coordinate on the grid. We use 2 seconds long temporal windows, with a 2/3 overlap. The result of the summation is a 2D map, representing the likelihood that the active speaker's mouth is located at a certain coordinate in the image, during the time interval W. The algorithm outputs the location of the detected active speaker as the (x,y) coordinates of the likelihood maximum:

$$L(x,y) = \sum_{t \in W} l(t,x,y) \quad (5.2)$$

$$(x_{speaker}, y_{speaker}) = \arg\max_{x,y}\left[L(x,y)\right] \quad (5.3)$$

Figure 5.11 shows the isocontours of such likelihood maps $L(x,y)$, superimposed on frames from the corresponding temporal windows. For the first image, the maximum likelihood point is emphasised by a cross, and, in this case, lies on the speaker's mouth, as expected. In the second image, both the left and the right person are simultaneously speaking, and, as can be seen, the two biggest local maxima of the likelihood function are on the speakers' mouths.

5.2.1.3 Speaker Localization Results

For our experiments, we use sequences from the CUAVE audio-visual database [171]. The video sequences are filmed at 30 fps, while the audio is sampled at

44kHz. Although the sampling rate of the audio is higher than the video frame rate, we need synchronised features. To this end, we compute the audio energy on short temporal windows, so as to obtain one audio feature value for every video frame.

The training sequence that we use belongs to the "individuals" part in the database. The speaker utters the English digits from "zero" to "nine" separately, for five times, with pauses between the repetitions. Testing sequences are from the "groups" section of the database. They consist of two speakers taking turns, and finally speaking simultaneously for a short time at the end. We ignored this final part of each sequence in testing.

For quantitative results, we use the frame-level ground truth established by Besson et al. [172]. Each frame is labelled as either *silence*, *left speaker* or *right speaker*.

The choice of detected speaker is based on where the maximum lies in the likelihood map. We compare the detected speaker with the frame-level ground truth on the center part of each 2-seconds long temporal window, obtaining in this way a quantitative result, comparable to others in the literature.

Nock et al. [164] obtain an average performance of 76% using a multimodal method. This result is lower than the 81% reported in the same paper using a visual-only method, so the multimodality did not improve performance in their case. In contrast, our multimodal method does increase performance. Taking only the last 11 sequences into account, we obtain an average accuracy of 93.7%. This confirms that we are able to profit from the extra information present in the audio.

In the next section, another application will be explored, showing how information from two modalities can be integrated with the goal of determining what has been said.

5.2.2 Audio-Visual Speech Recognition

The second multimodal application that we present is audio-visual speech recognition (AVSR). This is a more complex application than speaker localization since the temporal aspect of speech is taken into account. The sequence of speech sounds is modelled using Hidden Markov Models. Another difference is that the contribution of each modality can be quantified, and it can be easily shown that a system employing two modalities outperforms each of the two monomodal systems. Furthermore, the weight of each modality in the final result can be varied, showing directly its importance.

Humans use visual information subconsciously in order to understand speech, especially in noisy conditions, but also when the audio is clean. The movement of the speaker's lips offers hints about the place of articulation, which are automatically integrated by the brain. The McGurk effect [173] proves this by showing that, when presented with inconsistent audio and visual stimuli, the brain perceives a different sound from the one that was spoken.

The same integration can be done by computers to improve the performance of speech recognition systems, when dealing with difficult audio conditions. Audio-visual speech recognition (AVSR) can improve recognition rates beyond what is possible with only audio. An overview of audio-visual speech recognition can be found in [174].

While for audio speech recognition the types of features that are used are more or less established, with mel-frequency spectral coefficients being used in the majority of approaches, the same is not true for visual features. There are two major types of features that are used. The first is based on the image of the moving mouth, or some transform applied on this image. These transforms can be the same that are used for image compression, based on the assumption that a compressed image will also retain the information necessary for visual speech recognition. The second type of visual features is based on the shape of the mouth, its contour or some geometric attribute, like its height or width.

We will present a method to asses the quality of features based on their relevance, that is, their importance for a particular classification task. This method can be used in any multimodal context to find the best features for a certain task from each modality. We will also present our method of weighting modalities in a dynamic way, such that the system can quickly adapt to changing noise conditions, or even the complete interruption of one modality input.

5.2.2.1 Overview of AVSR

In this section we briefly present the structure of an audio-visual speech recognition system. While all such systems share common traits, they can differ in three major respects. The first one is the visual front-end; i.e., the part of the system that tracks the region of the mouth and extracts the visual features. The second one is the audio-visual integration strategy, that is, the way audio and visual information are put together in order to reach a decision about the recognised word. Finally, the type of speech recognition system can differ depending on the particular task (isolated-word recognition, continuous speech or large-vocabulary speech recognition). Our system recognizes sequences of words separated by silence, from a small-vocabulary database.

The majority of speech recognition systems use hidden Markov models [175] (HMMs) as the underlying classifiers used to represent and recognize the spoken words. Our audio-visual system also used HMMs, with two types of modality integration.

Visual Front-end

As commented in the Introduction, all audio-visual speech recognition systems require the identification and tracking of the region of interest (ROI), which can be either only the mouth, or a larger region, like the entire face. This typically begins with locating the face of the speaker, using a face detection algorithm. The second step is locating the mouth of the speaker and extracting the region of interest. This region can be scaled and rotated such that the mouth is centred and aligned.

Visual Feature Types

Once the ROI has been extracted, its useful information needs to be expressed using as few features as possible. This is because the high dimensionality of the ROI impairs its accurate statistical modelling. There are three main types of features that are used for visual speech recognition [174]:

- Appearance based features, which are extracted directly from the pixels of the ROI. They can be simple, like the pixels of the subsampled images, the difference between consecutive frames (delta images), or more complex, such as transforms of the ROI. Typical transforms are principal components analysis (PCA), linear discriminant analysis (LDA), or image transforms like the discrete cosine or wavelet transforms (DCT, DWT) [176]. Motion-based features like the optical flow can also be included in this category.
- Shape based features, which are extracted from the contour of the speaker's lips. They include simple geometric features, such as the lip height, width or area inside the lip contour. More complex features can be computed from a model of the lip's contour, such as the active shape model (see Section 1.2).
- Joint appearance and shape features, which are the result of combining both previous types. The reason for doing this is that the two feature types encode different kinds of information, so they complement each other.

In general, the use of shape features requires a good lip tracking algorithm and makes the limiting assumption that speech information is concentrated in the contour of the lips alone. Several articles report that DCT features outperform shape based ones [177, 178].

Audio-visual Integration

The integration of audio and visual information [174] can be performed in several ways. The simplest one is feature concatenation [179], where the audio and video feature vectors are simply concatenated before being presented to the classifier. Here, a single classifier is trained with combined data from the two modalities.

Although the feature concatenation method of integration does lead to an improved performance, it is impossible to model the reliability of each modality, depending on the changing conditions in the audio-visual environment.

Using decision fusion, separate audio and video classifiers are trained, and their output log-likelihoods are linearly combined with appropriate weights. There are three possible levels for combining individual modality likelihoods [174]:

5.2 Interaction

- Early integration, when likelihoods are combined at the state level, forcing the synchrony of the two streams. This leads to a multi-stream HMM classifier [180]. Practically, there is only one audio-visual HMM, whose state-level emission probabilities are estimated separately for the audio and visual streams.
- Late integration, which requires two separate HMMs. The final recognised word is selected based on the n-best hypotheses of the audio and visual HMMs.
- Intermediate integration, which uses models that force synchrony at the phone or word boundaries.

5.2.2.2 Overview of Information-theoretic Feature Selection

Feature Selection

Here we present previous work on feature selection algorithms [181]. The aim is to choose from a set F of n features, a subset S of m features, such that S retains most of the information in F relevant for the classification task. Since the number of possible subsets, $\binom{n}{m}$, is usually too high to allow the processing of each candidate, this leads to iterative "greedy" algorithms which choose features one by one according to some measure. In this respect, our focus is on information theoretic measures, as they can directly assess the relevance of a feature.

The mutual information (MI) $I(Y;C)$ between a feature Y and the class labels C is such a measure, extensively used for feature selection in classification. It represents the reduction in the uncertainty of C when Y is known. A high MI means that the feature is relevant for our particular classification problem.

Computing MI from data requires the estimation of probability densities, which cannot be accurately done in high dimensions. This is why a majority of feature selection algorithms use measures based on up to three variables (two features plus the class label).

We will now briefly present a few information-theoretic feature selection methods. Let $F = \{Y_1, Y_2 ... Y_n\}$ be the initial set of features. Let $\{\pi_1, \pi_2 ... \pi_m\}$ be a permutation on a subset of dimension m of the set of feature indices $\{1...n\}$. Then the set of selected features can be written as $S = \{Y_{\pi_1}, Y_{\pi_2} ... Y_{\pi_m}\} \subset F$.

The simplest information-theoretic criterion to select a feature at step $k+1$ is [178, 181]:

$$Y_{\pi_{k+1}} = \arg\max_{Y_i \in FS_k} I(Y_i; C) \qquad (5.4)$$

where $S_k = S_{k-1} \cup \{Y_{\pi_k}\}$ is the set of features selected at step k.

However, this method ranks each feature's relevance individually, irrespective of previous choices. In order to have a maximum of information with a small number of features, any redundancy should be eliminated.

A possible way of doing this is to penalize a feature's importance by a proportion of its summed redundancy with the already chosen features (the MIFS algorithm [182]):

$$Y_{\pi_{k+1}} = \arg\max_{Y_i \in FS_k} \left[I(Y_i;C) - \beta \sum_{Y_{\pi_j} \in S_k} I(Y_i;Y_{\pi_j}) \right] \quad (5.5)$$

A similar approach is to penalize the average redundancy [183]:

$$Y_{\pi_{k+1}} = \arg\max_{Y_i \in FS_k} \left[I(Y_i;C) - \frac{1}{|S_k|} \sum_{Y_{\pi_j} \in S_k} I(Y_i;Y_{\pi_j}) \right] \quad (5.6)$$

where $|S_k|$ is the size of set S_k.

Another family of information theoretic feature selection algorithms uses the conditional mutual information (CMI) as a measure [184], $I(X;C|Y) = I(X,Y;C) - I(Y;C)$. This shows how much the random variable X increases the information we have about C when Y is given. The selection criterion is the following:

$$Y_{\pi_{k+1}} = \arg\max_{Y_i \in FS_k} \left[\min_{Y_{\pi_j} \in S_k} I(Y_i;C|Y_{\pi_j}) \right] \quad (5.7)$$

$$= \arg\max_{Y_i \in FS_k} \left[I(Y_i;C) - \max_{Y_{\pi_j} \in S_k} I(Y_i;Y_{\pi_j};C) \right]$$

using $I(X;Y;C) = I(Y;C) - I(Y;C|X)$ [185]. For a certain Y_i, the particular Y_{π_j} is found with which Y_i is most redundant. By taking the maximum over this CMI, the feature that adds the most relevant information to the set S_k is found.

In the end, the goal of all these algorithms is to maximize the joint MI between the S and C, which could be expanded like this (chain rule [185]):

$$I(S;C) = I(Y_{\pi_1}, Y_{\pi_2}, \dots, Y_{\pi_m};C)$$

$$= \sum_{j=1}^{m} I(Y_{\pi_j};C|Y_{\pi_1},\dots,Y_{\pi_{j-1}}) \quad (5.8)$$

$$= \sum_{j=1}^{m} \left[I(Y_{\pi_j};C) - I(Y_{\pi_j};C;Y_{\pi_1},\dots,Y_{\pi_{j-1}}) \right]$$

Since not all subsets can be tested, an iterative algorithm could maximize the terms of this sum one by one. Since Y_{π_j} is the particular Y_i that maximizes the jth term of the sum, all previously mentioned criteria (Equations 5.2, 5.3, 5.4) can be interpreted as approximations of this general optimization. They all maximize the difference between $I(Y_i;C)$ and an approximation of the redundancy $I(Y_i;C;Y_{\pi_1},\dots,Y_{\pi_{j-1}})$ between Y_i, S_{j-1} and the class labels C.

Feature Selection with MI in Speech Recognition

Mutual information has been used before in speech recognition as a criterion for feature selection. In [186], conditional mutual information is used to find combinations of audio feature streams. In [187], mutual information is used to select visual features for AVSR from a set of DCT coefficients. Two criteria are used – either the maximum mutual information as shown in Equation 5.1, or the joint mutual information $I(Y_i, Y_j; C)$.

Our method uses two criteria, the one in Equation 5.4, and a new criterion based on clustering redundant features. We will present them in detail in the next section and then, in the results section, we will show the performance improvements that they bring over similar methods from the literature.

5.2.2.3 Our Feature Selection Method

MI Computation

To estimate the MI values, we opted for a histogram approach. We discretised the probability density function of each feature by finding its extreme values over the whole database and partitioning the interval into bins. Two and three-dimensional histograms (two features and the class label) were also computed. The number of bins was here empirically chosen as a trade-off between an adequately high number of bins for accurate estimation, and sufficient samples per bin. The class labels that we use for computing MI correspond to groups of HMM states representing speech phonemes.

Feature Selection Techniques

We propose two methods of feature selection. Both will maximize an individual feature's mutual information with the class labels, while at the same time removing redundancy between selected features. This is achieved with a greedy algorithm, selecting a new feature at each step, as shown previously. The particularity of our methods is the fact that only the "relevant" redundancy, that is tied to the class labels, is taken into account.

Our first method uses Equation 5.4. The most informative feature is chosen each time, provided that it also has little redundancy with the other chosen features.

Our second method, Selection by Redundant Features Clustering (SRFC), improves on this idea, reducing the redundancy even more. The algorithm is as follows. First, a feature is selected according to Equation 5.4. Then, it is assigned to the same group as $Y_{\pi_j} = \arg\max I(Y_i; Y_{\pi_j}; C)$ if the MI is positive. Here, a positive sign means there is some redundancy, while a negative one means there is none. If the max is negative, a new group is created containing just Y_i. However, in the case of positive MI, we can refuse the selection of a new feature if its assigned group is "full"; that is, there is too much redundancy inside it. This exclusion criterion is:

$$\sum_{Y_{\pi_j} \in Group} I(Y_i; Y_{\pi_j}; C) > I(Y_i; C) \tag{5.9}$$

In this way, if a feature satisfies the exclusion criterion, it is assumed that most of the information it conveys is already present in the group, distributed among its members. While our first technique uses a single feature for estimating redundancy, this method takes into account the whole selected subset redundancy by clustering it into internally redundant groups.

Multistream Classifier for AVSR

The classifiers that we use are multi-stream HMMs. They differ from traditional HMMs in the way that emission probabilities are computed. While in traditional HMMs each state has a mixture of Gaussians associated to it, in multi-stream HMMs there are multiple Gaussian mixtures per state, one for each modality. The emission likelihood b_j for state j and observation o_t at time t is the product of likelihoods from each modality s weighted by stream exponents λ_s [188]:

$$b_j(o_t) = \prod_{s=1}^{S} \left[\sum_{m=1}^{M_s} c_{jsm} N(o_{st}; \mu_{jsm}, \Sigma_{jsm}) \right]^{\lambda_s} \tag{5.10}$$

where $N(o; \mu, \Sigma)$ is the value in o of a multivariate Gaussian with mean μ and covariance matrix Σ. M_s Gaussians are used in a mixture, each weighed by c_{jsm}. The product in Equation 5.10 is in fact equivalent to a weighted sum in logarithmic domain.

In this context, the weights λ_s represent the individual importance of one stream at that particular time instant. The weights always sum to 1 $\sum_s \lambda_s = 1$.

If the noise level could be known in advance, the weights could be fixed to the ideal value for that particular noise level. This is one of the methods we used, selecting the best pair of weights for each noise level. We will call this method the "static weights" method, since the weights never change for a whole set of video sequences. All audio-visual results presented in the current section are obtained with this method. In contrast to the "static weights" method, there is a way to adjust the weights dynamically, which will be detailed in the "Dynamic stream weighting" section.

Implementation Details

For our experiments, we use sequences from the CUAVE audio-visual database [171]. It consists of 36 speakers uttering digit sequences. We use only the static

5.2 Interaction

part of the database; that is, 5 repetitions of the 10 digits. The region of interest (ROI) that we use is the mouth of the speaker, scaled and rotated, so that all the mouths have approximately the same size and position.

We use the HTK library [188] for the HMM implementation. Our word models have 8 states with one diagonal-covariance Gaussian per state. The silence model has 3 states with 3 Gaussians per state. Two streams are used, audio and video. The grammar consists of any combination of digits with silence in-between. The accuracy that we report is the number of correctly recognised words minus insertions, divided by the total number of test words.

The features that we extract from the audio stream are 13 Mel Frequency Cepstral Coefficients (MFCCs), together with their first and second temporal derivatives. As video features, we tested both Discrete Cosine Transform (DCT) and Principal Components Analysis (PCA) features computed on the ROI, with first and second temporal derivatives.

The testing method is as follows. The database is split into a training set of 30 sequences, and a testing set of 6. After recognition is performed, the process is repeated five times with different training and testing sets. In this way, all individual speakers are used once for testing, and five times for training. The 6 results are averaged at the end.

Feature Quality for AVSR

We now present a comparison in terms of relevant information brought by different features commonly used in AVSR. Figure 5.12 shows each individual fea-

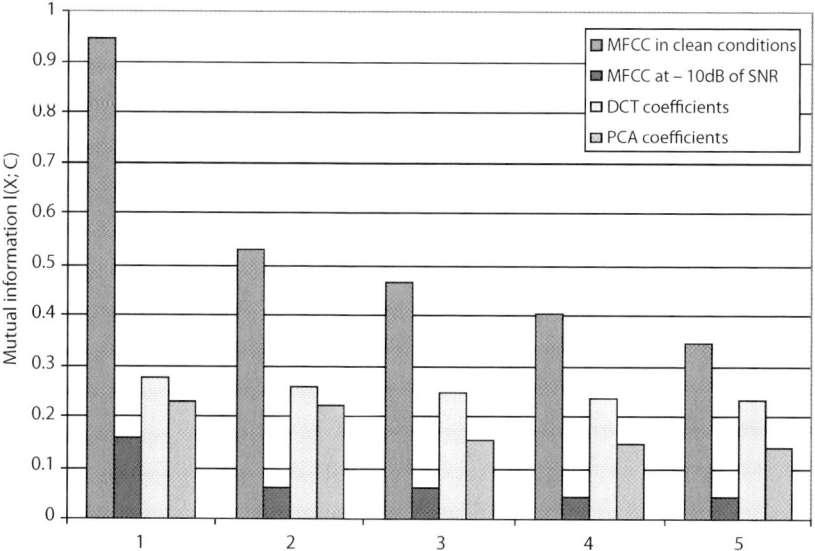

Fig. 5.12 Comparison of mutual information I(X;C) between different multimodal features

ture's mutual information with the class label for the five best features of four categories. The two first categories are MFCCs extracted on the audio signal for two noise levels: clean conditions and -10dB of SNR (with white noise added). A strong decrease of the audio features quality can be noted, justifying the need to complement the corrupted audio with visual information. The last categories are two of the commonly used visual features for AVSR: DCT and PCA, with DCT clearly outperforming PCA. However, note that Fig. 5.12 does not show the amount of redundancy among features inside a category, but just emphasizes the intrinsic relevant information of each feature.

The need for Feature Selection in AVSR

As discussed earlier, classifier performance degrades when dimensionality becomes too high. This degradation is even larger in the case of multimodal applications and AVSR in particular, since each modality increases the dimensionality that needs to be handled. Figure 5.13 shows the audio-visual word recognition rate with corrupted audio (-10db SNR) versus the visual features dimensionality for our AVSR system. Performance peaks and then decreases as the number of features is increased.

5.2.2.4 Feature Selection Results

The Performance of our Feature Selection Method

We compare here our feature selection method with two other methods from the literature, also described in the "Feature selection techniques" section (Equations 5.4 and 5.5). Results are shown in Fig. 5.14 for video-only speech recognition. Between our two proposed methods, SRFC is better performing by a small margin. The two methods from the literature perform worse, especially at lower dimensionality. The results clearly prove that computing relevant redundancy, as opposed to just redundancy, leads to better features.

5.2.2.5 Dynamic Stream Weighting

In the previous section we presented a method to select relevant features from each modality. Now our focus is the integration of the modalities, and in particular the way to make our system resistant to different types of noise, including noise that is dynamically varying.

In a real-life scenario, the noise may not be stationary, but varying, affecting in different degrees each modality, at different time instants. An extreme possibility is that one of the modalities becomes completely unavailable for a certain time, so it should be ignored; that is, only information from the other modality should be used.

5.2 Interaction

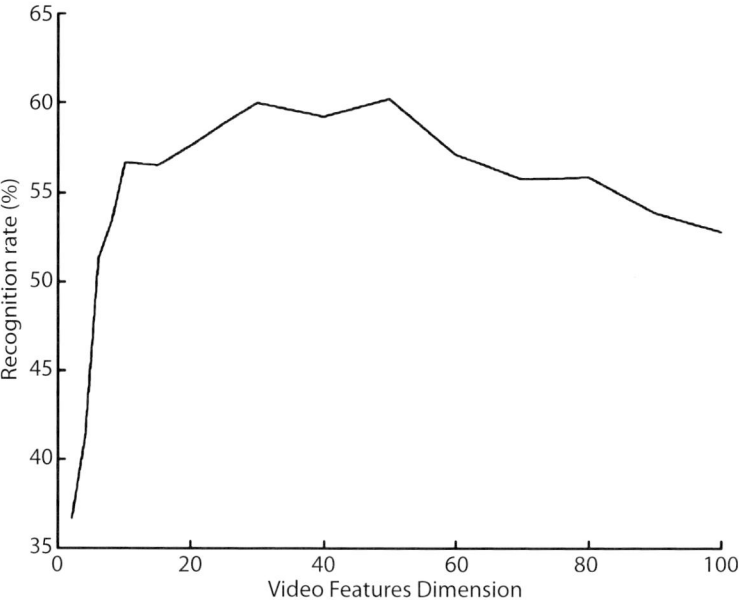

Fig. 5.13 Effect of dimensionality on a classifier performance for a multimodal application

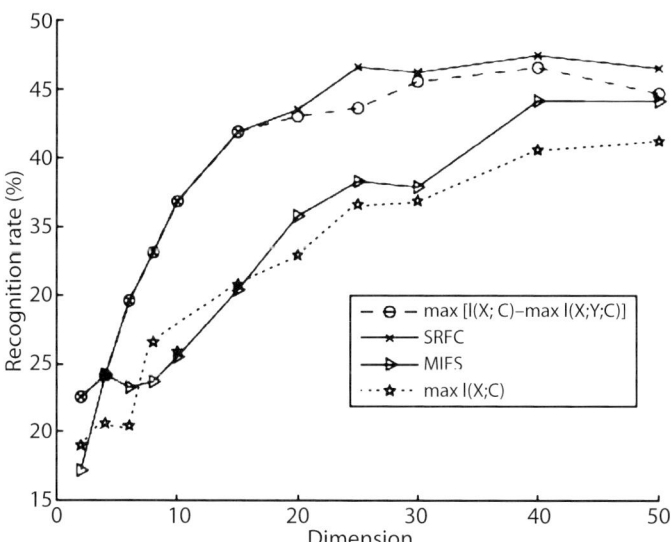

Fig. 5.14 Performance in video-only recognition according to dimensionality for different feature selection methods

Our method is based on determining the reliability of the classification for each time instant, or frame, and for each modality individually. Although used for speech recognition, this method could in theory be used in any case where the reliability of the modalities can be quantitatively determined.

Is Dynamic Weight Adjustment Necessary?

If the noise statistics never change, a fixed modality weight would perform just as well as an adaptive one. However, for different noise levels, the optimal weight varies considerably, as shown in Fig. 5.15. While for clean audio, the audio stream's weight should be 0.95, it decreases to 0.45 for a -10db SNR. This means that with changing noise conditions, the weights should change. In a scenario in which the noise varies in time, the "static weight" scheme would not work. This shows that an adaptive method which chooses the stream weight at each time instant based on the stream reliability at that particular moment is indeed required.

Estimating the Stream Reliability

In particular, for our application, AVSR, we are trying to classify frames as belonging to a certain phoneme class, or, more precisely, that it is associated to a certain HMM state. Each HMM state has two corresponding Gaussian mixtures, one for the audio and one for the video. For each frame and for each state we can compute the likelihood that the frame was generated by that particular state, that is, the emission likelihood.

The stream reliability can be considered to be tied to the dispersion of emission posterior probabilities for each stream, as computed through Baye's rule from the likelihoods. The more dispersed these posteriors are, the more reliable the classification. In other words, if the posterior probability distribution has a clear peak, the classification is clearly more reliable than if the same distribution is flat, i.e. no clear peak is present.

As a measure of this dispersion, we used the entropy of the posterior probabilities for each stream. Figure 5.16 shows the entropy value histograms for the visual stream, clean audio and audio at -10db SNR with either white or babble noise. As it can be seen from the figure, a large number of frames have very low entropy in the case of clean audio, showing that its reliability is high. The video stream frames have on average higher entropy values, showing that video is less reliable than clean audio, which confirms that there is more information in the audio than in the video, and also justifies why for clean audio the ideal static weight is 0.95. However, the entropy values for corrupted audio at -10db SNR are on average much higher than the ones corresponding to the video stream. This shows that the video contains more information than the corrupted audio.

5.2 Interaction

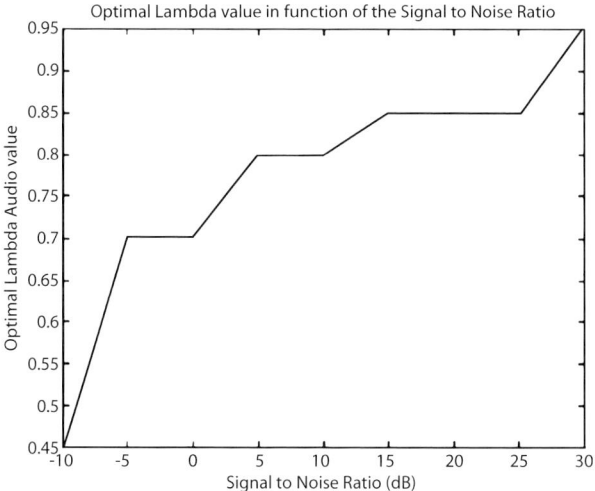

Fig. 5.15 Optimal audio stream weight for each audio SNR

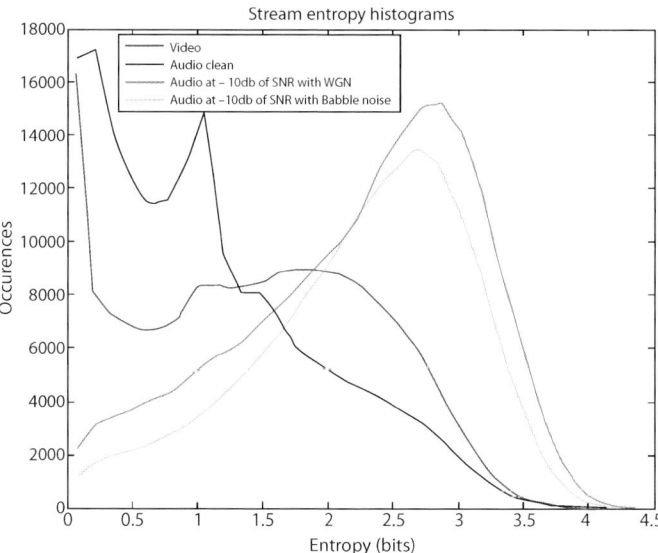

Fig. 5.16 Entropy value histograms for the audio and visual streams for different noise conditions

As high entropy means low dispersion and thus low reliability, the inverse entropy is actually used, as suggested in [189]. The actual formula of the weight for stream s is:

$$\lambda_s = \frac{1/H_s}{\sum_{i=1}^{S} 1/H_i} \tag{5.11}$$

where H_s is the entropy for stream s and λ_s its weight. In the extreme case, when H_s is zero, i.e. one of the posteriors is 1 and the others 0, the weight for stream s will be 1. If the entropies are equal, the weights are also equal.

The advantage of this method is that no matter how the noise changes, the weights adapt immediately to the change. Even if one of the modalities becomes completely corrupted, as for example if there is an interruption in the video stream, the weights would reflect this. In this case, the entropy of the posteriors corresponding to the corrupted stream would be very high, meaning that a very low weight would be assigned to it. The classification would be made based only on the posteriors of the other modality, which is the expected behaviour of such a multimodal system.

Dynamic Weighting Results

Figure 5.17 shows the results for both static and dynamic weights across a wide range of SNRs. The results show that the performance is comparable in all cases between the two methods.

However, this is a scenario where the SNR is kept constant for all test sequences in a certain test run. Obviously, in a different scenario where the SNR would vary in time, the dynamic weights method would outperform the static weights one.

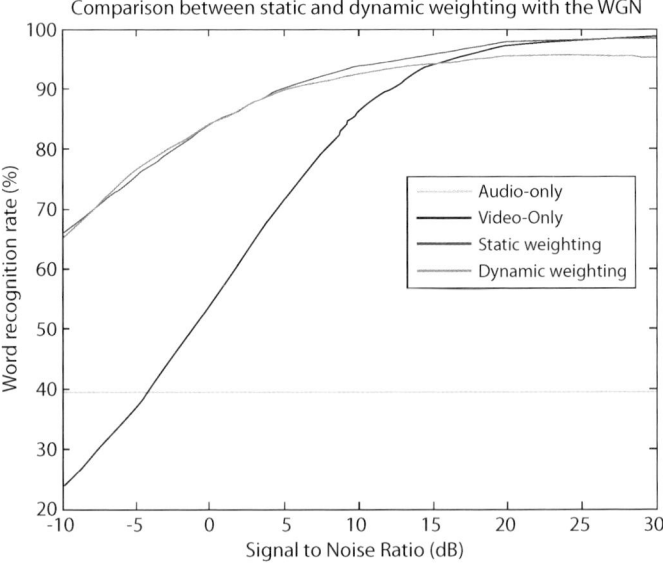

Fig. 5.17 Performance comparison between the static and the dynamic weights methods

5.3 Conclusions

In the first section of the chapter, we have reviewed the state of the art of face and facial feature detection. We have initially concentrated in face detection since most of facial feature detection algorithms rely on a previous face detection step. The various methods have been analysed from the perspective of the different models that they use to represent images and patterns. In this way, approaches have been classified into four different categories: pixel based, block based, transform coefficient based and region based techniques.

In the second section of the chapter, we have presented two examples of multimodal signal processing applications. The first one allows the localization of the speaker's mouth in a video sequence, using both the audio signal and the motion extracted from the video. This example shows how information from both the audio and the video streams can be integrated in a simple way, by using a joint probability density. The second application consists in recognizing the spoken words in a video sequence using both the audio and the images of moving lips. This application is more complex, and shows how the sequence of sounds can be modelled using Hidden Markov Models. Towards the end of the section, we show how the importance of the two streams can be varied dynamically, such that the system automatically adapts to any variation in the quality of any of the streams. This makes the system tolerant to varying noise or even to the complete loss of one modality stream.

Chapter 6
Recognition of Emotional States in Natural Human-Computer Interaction

R. Cowie[1], E. Douglas-Cowie[1], K. Karpouzis[2], G. Caridakis[2], M. Wallace[3] and S. Kollias[2]

[1] School of Psychology, Queen's University,University Road, Belfast, BT7 1NN, Northern Ireland, UK
[2] Image, Video and Multimedia Systems Laboratory, National Technical University of Athens, 15780, Zographou, Athens, Greece
[3] Department of Computer Science, University of Indianapolis, Athens Campus, 9 Ipitou St., GR-105 57 Athens, Greece
{r.cowie, e.douglas-cowie}@qub.ac.uk,
wallace@uindy.gr,
{kkarpou, gcari, skollias}@image.ntua.gr

6.1 Introduction

The introduction of the term 'affective computing' by R. Picard [190] epitomizes the fact that computing is no longer considered a 'number crunching' discipline, but should be thought of as an interfacing means between humans and machines and sometimes even between humans alone. To achieve this, application design must take into account the ability of humans to provide multimodal input to computers, thus moving away from the monolithic window-mouse-pointer interface paradigm and utilizing more intuitive concepts, closer to human niches ([191, 192]). A large part of this naturalistic interaction concept is expressivity [193], both in terms of interpreting the reaction of the user to a particular event or taking into account their emotional state and adapting presentation to it, since it alleviates the learning curve for conventional interfaces and makes less technology-savvy users feel more comfortable. In this framework, both speech and facial expressions are of great importance, since they usually provide a comprehensible view of users' reactions; actually, Cohen commented on the emergence and significance of multimodality, albeit in a slightly different human-computer interaction (HCI) domain, in [194, 195], while Oviatt [196] indicated that an interaction pattern constrained to mere 'speak-and-point' only makes up for a very small fraction of all spontaneous multimodal utterances in everyday HCI [197]. In the context of HCI, [198] defines a multimodal system as one that 'responds to inputs in more than one modality or communication channel' abundance, while Mehrabian [199] suggests that facial expressions and vocal intonations are the main means for someone to estimate a person's affective state

[200], with the face being more accurately judged, or correlating better with judgments based on full audiovisual input than on voice input ([198, 201]). This fact led to a number of approaches using video and audio to tackle emotion recognition in a multimodal manner ([202–205]), while recently the visual modality has been extended to include facial, head or body gesturing ([206, 207], extended in [208]).

Additional factors that contribute to the complexity of estimating expressivity in everyday HCI are the fusion of the information extracted from modalities ([196]), the interpretation of the data through time and the noise and uncertainty alleviation from the natural setting ([209, 210]). In the case of fusing multimodal information [211], systems can either integrate signals at the feature level ([212]) or, after coming up with a class decision at the feature level of each modality, by merging decisions at a semantic level (late identification, [212, 213]), possibly taking into account any confidence measures provided by each modality or, generally, a mixture of experts mechanism [214].

Regarding the dynamic nature of expressivity, Littlewort [215] states that while muscle-based techniques can describe the morphology of a facial expression, it is very difficult for them to illustrate in a measurable (and, therefore detectable) manner the dynamics, i.e. the temporal pattern of muscle activation and observable feature movement or deformation. She also makes a case of natural expressivity being inherently different in temporal terms than posed, presenting arguments from psychologists ([216, 217]), proving the dissimilarity of posed and natural data, in addition to the need to tackle expressivity using mechanisms that capture dynamic attributes. As a general rule, the naturalistic data chosen as input in this work, is closer to human reality since intercourse is not acted and expressivity is not guided by directives (e.g. Neutral expression → one of the six universal emotions → neutral). This amplifies the difficulty in discerning facial expressions and speech patterns [218]. Nevertheless it provides the perfect testbed for the combination of the conclusions drawn from each modality in one time unit and use as input in the following sequence of audio and visual events analysed.

The current work aims to interpret sequences of events by modelling the user's behaviour in a natural HCI setting through time. With the use of a recurrent neural network, the short term memory provided through its feedback connection, works as a memory buffer and the information remembered is taken under consideration in every next time cycle. Theory on this kind of network backs up the claim that it is suitable for learning to recognize and generate temporal patterns as well as spatial ones [190]. In addition to this, results show that this approach can capture the varying patterns of expressivity with a relatively low-scale network, which is not the case with other works operating on acted data.

The paper is structured as follows: in Section 6.2 we provide the fundamental notions upon which the remaining presentation is based. This includes the overall architecture of our approach as well as the running example which we will use throughout the paper in order to facilitate the presentation of our approach. In Section 0 we present our feature extraction methodologies, for both the visual and auditory modalities. In Section 6.4 we explain how the features extracted, although fundamentally different in nature, can be used to drive a recursive neural network in order to acquire an estimation of the human's state. In Section 6.5 we present results from the application of our methodology on naturalistic data and in Section 6.6 we list our concluding remarks.

6.2 Fundamentals

6.2.1 Emotion Representation

When it comes to recognizing emotions by computer, one of the key issues is the selection of appropriate ways to represent the user's emotional states. The most familiar and commonly used way of describing emotions is by using categorical labels, many of which are either drawn directly from everyday language, or adapted from it. This trend may be due to the great influence of the works of Ekman and Friesen who proposed that the archetypal emotions correspond to distinct facial expressions which are supposed to be universally recognizable across cultures [219, 220].

On the contrary psychological researchers have extensively investigated a broader variety of emotions. An extensive survey on emotion analysis can be found in [221]. The main problem with this approach is deciding which words qualify as genuinely emotional. There is, however, general agreement as to the large scale of the emotional lexicon, with most lists of descriptive terms numbering into the hundreds; the Semantic Atlas of Emotional Concepts lists 558 words with 'emotional connotations'. Of course, it is difficult to imagine an artificial systems being able to match the level of discrimination that is implied by the length of this list.

Although the labelling approach to emotion representation fits perfectly in some contexts and has thus been studied and used extensively in the literature, there are other cases in which a continuous, rather than discrete, approach to emotion representation is more suitable. At the opposite extreme from the list of categories are dimensional descriptions, which identify emotional states by associating them with points in a multidimensional space. The approach has a long history, dating from Wundt's [222] original proposal to Schlossberg's reintroduction of the idea in the modern era [222]. For example, activation-emotion space as a representation has great appeal as it is both simple, while at the same time makes it possible to capture a wide range of significant issues in emotion [223]. The concept is based on a simplified treatment of two key themes:

- Valence: The clearest common element of emotional states is that the person is materially influenced by feelings that are valenced, i.e., they are centrally concerned with positive or negative evaluations of people or things or events.
- Activation level: Research from Darwin forward has recognised that emotional states involve dispositions to act in certain ways. A basic way of reflecting that

theme turns out to be surprisingly useful. States are simply rated in terms of the associated activation level, i.e., the strength of the person's disposition to take some action rather than none.

There is general agreement on these two main dimensions. Still, in addition to these two, there are a number of other possible dimensions, such as power-control, or approach-avoidance. Dimensional representations are attractive mainly because they provide a way of describing emotional states that is more tractable than using words. This is of particular importance when dealing with naturalistic data, where a wide range of emotional states occur. Similarly, they are much more able to deal with non discrete emotions and variations in emotional state over time.

In this work we have focused on the general area in which the human emotion lies, rather than on the specific point on the diagram presented in Fig. 6.1. One of the reasons that has lead us to this decision is that it is not reasonable to expect human annotators to be able to discriminate between an extra pixel to the left or to the right as being an indication of a shift in observed emotional state, and therefore it does not make sense to construct a system that attempts to do so either. Thus, as is also displayed in Fig. 6.1, we have segmented the emotion representation space in broader areas.

As we can see in the figure, labels are typically given for emotions falling in areas where at least one of the two axes has a value considerably different than zero. On the other hand, the beginning of the axes (the center of the diagram) is typically considered as the neutral emotion. For the same reasons as mentioned above, we find it is not meaningful to define the neutral state so strictly. Therefore, we have added to the more conventional areas corresponding to the four quadrants a fifth one, corresponding to the neutral area of the diagram, as is depicted in Fig. 6.1.

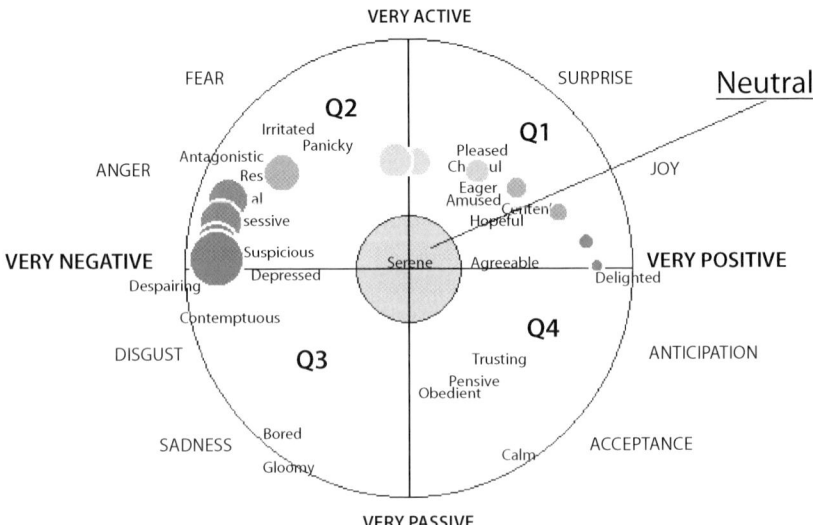

Fig. 6.1 The activation/valence dimensional representation [224]

6.2.2 Methodology Outline

As we have already mentioned, the overall approach is based on a multimodal processing of the input sequence. 'Multimodal processing' is a general term referring to the combination of multiple input queues in order to enhance the operation of a system. In fact, there are two different methodologies that fall under this general label; decision-level and feature-level.

In the first one, independent systems are developed, each one considering one of the available information queues. The results of the different systems are then considered as independent sources of evidence concerning the optimal result, and the overall output of the system is computed through some averaging approach. This approach has the benefit of being very easy to implement when the independent systems are already available in the literature.

In the second approach, a single system considers all input queues at the same time in order to reach a single conclusion. This approach has the drawback of being often difficult to implement, as different information queues are often different in nature and are thus difficult to incorporate in one uniform processing scheme. On the other hand, when successfully realised, the feature level approach produces systems that are able to achieve considerably better performances [226].

Our approach is of the latter type; the general architecture of our approach is depicted in Fig. 6.2.

The considered input sequence is split into the audio and visual sequences. The visual sequence is analysed frame by frame using the methodology presented in Section 0 and the audio sequence is analysed as outlined in Section 6.3.2 and further explained in [202]. Visual features of all corresponding frames are fed to a recurrent network as explained in Section 6.4 , where the dynamics in the visual channel are picked up and utilised in classifying the sequence to one of the five considered emotional classes mentioned in Table 6.1. Due to the fact that the fea-

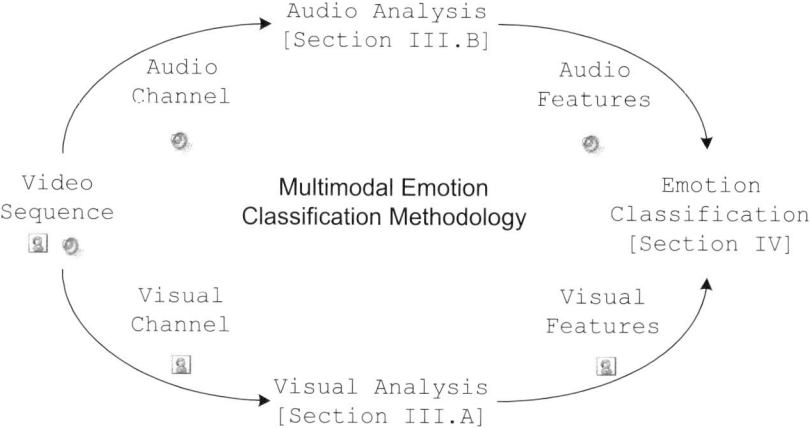

Fig. 6.2 Graphical outline of the proposed approach

Table 6.1 Emotion classes

Label	Location in FeelTrace [225] diagram
Q1	positive activation, positive evaluation (+/+)
Q2	positive activation, negative evaluation (+/-)
Q3	negative activation, negative evaluation (-/-)
Q4	negative activation, positive evaluation (-/+)
Neutral	close to the center

tures extracted from the audio channel are fundamentally different in nature than those extracted from the visual channel, the recurrent network structure is altered accordingly in order to allow both inputs to be fed to the network at the same time, thus allowing for a truly multimodal classification scheme.

The evaluation of the performance of our methodology includes statistical analysis of application results, quantitative comparisons with other approaches focusing on naturalistic data and qualitative comparisons with other known approaches to emotion recognition, all listed in Section 6.5 .

6.2.3 Running Example

In developing a multimodal system one needs to integrate diverse components which are meant to deal with the different modalities. As a result, the overall ar-

Frame: 00815

Frame: 00820

Frame: 00825

Frame: 00830

Frame: 00835

Fig. 6.3 Frames from the running example

chitecture comprises a wealth of methodologies and technologies and can often be difficult to grasp in full detail. In order to facilitate the presentation of the multi-modal approach proposed herein for the estimation of human emotional state we will use the concept of a running example.

Our example is a sample from the dataset on which we will apply our overall methodology in section 6.5 . In Fig. 6.3 we present some frames from the sequence of the running example.

6.3 Feature Extraction

6.3.1 Visual Modality

6.3.1.1 State of the Art

Automatic estimation of facial model parameters is a difficult problem and although a lot of work has been done on selection and tracking of features [227], relatively little wp1p1ork has been reported [228] on the necessary initialization step of tracking algorithms, which is required in the context of facial feature extraction and expression recognition. Most facial expression recognition systems use the Facial Action Coding System (FACS) model introduced by Ekman and Friesen [219] for describing facial expressions. FACS describes expressions using 44 Action Units (AU) which relate to the contractions of specific facial muscles.

Additionally to FACS, MPEG-4 metrics [229] are commonly used to model facial expressions and underlying emotions. They define an alternative way of modelling facial expressions and the underlying emotions, which is strongly influenced by neurophysiologic and psychological studies. MPEG-4, mainly focusing on facial expression synthesis and animation, defines the Facial Animation Parameters (FAPs) that are strongly related to the Action Units (AUs), the core of the FACS. A comparison and mapping between FAPs and AUs can be found in [230].

Most existing approaches in facial feature extraction are either designed to cope with limited diversity of video characteristics or require manual initialization or intervention. Specifically [228] depends on optical flow, [231] depends on high resolution or noise-free input video, [232] depends on colour information, [233] requires two head-mounted cameras and [234] requires manual selection of feature points on the first frame. Additionally very few approaches can perform in near-real time. In this work we combine a variety of feature detection methodologies in order to produce a robust FAP estimator, as outlined in the following.

6.3.1.2 Face Localization

The first step in the process of detecting facial feature is that of face detection. In this step the goal is to determine whether or not there are faces in the image and, if yes, to return the image location and extent of each face [235]. Face detection can be performed with a variety of methods [236–238]. In this paper we have chosen to use nonparametric discriminant analysis with a Support Vector Machine (SVM) which classifies face and non-face areas, thus reducing the training problem dimension to a fraction of the original with negligible loss of classification performance [239].

In order to train the SVM and fine-tune the procedure we used 800 face examples from the NIST Special Database 18. All these examples were first aligned with respect to the coordinates of the eyes and mouth and rescaled to the required size and then the set was extended by applying small scale, translation and rotation perturbations to all samples, resulting in a final training set consisting of 16695 examples.

The accuracy of the feature extraction step that will follow greatly depends on head pose, and thus rotations of the face need to be removed before the image is further processed. In this work we choose to focus on roll rotation, since it is the most frequent rotation encountered in real life video sequences. So, we need to first estimate the head pose and then eliminate it by rotating the facial image accordingly. In order to estimate the head pose we start by locating the two eyes in the detected head location.

For this tack we utilize a multi-layer perceptron (MLP). As activation function we choose a sigmoidal function and for learning we employ the Marquardt-Levenberg learning algorithm [240]. In order to train the network we have used approximately 100 random images of diverse quality, resolution and lighting conditions from the ERMIS database [241], in which eye masks were manually specified. The network has 13 input neurons; the 13 inputs are the luminance Y, the Cr & Cb chrominance values and the 10 most important DCT coefficients (with zigzag selection) of the neighbouring 8x8 pixel area. The outputs are 2, one for eye and one for non eye regions. Through pruning, the remaining architecture of the MLP has been trimmed and optimised to comprise two hidden layers of 20 neurons each.

The locations of the two eyes on the face are initially estimated roughly using the approximate anthropometric rules presented in Table 6.2 and then the MLP is applied separately for each pixel in the two selected regions of interest. For rotations up to 30 degrees, this methodology is successfully at a rate close to 100% in locating the eye pupils accurately.

In Fig. 6.4 and Fig. 6.5 we see the result of applying the MLP in the first frame of the running example. Once we have located the eye pupils, we can estimate the head roll rotation by calculating the angle between the horizontal plane and the line defined by the eye centres. We can then rotate the input frame in order to bring the head in the upright position. Finally, we can then roughly segment the

Table 6.2 Anthropometric rules for feature-candidate facial areas [202]

Segment	Location	Width	Height
Left eye and eyebrow	Top left	$0.6 \times$ (width of face)	$0.5 \times$ (height of face)
Left eye and eyebrow	Top right	$0.6 \times$ (width of face)	$0.5 \times$ (height of face)
Mouth and nose	Bottom center	width of face	$0.5 \times$ (height of face)

6.3 Feature Extraction

Fig. 6.4 Eye location using the MLP

Fig. 6.5 Detail from Fig. 6.4

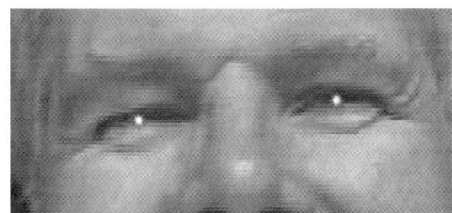

Fig. 6.6 Frame rotation based on eye locations

Fig. 6.7 Regions of interest for facial feature extraction

rotated frame into three overlapping rectangle regions of interest which include both facial features and facial background; these three feature-candidate areas are the left eye/eyebrow, the right eye/eyebrow and the mouth. The segmentation is once more based on the approximate anthropometric rules presented in Table 6.2.

6.3.1.3 Nose Localization

The nose is not used for expression estimation by itself, but is a fixed point that facilitates distance measurements for FAP estimation (see Fig. 6.21). Thus, it is sufficient to locate the tip of the nose and it is not required to precisely locate its boundaries. The most common approach to nose localization is starting from nostril localization; nostrils are easily detected based on their low intensity. In order to identify candidate nostril locations we apply the threshold t_n on the luminance channel of the area above the mouth region

$$t_n = \frac{\overline{L^n} + 2\min(L^n)}{3}$$

where L^n is the luminance matrix for the examined area and $\overline{L^n}$ is the average luminance in the area. The result of this thresholding is presented in Fig. 6.8. Connected objects in this binary map are labelled and considered as nostril candidates. In poor lighting conditions, long shadows may exist along either side of the nose, resulting in more than two nostril candidates appearing in the mask. Using statistical anthropometric data about the distance of left and right eyes (bipupil breadth, D_{bp}) we can remove these invalid candidate objects and identify the true nostrils. The nose centre is defined as the midpoint of the nostrils.

6.3 Feature Extraction

Fig. 6.8 Candidate nostril locations

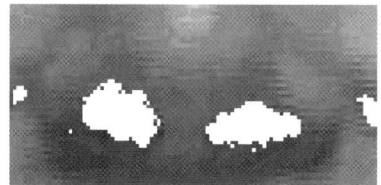

6.3.1.4 Eyebrow Localization

Eyebrows are extracted based on the fact that they have a simple directional shape and that they are located on the forehead, which due to its protrusion, has a mostly uniform illumination.

The first step in eyebrow detection is the construction of an edge map of the greyscale eye and eyebrow region of interest. This map is constructed by subtracting the dilation and erosion of the greyscale image using a line structuring element. The selected edge detection mechanism is appropriate for eyebrows because it can be directional, preserves the feature's original size and can be combined with a threshold to remove smaller skin anomalies such as wrinkles. This procedure can be considered as a special case of a non-linear high-pass filter.

Each connected component on the edge map is then tested against a set of filtering criteria that have been formed through statistical analysis of the eyebrow lengths and positions on 20 persons of the ERMIS database [223]. The results of this procedure for the left eyebrow are presented in Fig. 6.9.

The same procedure is also applied for the right eyebrow.

6.3.1.5 Eye Localization

A wide variety of methodologies have been proposed in the literature for the extraction of different facial characteristics and especially for the eyes, in both controlled and uncontrolled environments. What is common among them is that, re-

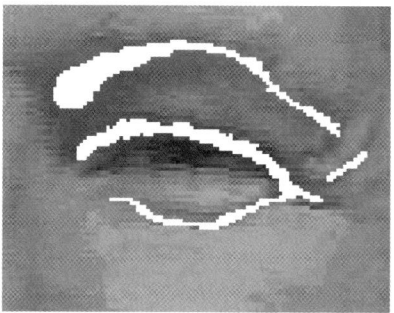 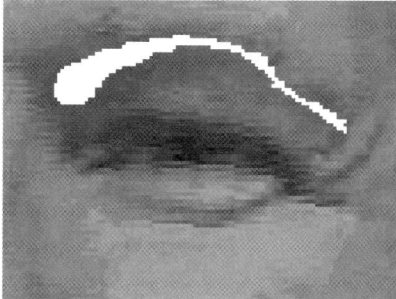

Fig. 6.9 Eyebrow detection steps

gardless of the overall success rate that they have, they all fail in some set of cases, due to the inherent difficulties and external problems that are associated with the task. As a result, it is not reasonable to select a single methodology and expect it to work optimally in all cases. In order to overcome this, in this work we choose to utilize multiple different techniques in order to locate the most difficult facial features, i. e. the eyes and the mouth.

- MLP based mask

This approach refines eye locations extracted by the MLP network that was used in order to identify the eye pupils in the eye detection phase. It builds on the fact that eyelids usually appear darker than skin due to eyelashes and are almost always adjacent to the iris. Thus, by including dark objects near the eye centre, we add the eyelashes and the iris in the eye mask. The result is depicted in Fig. 6.10.

- Edge based mask

This is a mask describing the area between the upper and lower eyelids. Since the eye-center is almost always detected correctly from the MLP, the horizontal edges of the eyelids in the eye area around it are used to limit the eye mask in the vertical direction. For the detection of horizontal edges we utilize the Canny edge operator due to its property of providing good localization. Out of all edges detected in the image we choose the ones right above and below the detected eye center and fill the area between them in order to get the final eye mask. The result is depicted in Fig. 6.11.

- Region growing based mask

This mask is created using a region growing technique; the latter usually gives very good segmentation results corresponding well to the observed edges. The construction of this mask relies on the fact that facial texture is more complex and darker inside the eye area and especially in the eyelid-sclera-iris borders, than in the areas around them. Instead of using an edge density criterion, we utilize a simple yet effective new method to estimate both the eye centre and eye mask.

For each pixel in the area of the center of the eye we calculate the standard deviation of the luminance channel in its 3x3 neighbourhood and then threshold the result by the luminance of the pixel itself. This process actually results in the area

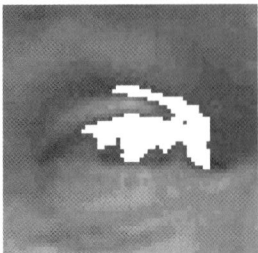

Fig. 6.10 The MLP based eye mask

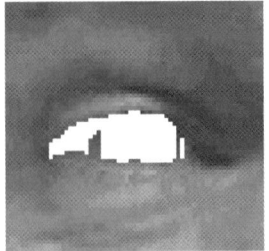

Fig. 6.11 The edge based eye mask

6.3 Feature Extraction

of the center of the eye being extended in order to include some of its adjacent facial characteristics. The same procedure is also repeated for 5x5 neighbourhoods; by using different block sizes we enhance the procedure's robustness against variations of image resolution and eye detail information. The two results are then merged in order to produce the final mask depicted in Fig.6.12. The process is found to fail more often than the other approaches we utilize, but it is found to perform very well for images of very-low resolution and low colour quality. The overall procedure is quite similar to that of a morphological bottom hat operation, with the difference that the latter is rather sensitive to the structuring element size.

- Luminance based mask

Finally, a second luminance based mask is constructed for eye and eyelid border extraction, using the normal probability of luminance using a simple adaptive threshold on the eye area. The result is usually a blob depicting the boundaries of the eye. In some cases, though, the luminance values around the eye are very low due to shadows from the eyebrows and the upper part of the nose. To improve the outcome in such cases, the detected blob is cut vertically at its thinnest points on either side of the eye centre; the resulting mask's convex hull is depicted in Fig. 6.13.

- Mask fusion

The reason we have chosen to utilize four different masks is that there is no standard way in the literature based on which to select the ideal eye localization methodology for a given facial image. Consequently, having the four detected masks it is not easy to judge which one is the most correct and select it as the output of the overall eye localization module. Instead, we choose to combine the different masks using a committee machine.

Given the fact that each one of the different methodologies that we have utilised has some known strong and weak points, the committee machine that is most suitable for the task of mask fusion is the mixture of experts dynamic structure, properly modified to match our application requirements [214]. The general structure of this methodology is presented in Fig. 6.14. It consists of k supervised modules called the experts and a gating network that performs the function of a media-

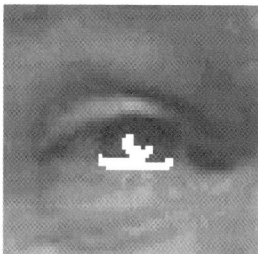

Fig. 6.12 The standard deviation based eye mask

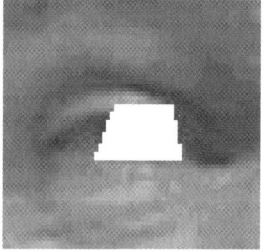

Fig. 6.13 The luminance based eye mask

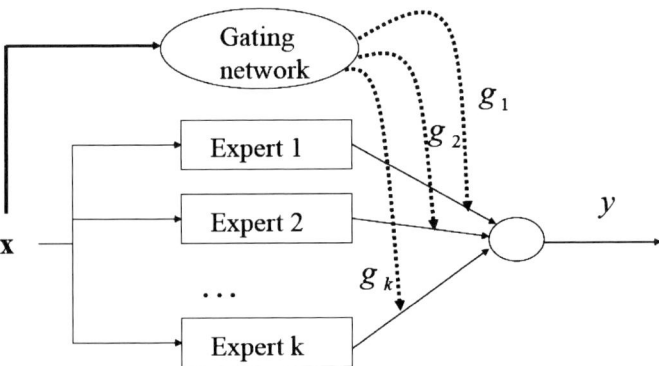

Fig. 6.14 Mixture of experts architecture

tor among the experts. The main assumption is that each one of the experts operates best in different regions of the input space in accordance with a probabilistic model that is known a priori, hence the need of the gating network.

The role of the gating network is to estimate, based on the input, the probability g_i that each individual expert i operates correctly, and to provide these estimations to the output combiner module. The gating network consists of a single layer of *softmax* neurons; the choice of *softmax* as the activation function for the neurons has the important properties of

$$0 \leq g_i \leq 1, \forall i \in 1..k$$

$$\sum_{i=1}^{k} g_i = 1 \quad (6.1)$$

i.e. it allows for the estimations to be interpreted as probabilities. In our work we have $k = 4$ experts; the implementations of the eye detection methodologies presented earlier in the section. The gating network favours the colour based feature extraction methods in images of high colour and resolution, thus incorporating the a priori known probabilities of success for our experts in the fusion process.

Additionally, the output combiner module which normally operates as $y = \bar{g} \cdot \bar{e}$, where $\bar{e}$ is the vector of expert estimations, is modified in our work to operate as

$$y = \frac{\bar{g} \cdot \bar{f} \cdot \bar{e}}{|\bar{f}|} \quad (6.2)$$

where $\bar{f}$ is the vector of confidence values associated with the output of each expert, thus further enhancing the quality of the mask fusion procedure. Confi-

6.3 Feature Extraction

Fig. 6.15 The final mask for the left eye

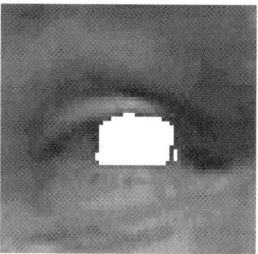

dence values are computed by comparing the location, shape and size of the detected masks to those acquired from anthropometric statistical studies.

The modified combiner module fuses the four masks together by making pixel by pixel decisions. The result of the procedure for the left eye in the frame of the running example is depicted in Fig. 6.15.

6.3.1.6 Mouth Localization

Similarly to the eyes, the mouth is a facial feature that is not always detected and localised successfully, mainly due to the wide range of deformations that are observed in it in sequences where the human is talking, which is the typical case in our chosen field of application. In this work we utilize the following methodologies in order to estimate the location and boundaries of the mouth:

- MLP based mask

An MLP neural network is trained to identify the mouth region using the neutral image. The network has similar architecture as the one used for the eyes. The train data are acquired from the neutral image. Since the mouth is closed in the neutral image, a long region of low luminance region exists between the lips. Thus, the mouth-candidate region of interest is first filtered with Alternating Sequential Filtering by Reconstruction (ASFR) to simplify and create connected areas of similar luminance. Luminance thresholding is then used to find the area between the lips.

This area is dilated vertically and the data depicted by this area are used to train the network.

Fig. 6.16 The MLP based mouth mask

The MLP network that has been trained on the neutral expression frame is the one used to produce an estimate of the mouth area in all other frames. The output of the neural network on the mouth region of interest is thresholded in order to form a binary map containing several small sub-areas. The convex hull of these areas is calculated to generate the final mask for the mouth. The result of this procedure is depicted in Fig. 6.16.

- Edge based mask

In this second approach, the mouth luminance channel is again filtered using ASFR for image simplification. The horizontal morphological gradient of the mouth region of interest is then calculated. Since the position of the nose has already been detected, and, as we have already explained, the procedure of nose detection rarely fails, we can use the position of the nose to drive the procedure of mouth detection. Thus, the connected elements that are too close to the nose center to be a part of the mouth are removed. From the rest of the mask, very small objects are also removed. A morphological closing is then performed and the longest of the remaining objects in the horizontal sense is selected as the final mouth mask. The result of this procedure is depicted in

- Lip corner based mask

The main problem of most intensity based methods for the detection of the mouth is the existence of the upper teeth, which tend to alter the saturation and intensity uniformity of the region. Our final approach to mouth detection takes advantage of the relative low luminance of the lip corners and contributes to the correct identification of horizontal mouth extent which is not always detected by the previous methods.

The image is first thresholded providing an estimate of the mouth interior area, or the area between the lips in case of a closed mouth. Then, we discriminate between two different cases:

1. there are no apparent teeth and the mouth area is denoted by a cohesive dark area
2. there are teeth and thus two dark areas appear at both sides of the teeth

In the first case mouth extend is straightforward to detect; in the latter mouth centre proximity of each object is assessed and the appropriate objects are selected. The convex hull of the result is then merged through morphological reconstruction with an horizontal edge map to include the upper and bottom lips.

In order to classify the mouth region in one of the two cases and apply the corresponding mouth detection methodology we start by selecting the largest connected object in the thresholded input image and finding its centroid. If the horizontal position of the centroid is close to the horizontal position of the tip of the nose, then we assume that the object is actually the interior of the mouth and we have the first case where there are no apparent teeth. If the centroid is not close to the horizontal position of the nose then we assume that we have the second case

6.3 Feature Extraction

Fig. 6.17 Edge based mouth mask

Fig. 6.18 The final mask for the mouth

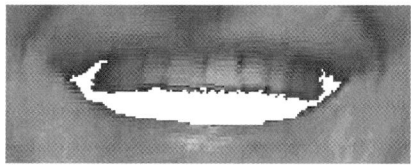

Fig. 6.19 The lip corner based mouth mask

where there are apparent teeth and the object examined is the dark area on one of the two sides of the teeth.

The result from the application of this methodology on the running example frame is depicted in Fig. 6.19.

- Mask fusion

The fusion of the masks is performed using a modified mixture of experts model similar to the one used for the fusion of the different masks for the eyes. The main difference here is that we cannot assess the probability of success of either of the methods using information readily available in the input, such as resolution or colour depth, and therefore the gating network has the trivial role of weighting the three experts equally.

This does not mean that the output combiner module is also trivial. Quite the contrary, we still utilize the modified version of the module, where anthropometric statistics are used to validate the three masks and the degree of validation is utilised in the process of mask fusion. The resulting mask for the mouth for the examined frame is depicted in Fig. 6.18.

6.3.1.7 Feature Points and FAPs

The facial feature masks detected in the previous section are not used directly for the emotion recognition procedure. They are merely the basis from which other, more refined, information elements will be drawn. Specifically, we utilize the masks in order to detect the marginal points of the studied elements on the face. Table 6.3 presents the complete list of points detected on the human face; these are a subset of the complete list of facial feature points defined in the MPEG-4 standard [229]. For example, Fig. 6.20 depicts the feature points detected on the frame of the running example.

As people change their facial expression their face is altered and the position of some of these points is changed (see Fig. 6.22). Therefore, the main information unit we will consider during the emotion classification stage will be the set of FAPs that describe a frame.

In order to produce this set we start by computing a 25-dimensional distance vector $\bar{d}$ containing vertical and horizontal distances between the 19 extracted FPs, as shown in Fig. 6.21. Distances are not measured in pixels, but in normalised scale-invariant MPEG-4 units, i.e. ENS, MNS, MW, IRISD and ES [229];

Table 6.3 Considered feature points

Feature Point	MPEG-4	Description
1	4.5	Outer point of Left eyebrow
2	4.3	Middle point of Left eyebrow
3	4.1	Inner point of Left eyebrow
4	4.6	Outer point of Right eyebrow
5	4.4	Middle point of Right eyebrow
6	4.2	Inner point of Right eyebrow
7	3.7	Outer point of Left eye
8	3.11	Inner point of Left eye
9	3.13	Upper point of Left eyelid
10	3.9	Lower point of Left eyelid
11	3.12	Outer point of Right eye
12	3.8	Inner point of Right eye
13	3.14	Upper point of Right eyelid
14	3.10	Lower point of Right eyelid
15	9.15	Nose point
16	8.3	Left corner of mouth
17	8.4	Right corner of mouth
18	8.1	Upper point of mouth
19	8.2	Lower point of mouth

6.3 Feature Extraction

unit bases are measured directly from FP distances on the neutral image, for example ES is calculated as the distance between FP_9 and FP_{13} (distance between eye pupils). The first step is to create the reference distance vector $\overline{d}_n$ by processing the neutral frame and calculating the distances described in Fig. 6.21 and then a similar distance vector $\overline{d}_i$ is created for each examined frame i. FAPs are calculated by comparing $\overline{d}_n$ and $\overline{d}_i$.

Fig. 6.20 Feature points detected on the input frame

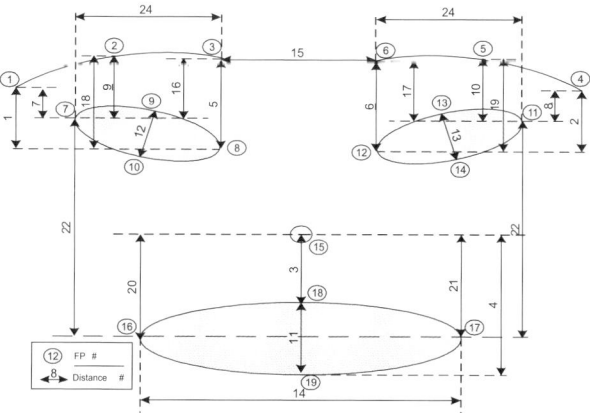

Fig. 6.21 Feature Point Distances

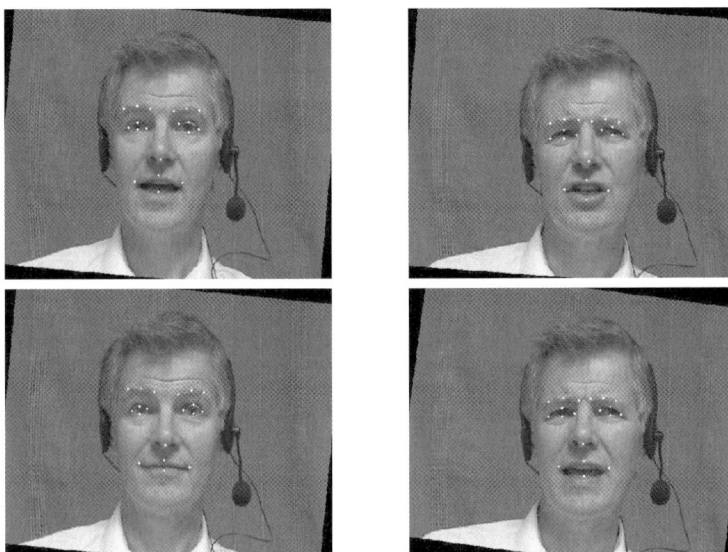

Fig. 6.22 Feature points detected from frames belonging to different sequences

6.3.2 Auditory Modality

6.3.2.1 State of the Art

Figure 6.23 presents classification results from a selection of recent studies that it seems fair to take as representing the state of the art. They show a complex picture, key parts of which have not been formulated clearly.

The horizontal axis shows one variable that clearly affects classification rate, that is, the number of categories considered. Under some circumstances, performance approaches 100% with two-choice classification. However, when techniques that achieved 90% in pair wise discriminations were used to assign samples to a set of 15 possible emotions [246], recognition rate falls to 8.7% (though note that this is still above chance). Intermediate points suggest an almost linear change between the two extremes.

The implication of the fall is that the techniques used in contemporary systems have limitations are concealed by using pair wise discrimination as a test. Yacoub et al. [246] clarified the issue in elegant follow-up studies, showing that their system effectively discriminated two emotion clusters, happiness/hot anger and sadness/boredom. These strongly suggest that that the information available may be more strongly related to dimensions than to categories: they appear to be a high- and a low-activation group. Of course, when analyzers that function as activation detectors are applied to stimuli which are either neutral or irritated, they will appear to detect irritation with high reliability: but the appearance is quite misleading.

6.3 Feature Extraction

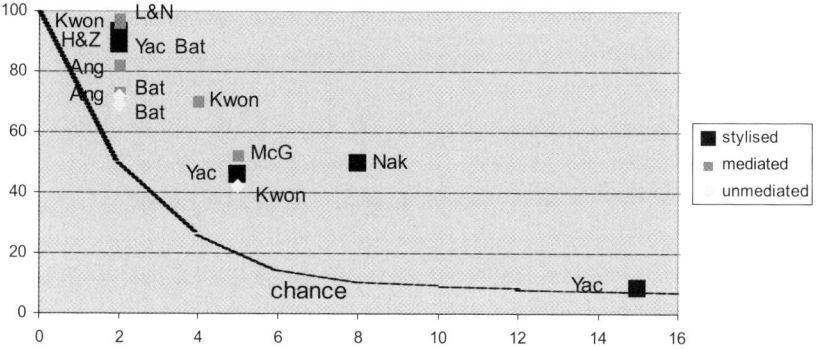

Fig. 6.23 Plot of discrimination results from key recent studies of emotion recognition – [242–249] against number of categories to be discriminated (horizontal axis)

The other major dimension, represented by the different types of symbol on the graph, is the extent to which the material has been stylised to simplify the task. The graph uses a simplified classification into three broad levels. Fully stylised speech is produced by competent actors, often in a carefully structured format. The second level, mediated speech, includes two main types: emotion simulated by people without particular acting skill or direction; and samples selected from a naturalistic database as clear examples of the category being considered. The third level includes speech that arises spontaneously from the speaker's emotional state, and which includes naturally occurring shades, not only well-defined examples.

6.3.2.2 Feature Extraction

An important difference between the visual and audio modalities is related to the duration of the sequence that we need to observe in order to be able to gain an understanding of the sequence's content. In case of video, a single frame is often enough in order for us to understand what the video displays and always enough for us to be able to process it and extract information. On the other hand, an audio signal needs to have a minimum duration for any kind of processing to be able to be made.

Therefore, instead of processing different moments of the audio signal, as we did with the visual modality, we need to process sound recordings in groups. Obviously, the meaningful definition of these groups will have a major role in the overall performance of the resulting system. In this work we consider sound samples grouped as tunes, i.e. as sequences demarcated by pauses. The basis behind this is that that although expressions may change within a single tune, the underlying human emotion does not change dramatically enough to shift from one quadrant to another. For this reason, the tune is not only the audio unit upon which we apply our audio feature detection techniques but also the unit considered during the operation of the overall emotion classification system.

From Section 6.3.2.1 it is quite obvious that selecting the right set of audio features to consider for classification is far from a trivial procedure. Batliner et al. [250]

classify features in categories and present a systematic comparison of different sets of features and combination strategies in the presence of natural, spontaneous speech. In order to overcome this in our work, we start by extracting an extensive set of 377 audio features. This comprises features based on intensity, pitch, MFCC (Mel Frequency Cepstral Coefficient), Bark spectral bands, voiced segment characteristics and pause length.

We analysed each tune with a method employing prosodic representation based on perception called Prosogram [251]. Prosogram is based on a stylization of the fundamental frequency data (contour) for vocalic (or syllabic) nuclei. It gives globally for each voiced nucleus a pitch and a length. According to a 'glissando threshold' in some cases we don't get a fixed pitch but one or more lines to define the evolution of pitch for this nucleus. This representation is in a way similar to the 'piano roll' representation used in music sequencers. This method, based on the Praat environment, offers the possibility of automatic segmentation based both on voiced part and energy maxima. From this model – representation stylization we extracted several types of features: pitch interval based features, nucleus length features and distances between nuclei.

Given that the classification model used in this work, as we will see in Section 6.4 , is based on a neural network, using such a wide range of features as input to the classifies means that the size of the annotated data set as well as the time required for training will be huge. In order to overcome this we need to statistically process the acoustic feature, so as to discriminate the more prominent ones, thus performing feature reduction. In our work we achieve this by combining two well known techniques: analysis of variance (ANOVA) and Pearson product-moment correlation coefficient (PMCC). ANOVA is used first to test the discriminative ability of each feature. This resulting in a reduced feature set, containing about half of the features tested. To further reduce the feature space we continued by calculating the PMCC for all of the remaining feature pairs; PMCC is a measure of the tendency of two variables measured on the same object to increase or decrease together. Groups of highly correlated (>90%) features were formed, and a single feature from each group was selected.

The overall process results in reducing the number of audio features considered during classification from 377 to only 32 [208]; all selected features are numerical and continuous.

6.4 Multimodal Expression Classification

6.4.1 The Elman Net

In order to consider the dynamics of displayed expressions we need to utilize a classification model that is able to model and learn dynamics, such as a Hidden Markov Model or a recursive neural network In this work we are using a recursive

6.4 Multimodal Expression Classification

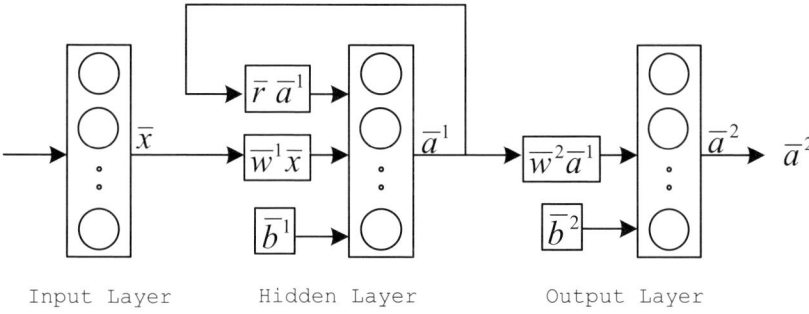

Fig. 6.24 The recursive neural network

neural network; see Fig. 6.24. This type of network differs from conventional feed-forward networks in that the first layer has a recurrent connection. The delay in this connection stores values from the previous time step which can be used in the current time step, thus providing the element of memory.

Although we are following an approach that only comprises a single layer of recurrent connections, in reality the network has the ability to learn patterns of a greater length as well, as current values are affected by all previous values and not only by the last one.

Out of all possible recurrent implementations we have chosen the Elman net for our work [190, 252]. This is a two-layer network with feedback from the first layer output to the first layer input. This recurrent connection allows the Elman network to both detect and generate time-varying patterns.

The transfer functions of the neurons used in the Elman net are tan-sigmoid for the hidden (recurrent) layer and purely linear for the output layer. More formally

$$a_i^1 = \tan sig(k_i^1) = \frac{2}{1+e^{-2k_i^1}} - 1, \ a_j^2 = k_j^2 \quad (6.3)$$

where a_i^1 is the activation of the i-th neuron in the first (hidden) layer, k_i^1 is the induced local field or activation potential of the i-th neuron in the first layer, a_j^2 is the activation of the j-th neuron in the second (output) layer and k_j^2 is the induced local field or activation potential of the j-th neuron in the second layer.

The induced local field in the first layer is computed as:

$$k_i^1 = \overline{w}_i^1 \cdot \overline{x} + \overline{r}_i \cdot \overline{a}^1 + b_i^1 \quad (6.4)$$

where $\overline{x}$ is the input vector, $\overline{w}_i^1$ is the input weight vector for the i-th neuron, $\overline{a}^1$ is the first layer's output vector for the previous time step, $\overline{r}_i$ is the recurrent weight vector and b_i^1 is the bias. The local field in the second layer is computed in the conventional way as:

$$k_j^2 = \overline{w}_j^2 \cdot \overline{a}^1 + b_j^2 \quad (6.5)$$

where $\overline{w}_i^2$ is the input weight and b_j^2 is the bias.

This combination of activation functions is special in that two-layer networks with these transfer functions can approximate any function (with a finite number of discontinuities) with arbitrary accuracy. The only requirement is that the hidden layer must have enough neurons [253, 254].

As far as training is concerned, the truncated back-propagation through time (truncated BPTT) algorithm is used [214].

The input layer of the utilised network has 57 neurons (25 for the FAPs and 32 for the audio features). The hidden layer has 20 neurons and the output layer has 5 neurons, one for each one of five possible classes: Neutral, Q1 (first quadrant of the Feeltrace [225] plane), Q2, Q3 and Q4. The network is trained to produce a level of 1 at the output that corresponds to the quadrant of the examined tune and levels of 0 at the other outputs.

6.4.1.1 Dynamic and Non Dynamic Inputs

In order for the network to operate we need to provide as inputs the values of the considered features for each frame. As the network moves from one frame to the next it picks up the dynamics described by the way these features are changed and thus manages to provide a correct classification in its output.

One issue that we need to consider, though, is that not all of the considered inputs are dynamic. Specifically, as we have already seen in section 6.3.2, as far as the auditory modality is concerned the tune is seen and processed as a single unit. Thus, the acquired feature values are referring to the whole tune and cannot be allocated to specific frames. As a result, a recurrent neural network cannot be used directly and unchanged in order to process our data.

In order to overcome this, we modify the simple network structure of Fig. 6.24 as shown in Fig. 6.25. In this modified version input nodes of two different types are utilised:

1. For the visual modality features we maintain the conventional input neurons that are met in all neural networks.
2. For the auditory modality features we use static value neurons. These maintain the same value throughout the operation of the neural network.

The auditory feature values that have been computed for a tune are fed to the network as the values that correspond to the first frame. In the next time steps, while visual features corresponding to the next frames are fed to the first input neurons of the network, the static input neurons maintain the original values for the auditory modality features, thus allowing the network to operate normally.

One can easily notice that although the network has the ability to pick up the dynamics that exist in its input, it cannot learn how to detect the dynamics in the auditory modality since it is only fed with static values. Still, we should comment that the dynamics of this modality are not ignored. Quite the contrary, the static feature values computed for this modality, as has been explained in section 6.3.2, are all based on the dynamics of the audio channel of the recording.

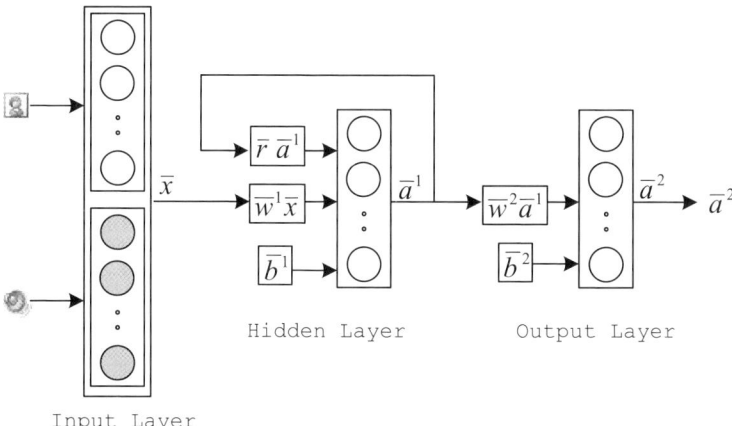

Fig. 6.25 The modified Elman net

6.4.2 Classification

The most common applications of recurrent neural networks include complex tasks such as modelling, approximating, generating and predicting dynamic sequences of known or unknown statistical characteristics. In contrast to simpler neural network structures, using them for the seemingly easier task of input classification is not equally simple or straight forward.

The reason is that where simple neural networks provide one response in the form of a value or vector of values at their output after considering a given input, recurrent neural networks provide such inputs after each different time step. So, one question to answer is at which time step the network's output should be read for the best classification decision to be reached.

As a general rule of thumb, the very first outputs of a recurrent neural network are not very reliable. The reason is that a recurrent neural network is typically trained to pick up the dynamics that exist in sequential data and therefore needs to see an adequate length of the data in order to be able to detect and classify these dynamics. On the other hand, it is not always safe to utilize the output of the very last time step as the classification result of the network because:

1. the duration of the input data may be a few time steps longer than the duration of the dominating dynamic behaviour and thus the operation of the network during the last time steps may be random
2. a temporary error may occur at any time step of the operation of the network

For example, in Fig. 6.24 we present the output levels of the network after each frame when processing the tune of the running example. We can see that during the first frames the output of the network is quite random and changes swiftly. When enough length of the sequence has been seen by the network so that the dynamics can be picked up, the outputs start to converge to their final values. But even then small changes to the output levels can be observed between consecutive frames.

Although these are not enough to change the classification decision (see Fig. 6.26) for this example where the classification to Q1 is clear, there are cases in which the classification margin is smaller and these changes also lead to temporary classification decision changes.

In order to arm our classification model with robustness we have added a weighting integrating module to the output of the neural network which increases its stability. Specifically, the final outputs of the model are computed as:

$$o_j(t) = c \cdot a_j^2 + (1-c) \cdot o_j(t-1) \tag{6.6}$$

where $o_j(t)$ is the value computed for the j-th output after time step t, $o_j(t-1)$ is the output value computed at the previous time step and c is a parameter taken from the $(0,1]$ range that controls the sensitivity/stability of the classification model. When c is closer to zero the model becomes very stable and a large sequence of changed values of k_j^2 is required to affect the classification results while as c approaches one the model becomes more sensitive to changes in the output of the network. When $c = 1$ the integrating module is disabled and the network output is acquired as overall classification result. In our work, after observing the models performance for different values of c, we have chosen $c = 0.5$.

In Fig. 6.29 we can see the decision margin when using the weighting integration module at the output of the network. When comparing to Fig. 6.27 we can clearly see that the progress of the margin is more smooth, which indicates that we have indeed succeeded in making the classification performance of the network more stable and less dependent on frame that is chosen as the end of a tune.

Of course, in order for this weighted integrator to operate, we need to define output values for the network for time step 0, i.e. before the first frame. It is easy to see that due to the way that the effect of previous outputs wares off as time steps elapse due to c, this initialization is practically indifferent for tunes of adequate length. On the other hand, this value may have an important affect on tunes that are very short. In this work, we have chosen to initialize all initial outputs at

$$\overline{o}(0) = 0 \tag{6.7}$$

Another meaningful alternative would be to initialize $\overline{o}(0)$ based on the percentages of the different output classes in the ground truth data used to train the classifier. We have avoided doing this in order not to add a bias towards any of the outputs, as we wanted to be sure that the performance acquired during testing is due solely to the dynamic and multimodal approach proposed in this work.

It is worth noting that from a modelling point of view it was feasible to include this integrator in the structure of the network rather than having it as an external module, simply by adding a recurrent loop at the output layer as well. We have decided to avoid doing so, in order not to also affect the training behaviour of the network, as an additional recurrent loop would greatly augment the training time and size and average length of training data required.

6.4 Multimodal Expression Classification

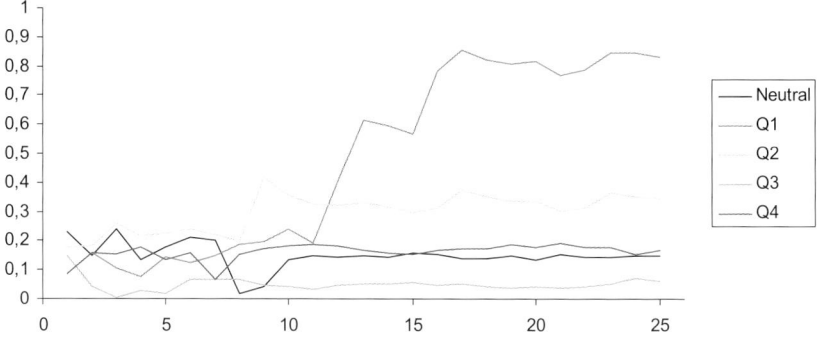

Fig. 6.26 Individual network outputs after each frame

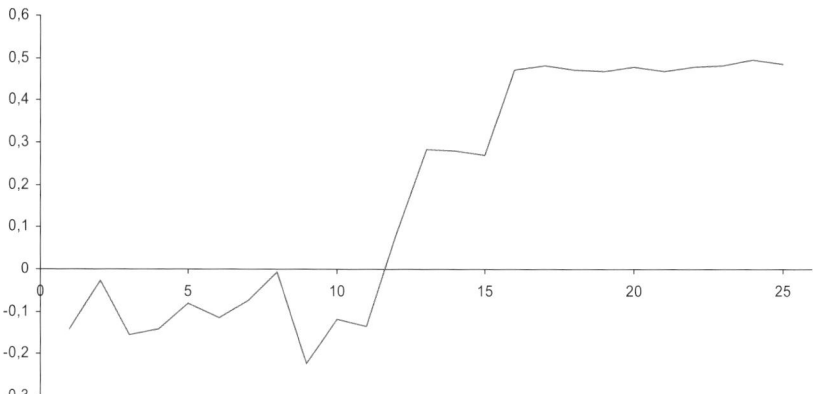

Fig. 6.27 Margin between correct and next best output

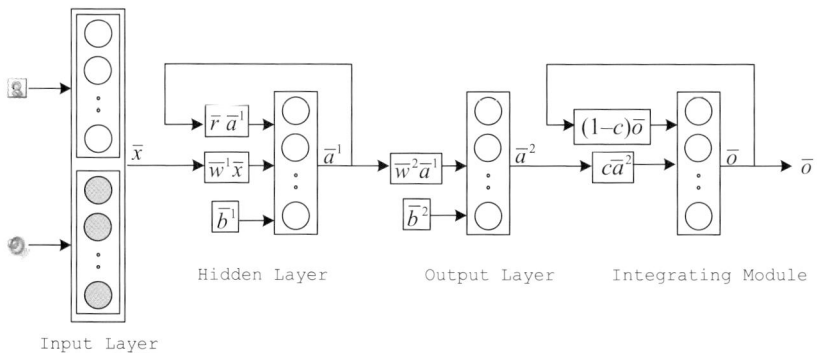

Fig. 6.28 The Elman net with the output integrator

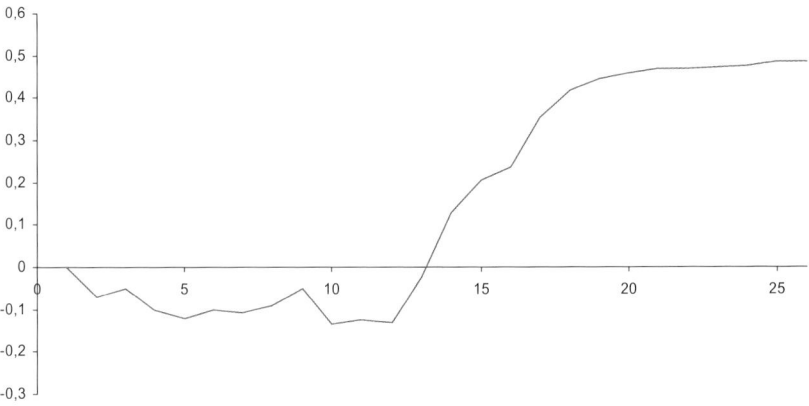

Fig. 6.29 Decision margin when using the integrator

6.5 Experimental Results

6.5.1 The Case for Naturalistic Data

As Douglas-Cowie mentions in [255], the easiest way to collect emotional data is to have actors simulate it; the data produced in such experiments lends itself to extremely high recognition rates possibly in the high 70s [256]. However, this kind of intercourse is very rare in everyday human-human or human-computer interaction contexts and this is attributed to a number of reasons:

- acted experiments usually involve reading a passage or uttering a specific phrase, which produces speech recordings with particular qualities. Older theoretical [257, 258], as well as some recent works [259] deal with picking out spontaneous or posed behaviour
- turn-based interaction results in associations between expressivity and stimulus: e.g. participants express anger at a particular cue or their computer freezing, thus catering for annotation of context, in addition to expressivity. An initial view of context can be the relation of an expressive utterance to answers to the W-questions ('who?', 'when?', 'where?', 'why?')
- speech and facial expressivity hardly follow a specific pattern; in addition to this, some parameters may be stable, e.g. the *valence* of the observed emotion, while *activation* may change to indicate semantic emphasis. Most 'regular' databases provide samples of participants moving from neutral to highly expressive and back

Since the aim of this work is to emphasize on the ability to classify sequences with naturalistic expressions, we have chosen to utilize the SAL database for training and testing purposes [260]. Recordings were based on the notion of the "Sensitive Artificial Listener", where the SAL simulates what some interviewers and

6.5 Experimental Results

Table 6.4 Class distribution in the SAL dataset

	Neutral	Q1	Q2	Q3	Q4	Totals
Tunes	47	205	90	63	72	477
Percentages	9,85%	42,98%	18,87%	13,21%	15,09%	100,00%

good listeners do, i.e. engages a willing person in emotionally coloured interaction on the basis of stock lines keyed in a broad way to the speaker's emotions. Although the final goal is to let the SAL automatically assess the content of the interaction and select the line with which to respond, this had not yet been fully implemented at the time of the creation of the SAL database and thus a "Wizard of Oz" approach was used for the selection of the SAL's answers [261].

A point to consider in natural human interaction is that each individual's character has an important role on the human's emotional state; different individuals may have different emotional responses to similar stimuli. Therefore, the annotation of the recordings should not be based on the intended induced emotion but on the actual result of the interaction with the SAL. Towards this end, FeelTrace was used for the annotation of recordings in SAL [222]. This is a descriptive tool that has been developed at Queen's University Belfast using dimensional representations, which provides time-sensitive dimensional representations. It lets observers track the emotional content of a time-varying stimulus as they perceive it. Table 6.1, illustrates the kind of display that FeelTrace users see.

The space is represented by a circle on a computer screen, split into four quadrants by the two main axes. The vertical axis represents activation, running from very active to very passive and the horizontal axis represents evaluation, running from very positive to very negative. It reflects the popular view that emotional space is roughly circular. The centre of the circle marks a sort of neutral default state, and putting the cursor in this area indicates that there is no real emotion being expressed. A user uses the mouse to move the cursor through the emotional space, so that its position signals the levels of activation and evaluation perceived by her/him, and the system automatically records the co-ordinates of the cursor at any time.

For reasons outlined in Section 6.2.1 the x-y coordinates of the mouse movements on the two-dimensional user interface are mapped to the five emotional categories presented in Table 6.1. Applying a standard pause detection algorithm on the audio channel of the recordings in examination, the database has been split into 477 tunes, with lengths ranging from 1 frame up to 174 frames. A bias towards Q1 exists in the database, as 42,98% of the tunes are classified to Q1, as shown in Table 6.4.

6.5.2 Statistical Results

From the application of the proposed methodology on the data set annotated as ground truth we acquire a measurement of 81,55% for the system's accuracy.

Table 6.5 Overall confusion matrix

	Neutral	Q1	Q2	Q3	Q4	Totals
Neutral	**34**	1	5	3	0	43
Q1	1	**189**	9	12	6	217
Q2	4	3	**65**	2	1	75
Q3	4	6	7	**39**	3	59
Q4	4	6	4	7	**62**	83
Totals	47	205	90	63	72	477

Table 6.6 Overall confusion matrix expressed in percentages

	Neutral	Q1	Q2	Q3	Q4	Totals
Neutral	**79,07%**	2,33%	11,63%	6,98%	0,00%	100,00%
Q1	0,46%	**87,10%**	4,15%	5,53%	2,76%	100,00%
Q2	5,33%	4,00%	**86,67%**	2,67%	1,33%	100,00%
Q3	6,78%	10,17%	11,86%	**66,10%**	5,08%	100,00%
Q4	4,82%	7,23%	4,82%	8,43%	**74,70%**	100,00%
Totals	9,85%	42,98%	18,87%	13,21%	15,09%	100,00%

Specifically, 389 tunes were classified correctly, while 88 were misclassified. Clearly, this kind of information, although indicative, is not sufficient to fully comprehend and assess the performance of our methodology.

Towards this end, we provide in Table 6.5 the confusion matrix for the experiment. In the table rows correspond to the ground truth and columns to the system's response. Thus, for example, there were 5 tunes that were labelled as neutral in the ground truth but were misclassified as belonging to Q2 by our system.

Given the fact that our ground truth is biased towards Q1, we also provide in Table 6.6 the confusion matrix in the form of percentages so that the bias is removed from the numbers. There we can see that the proposed methodology performs reasonably well for most cases, with the exception of Q3, for which the classification rate is very low. What is more alarming is that more than 10% of the tunes of Q3 have been classified as belonging to the exactly opposite quadrant, which is certainly a major mistake.

Still, in our analysis of the experimental results so far we have not taken into consideration a very important factor: that of the length of the tunes. As we have explained in section 6.5, in order for the Elman net to pick up the expression dynamics of the tune an adequate number of frames needs to be available as input. Still, there is a number of tunes in the ground truth that are too short for the network to reach a point where its output can be read with high confidence.

In order to see how this may have influence our results we present in the following separate confusion matrices for short and normal length tunes. In this con-

6.5 Experimental Results

Table 6.7 Confusion matrix for normal tunes

	Neutral	Q1	Q2	Q3	Q4	Totals
Neutral	**29**	0	0	0	0	29
Q1	0	**172**	3	0	0	175
Q2	1	1	**54**	0	0	56
Q3	0	0	0	**30**	0	30
Q4	0	0	0	0	**56**	56
Totals	30	173	57	30	56	346

Table 6.8 Confusion matrix for normal tunes expressed in percentages

	Neutral	Q1	Q2	Q3	Q4	Totals
Neutral	**100,00%**	0,00%	0,00%	0,00%	0,00%	100,00%
Q1	0,00%	**98,29%**	1,71%	0,00%	0,00%	100,00%
Q2	1,79%	1,79%	**96,43%**	0,00%	0,00%	100,00%
Q3	0,00%	0,00%	0,00%	**100,00%**	0,00%	100,00%
Q4	0,00%	0,00%	0,00%	0,00%	**100,00%**	100,00%
Totals	8,67%	50,00%	16,47%	8,67%	16,18%	100,00%

text we consider as normal tunes that comprise at least 10 frames and as short tunes with length from 1 up to 9 frames.

First of all, we can see right away that the performance of the system, as was expected is quite different in these two cases. Specifically, there are 83 errors in just 131 short tunes while there are only 5 errors in 346 normal tunes. Moreover, there are no severe errors in the case of long tunes, i. e. there are no cases in which a tune is classified in the exact opposite quadrant than in the ground truth.

Overall, the operation of our system in normal operating conditions (as such we consider the case in which tunes have a length of at least 10 frames) is accompanied by a classification rate of 98,55%, which is certainly very high, even for controlled data, let alone for naturalistic data.

6.5.3 Quantitative Comparative Study

In a previous work we have proposed a different methodology to process naturalistic data with the goal of estimating the human's emotional state [202]. In that work a very similar approach is followed in the analysis of the visual component of the video with the aim of locating facial features. FAP values are then fed into a rule based system which provides a response concerning the human's emotional state.

Table 6.9 Classification rates on parts of the naturalistic data set

Methodology	Classification rate
Rule based	78,4%
Possibilistic rule based	65,1%
Dynamic and multimodal	98,55%

Table 6.10 Classification rates on the naturalistic data set

Methodology	Classification rate
Rule based	27,8%
Possibilistic rule based	38,5%
Dynamic and multimodal	98,55%

In a later version of this work, we evaluate the likelihood of the detected regions being indeed the desired facial features with the help of anthropometric statistics acquired from [262] and produce degrees of confidence which are associated with the FAPs; the rule evaluation model is also altered and equipped with the ability to consider confidence degrees associated with each FAP in order to minimize the propagation of feature extraction errors in the overall result [209].

When compared to our current work, these systems have the extra advantages of

1. considering expert knowledge in the form of rules in the classification process
2. being able to cope with feature detection deficiencies and

On the other hand, they are lacking in the sense that

3. they do not consider the dynamics of the displayed expression and
4. they do not consider other modalities besides the visual one.

Thus, they make excellent candidates to compare our current work against in order to evaluate the practical gain from the proposed dynamic and multimodal approach. In Table 6.10 we present the results from the two former and the current approach. Since dynamics are not considered, each frame is treated independently in the pre-existing systems. Therefore, statistics are calculated by estimating the number of correctly classified frames; each frame is considered to belong to the same quadrant as the whole tune.

It is worth mentioning that the results are from the parts of the data set that were selected as expressive for each methodology. But, whilst for the current work this refers to 72,54% of the data set and the selection criterion is the length of the tune, in the previous works only about 20% of the frames was selected with a criterion of the clarity with which the expression is observed, since frames close to the beginning or the end of the tune are often too close to neutral to provide meaningful visual input to a system.

6.5.4 Qualitative Comparative Study

As we have already mentioned, during the recent years we have seen a very large number of publications in the field of the estimation of human expression and/or emotion. Although the vast majority of these works is focused on the six universal expressions and use sequences where extreme expressions are posed by actors, it would be an omission if not even a qualitative comparison was made to the broader state of the art.

In Table 6.10 we present the classification rates reported in some of the most promising and well known works in the current state of the art. Certainly, it is not possible or fair to compare numbers directly, since they come from the application on different data sets. Still, it is possible to make qualitative comparisons base on the following information:

1. The Cohen2003 is a database collected of subjects that were instructed to display facial expressions corresponding to the six types of emotions. In the Cohn–Kanade database subjects were instructed by an experimenter to perform a series of 23 facial displays that included single action units and combinations of action units.
2. In the MMI database subjects were asked to display 79 series of expressions that included either a single AU or a combination of a minimal number of AUs or a prototypic combination of AUs (such as in expressions of emotion). They were instructed by an expert (a FACS coder) on how to display the required facial expressions, and they were asked to include a short neutral state at the beginning and at the end of each expression. The subjects were asked to display the required expressions while minimizing out-of-plane head motions.
3. The original instruction given to the actors has been taken as the actual displayed expression in all abovementioned databases, which means that there is an underlying assumption is that there is no difference between natural and acted expression.

As we can see, what is common among the datasets most commonly used in the literature for the evaluation of facial expression and/or emotion recognition is that expressions are solicited and acted. As a result, they are generally displayed clearly and to their extremes. In the case of natural human interaction, on the other hand, expressions are typically more subtle and often different expressions are mixed. Also, the element of speech adds an important degree of deformation to facial features which is not associated with the displayed expression and can be misleading for an automated expression analysis system.

Consequently, we can argue that the fact that the performance of the proposed methodology when applied to a naturalistic dataset is comparable to the performance of other works in the state of the art when applied to acted sequences is an indication of its success. Additionally, we can observe that when extremely short tunes are removed from the data set the classification performance of the proposed

Table 6.11 Classification rates reported in the broader state of the art [263, 264]

Methodology	Classification rate	Data set
TAN	83,31%	Cohen2003
Multi-level HMM	82,46%	Cohen2003
TAN	73,22%	Cohn–Kanade
PanticPatras2006	86,6%	MMI
Proposed methodology	81,55%	SAL Database
Proposed methodology	98,55%	Normal section of the SAL database

approach exceeds 98%, which, in current standards, is very high for an emotion recognition system.

6.6 Conclusions

In this work we have focused on the problem of human emotion recognition in the case of naturalistic, rather than acted and extreme, expressions. The main elements of our approach are that i) we use multiple algorithms for the extraction of the "difficult" facial features in order to make the overall approach more robust to image processing errors, ii) we focus on the dynamics of facial expressions rather that on the exact facial deformations they are associated with, thus being able to handle sequences in which the interaction is natural or naturalistic rather than posed or extreme and iii) we follow a multimodal approach where audio and visual modalities are combined, thus enhancing both performance and stability of the system.

From a more technical point of view, our contributions include: i) A modified input layer that allows the Elman net to process both dynamic and static inputs at the same time. This is used to fuse the fundamentally different visual and audio inputs in order to provide for a truly multimodal classification scheme. ii) A modified output scheme that allows the Elman that integrates previous values, with value significance decreasing exponentially through time. This allows the network to display augmented stability. iii) a modified mixture of experts module that, additionally to characteristics drawn from the experts' input, can also draw information from the experts' output in order to drive the output mediation step. This is used in order to incorporate the results from the statistical anthropometric evaluation of the acquired masks in the operation of the output combiner module.

Practical application of our methodology in a ground truth data set of naturalistic sequences has given a performance of 98,55% for tunes that are long enough for dynamics to be able to be picked up in both the visual and the audio channel.

For our future work, we intend to further extend our work in multimodal naturalistic expression recognition by considering more modalities such as posture and gestures and by incorporating uncertainty measuring and handling modules in order to maximize the system's performance and stability in difficult and uncontrolled environments.

6.7 Acknowledgements

The authors would like to thank all collaborators within the Humaine Network of Excellence, as well as all those that participated in the SAL data collection and annotation process. This work has been funded by the FP6 Network of Excellence Humaine: Human-Machine Interaction Network on Emotion, http://www.emotion-research.net.

Chapter 7
Two SIMILAR Different Speech and Gestures Multimodal Interfaces

Alexey Karpov[1], Sebastien Carbini[2], Andrey Ronzhin[1], Jean Emmanuel Viallet[3]

[1] St. Petersburg Institute for Informatics and Automation of the Russian Academy of Sciences, 39, 14-th line, 199178, St. Petersburg, Russia
[2] Ifremer/LASAA, Technopôle Brest-Iroise, BP 70, 29280 Plouzané, France
[3] Orange Labs France Telecom, 2 avenue Pierre Marzin, BP40, 22307 Lannion, France

7.1 Introduction and State-of-the-art

Mouse and keyboard are the most common input interfaces of human-computer interaction within the windows, icons, menus, pointer (WIMP) paradigm. However during last 25 years, scientists and engineers have been actively looking for and developing more intuitive, natural and ergonomic interfaces. Of particular interest are multimodal interfaces that use speech and gestures to create human-machine interfaces similar to human-human communication. This new class of interfaces aims to recognize naturally occurring forms of human language and behaviour, which incorporate several recognition-based technologies into one intelligent multimodal system.

Speech and gesture are often associated in multimodal interfaces, even if it can be argued that other modalities such as gaze (or even speech) involve a gesture of the body. Thus by speech, we refer to the information held in the acoustical signal, even if audiovisual automatic speech recognition can be carried out by a human or a machine. Speech, the oral component of language corresponds to a code and different languages lead to different codes that must be learned by automatic speech recognition engines.

Speech or speech communication refers to the processes associated with the production and perception of sounds used in spoken language, and automatic speech recognition (ASR) is a process of converting a speech signal to a sequence of words, by means of an algorithm implemented as a software or hardware module. Several kinds of speech are identified: spelled speech (with pauses between phonemes), isolated speech (with pauses between words), continuous speech (when a speaker does not make any pauses between words) and spontaneous natural speech. The most common classification of ASR by recognition vocabulary is following [265]:

- small vocabulary (10-1000 words);
- medium vocabulary (up to 10 000 words);

- large vocabulary (up to 100 000 words);
- extra large vocabulary (above 1 million words that can be applied for inflective or agglutinative languages only)

Recent automatic speech recognizers exploit mathematical techniques such as Hidden Markov Models (HMMs), Artificial Neural Networks (ANN), Bayesian Networks or Dynamic Time Warping (dynamic programming) methods. The most popular ASR models apply speaker-independent speech recognition though in some cases (for instance, personalised systems that have to recognize their owner only) speaker-dependant systems are more adequate.

Gesture is an expressive, meaningful body motion i.e. physical movement of the fingers, hands, arms, head, face, or body with the intent to convey information or interact with the environment. As speech, gesture can correspond to a code as in sign language. It can also relate to ordinary life experience for example when indicating a direction in space with one hand, rotating a car wheel or expanding a rubber band with two hands. In these latter cases, gestures can be interpreted by other persons, sharing a common collaborative environment, without explicit knowledge of a code as with native or foreign languages. State-of-the-art gesture recognition technologies can be split into three main categories [266]:

1. *Pen-based recognition*: This kind of processing involves recognizing gestures from 2D input devices (for instance, a pen or a touch screen). However this interaction is quite constraining and techniques that allow the user to move around and interact in more natural ways are more compelling. This kind of interface is preferable for personal digital assistants (PDAs) that have a touch screen and low computational resources (Fig. 7.1).
2. *Hardware tracker-based recognition*: There are hardware devices that can process input gestures primarily for tracking user's head, eye gaze, hand or body position. Head helmet, body suits or instrumented gloves include different sensors to recognize natural gestures. For instance, there exist several commercially available head trackers. NaturalPoint manufactures the SmartNAV hands-free mouse that consists of a special transmitter-receiver device working

Fig. 7.1 An example of pen-based personal digital assistant device.

7.1 Introduction and State-of-the-art

in the infrared mode and of several reflective marks, which should be attached to the face of a user or to a special hat (Fig. 7.2, left). The company InterSence produces InterTrax, a professional tracker for helmets dedicated to virtual reality applications and computer stereo glasses. There is a gyroscope inside the device, which allows determining the orientation and position of the head (Fig. 7.2, right).

3. *Vision-based recognition:* This kind of gesture recognition in contrast to the previous ones does not require the user to wear any special markers or to use hardware equipment. Vision-based recognition systems employ one or several video-cameras to capture images as well as computer algorithms to interpret video frames and recognize human's activity and natural gestures (Fig. 7.3). Hand, head or body gestures and eyes gaze can be interpreted by a vision-based recognition system.

Both speech and vision-based gesture recognition allow building contactless interaction devices that offer the advantage of distant interaction without having to be within reach or in contact with a physical device in order to interact.

Although pointing and selection are very common interactions with a computer, more complex interactions, either monomodal or multimodal, can be performed with speech and vision-based gesture.

Speech is a very useful modality to reference objects and actions on objects whereas pointing gesture is a very powerful modality to indicate spatial locations, even if it is possible for a disabled person to specify a location with words or to

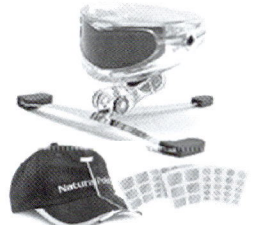

Fig. 7.2 Two examples of hardware-based head trackers.

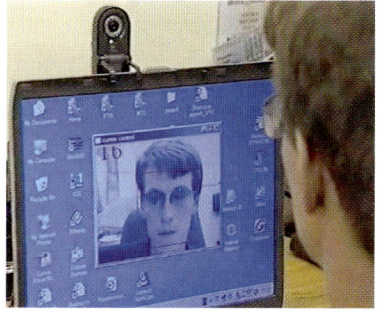

Fig. 7.3 An example of vision-based head gesture recognition system.

use sign language to communicate. Both modalities benefit from a long learning process which begins in early childhood. Pointing gesture for communication is clearly established at the age of 14 months [267], and according to Poggi [268], not only hands or fingers can be used to designate the direction of an object but also the tip of the nose, the chin, the lips or the gaze.

This is why, our goal is to build multimodal interfaces that benefits from the advantages of both modalities to interact as easily and naturally as possible using a representation of the world with which we are familiar where objects are described by their name and attributes and locations indicated with hand gesture or nose gesture in case of hands or arms disabilities.

The general architecture of speech & gestures multimodal user interface is presented on the Fig. 7.4 [269]. Multimodal information processing uses late-stage semantic architecture and two input modes are recognised in parallel and processed by understanding components. The results involve partial meaning representations that are fused by the multimodal fusion component, which also is influenced by the dialogue management and interpretation of current context. The best-ranked multimodal interpretation is transformed into control commands which are sent to a user application. System's feedback includes multimedia output, which may incorporate text-to-speech and non-speech audio, graphics and animation.

Since the "Put-That-There" system described by Bolt [270], it is known that among gestures naturally performed by humans as they communicate, pointing gestures associated to speech recognition, lead to powerful and more natural man machine interfaces.

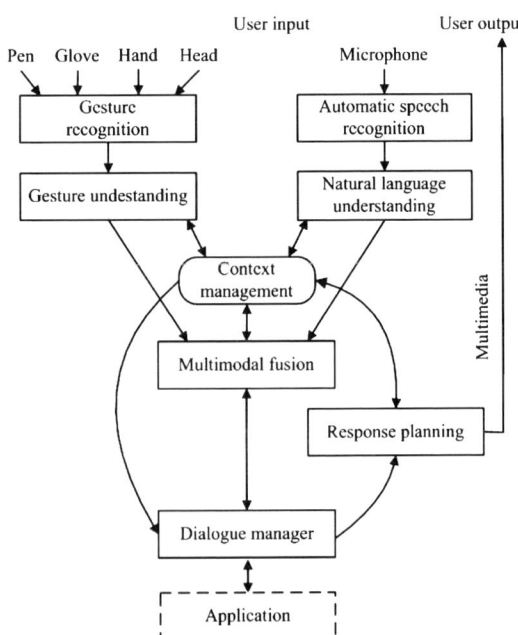

Fig. 7.4 General architecture of gesture and speech-based multimodal interface.

Pointing alone is a very efficient interaction mechanism for exploring a visual scene. Computer-based pointing gesture recognition [271] allows to explore digitised artworks.

The FlowMouse device described in [272] is an optical flow vision-based system where a camera, located on top of a laptop screen overlooks a hand above the laptop keyboard. Optical flow based recognition is triggered when a user touches a mouse button. Therefore, this system has a limited field of view, which can accommodate only one user at a sufficiently close distance with the interface so that the user can be in physical contact.

A facial gesture interface [273], allows cursor control with head movements analysed by a camera on top of a PC display. Since the nose movement used to control the cursor cannot directly map the surface of the display, a displacement (corresponding to head rotation) gain must be introduced to cover without resulting uncomfortable movements the entire horizontal display. Recognition of mouth opening gesture is interpreted as a mouse click. But the system does not include speech recognition, which allows a richer interaction than a simple opening or closing of the mouth.

Kaiser et al. [274] propose a gaze, gesture and speech multimodal interaction, for virtual and augmented reality, allowing mutual disambiguation of modalities. But gaze and gesture involve four 6DOF trackers worn by a user. Stiefelhagen et al. [275] use stereo computer vision for gesture and head pose recognition for multimodal interaction with a robot. Head pose improves to decide if the user is pointing and where with an 80% recall, 87% precision. Pointing gesture (of average length 1.75 s) has error below 20%, sufficient for interacting with a robot in an open scene, where no available display feedback cursor allows a user to correct the estimated pointing direction.

In Sharma et al. [276], for large plasma display multimodal applications, due to the statistical methods employed for continuous recognition of speech and gesture (point, contour and circle), results are emitted with a time delay of typically 1 s. Deictic words are found to occur during or after gesture in 97% of the case. With a single camera, hand detection integrates evidence over a sequence of 60 frames, and the hand tracker is a continuous version of the hand detector geared towards skin coloured moving objects. Although this tracking requires less than 15% of CPU, the placement of the camera and microphone require a user to interact with the system from a certain location to ensure optimal performance which restricts users from moving freely in a room and sharing space for interactive and collaborative purposes.

In the study of Schapira and Sharma [277], pointing gestures are recognised by computer vision with a camera on top of a plasma display whereas selection is performed either by a Point & Wait during 300 ms, a Point & Shake near the target, or by a Point & Speak interaction. However, the 20 ms time delay necessary to trigger speech detection is far longer than the 1 s delay, necessary for gesture or speech recognition delay involve in their operational multimodal system reported in (Sharma et al. 2003).

In [278], with a giant screen (2x1.5 m), for a transport management multimodal application, gesture selection is achieved with a "point and wait" (500 ms) strategy. A low-cost camera and a light, a meter away from the user, are used to track only one hand. This set-up is an obstacle in the room and grounds the user to a specific location, although allowing pointing anywhere on the giant screen with a good precision (minimum effective target of 80 pixels).

In [279], with a stereo camera on top of a giant screen, body pose of a user freely moving in a large field of view is estimated with a 3D articulated model and gestures recognize with hidden Markov models. Inferring body pose with speech, decreases pose error rate from 20% to 12%, and average pointing error from 13% to 7% using speech with a word error rate of 37%.

Examples of two multimodal interfaces, allowing contactless interaction using speech and gesture, developed within the framework of the SIMILAR Network of Excellence [280] are presented below in sections 2 and 3.

7.2 ICANDO Multimodal Interface

This section presents ICANDO (Intellectual Computer AssistaNt for Disabled Operators) multimodal user interface intended mainly for assistance to the persons without hands or with disabilities of their hands or arms, but it could be useful for contactless human-computer interaction by ordinary users too. Users can manipulate screen cursor by moving their head and giving the speech commands instead of using such input devices as mouse, keyboard, trackball, touchpad or joystick. It combines the module for automatic recognition of voice commands in English, French and Russian as well as the head tracking module in one multimodal interface. Video and audio signals are captured by a low cost web-camera and recognised automatically.

ICANDO provides contactless access to a Graphical User Interface (GUI) of an operational system and peripheral devices hands-free. Possible application of the interface for the rehabilitation process allows a person with special needs to enter into information society, to perform any computer work, to communicate with other people via Internet and to make a successful professional career and personal life as the result.

7.2.1 Objectives

Nowadays the European Society pays especial attention to the problems of the physically handicapped persons with partial or full dysfunctions of their body parts and organs. There are several kinds of physical disabilities of: voice, ears, eyes, hands, legs, besides millions of people all over the world are not able to speak, hear or move their hands. Many governmental programs are elaborated for social and professional rehabilitation of disabled people. Informational technologies have made significant contribution in this area too. For instance, there are talking books for blind persons, automatic recognition of sign language and cued speech, multi-modal systems for visually and speaking impaired people [281]. Such systems allow supporting the equal participation and socio-economic integration of people with disabilities in the information society and increase their independence from other people. Some people cannot operate a personal computer with a standard mouse-manipulator or a keyboard because of hands or arms disabilities. It leads to restriction of the kinds of activities and strong limitation of their socio-economic activities and life status. We propose an alternative for these persons: a multimodal interface, which allows one to control a computer without input devices, by means of: (1) head or face movements controlling the mouse pointer (cursor) on the monitor screen; (2) speech input for giving control commands.

It is clear that disability may affect also the person's neck along with the hands or the arms. A human could suffer from problems connected with activity of neck and reduced ability of moving the head in one or more directions. In some of such instances an eye tracking system could be successfully applied [282]. In this case gaze direction of a user can point to objects on the monitor screen and navigate mouse pointer through the menu items. Now there are some researches in the area of human's gaze tracking, for instance, Eyegaze System [283] or Visual Mouse [284] projects. Moreover eye blinks or eyebrow raises can be processed as a signal for clicking mouse buttons (left eye double blink for left button and right eye double blink for right one) [285]. However some researches have shown that usage of eye tracking is worse than that of head tracking in such parameters as: task performance, human's workload and comfort both for novices and experienced users [286]. The hardware required for the gaze tracking system (digital camera with high resolution) is much more expensive as well. Meanwhile the speech input is only one acceptable alternative to the keyboard for hand-impaired operators.

7.2.2 System's Description

7.2.2.1 The Architecture of the ICANDO Multimodal Interface

Two human's natural input modalities: speech and head motions are processed by ICANDO. Both modalities are active [269], so their input must be controlled constantly (non-stop) by the system. It processes human's speech and head motions in parallel and then combines both informational streams in a joint multimodal command, which is used for operating with GUI of a computer. Figure 7.5 shows the general architecture of ICANDO multimodal interface. Each of the modalities transmits own semantic information: head position indicates the coordinates of the mouse pointer at the current time moment, while speech signal transmits the information about the meaning of the action, which must be performed with an object selected with the cursor (or irrespectively to the cursor position) [287].

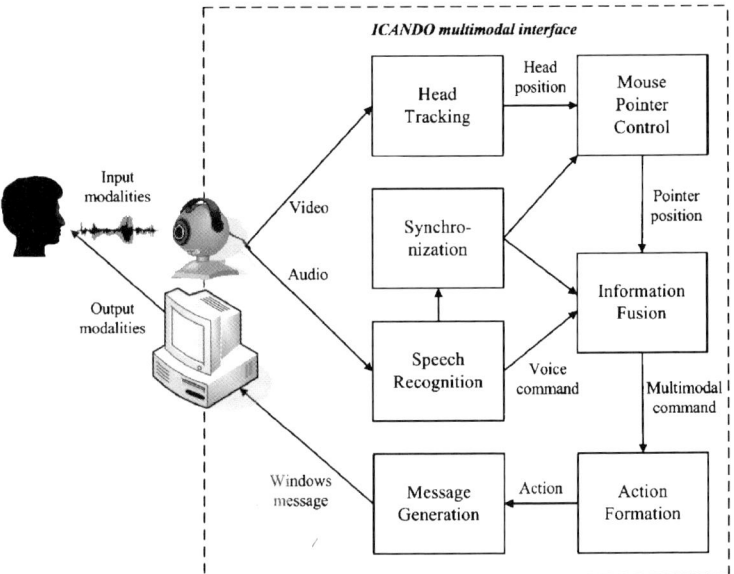

Fig. 7.5 The general architecture of the ICANDO multimodal user interface

The audio signal captured via a microphone is processed constantly by the automatic speech recognition module. The speech recognition process is started at the moment of triggering an algorithm for speech endpoint detection. The last finds presence of some useful signal (speech) different from silence (or permanent background noise). The speech recognition process is finished after finding a best hypothesis of voice command recognition. The synchronization of two information streams is performed by the speech recognition module, which gives the special signals for storing of the mouse cursor coordinates, calculated by the head tracking module, and for multimodal information fusion. For organization of the hands-free computer controlling, the standard means of the operating system MS Windows are used, namely the mechanism of message queues [288]. After the stage of multimodal information fusion a Windows message, corresponding to the required action, is generated and then sent to the queue of an active window. Each of the Windows-compatible programs has its own message queue and ICANDO sends the messages into this queue. When any such message is received, an applied program identifies it and performs corresponding actions. After the completion of the actions the computer produces a multimodal output (feedback) to the operator in the form of graphical interface, sound, speech synthesis, etc.

USB web-camera Logitech QuickCam for Notebooks Pro is applied as the hardware of the multimodal interface. The camera provides both video signal with the resolution 640x480 and 25 fps and audio signals with 16 KHz sampling via the built-in microphone with acceptable SNR level. Usage of a professional digital camera (Sony DCR-PC1000E was tested in some experiments) provides better precision of video processing and quality of audio signal. However ICANDO is positioned as a low-cost multimodal interface and should be available for the most of potential users so the web-camera with the price under 50 € is employed.

7.2.2.2 Automatic Speech Recognition

ICANDO system is able to recognize voice commands of a user in the three languages: English, French and Russian. The audio signal is captured by the microphone built-in the web-camera and the SIRIUS system (SPIIRAS Interface for Recognition and Integral Understanding of Speech) is applied for automatic speech recognition [289]. The SIRIUS system has also been used for automatic speech recognition in other multimodal interfaces in medical (neurosurgical) applications [290]. The voice command list contains 23 commands, which are similar to the mouse buttons ("Left", "Right", "Double click", "Scroll up", etc.) or the keyboard shortcuts ("Print", "Open", "Enter", "Start", etc.). All the voice commands can be divided into four classes according to their functional purpose: mouse manipulation commands, keyboard commands, Windows Graphical User Interface commands, as well as the special command "Calibration" in order to start the process of the head tracker tuning. However the mouse manipulation commands only have multimodal nature. They use the information on the coordinates of the mouse cursor at a current time moment. All the other commands are purely speech ones (unimodal) and pointer position is not taken into account for them. Theoretically, two voice commands ("Left" and "Right") could be enough to work with a PC (or a PDA), but introduction of the additional commands, which are often used by a user, increases the speed of human-computer interaction essentially.

The captured audio signal is digitised with the sampling rate 16 KHz, then and the samples are combined in the segments (100 segments per second) and the features are extracted. For the speech parameterization the mel-frequency cepstral coefficients (MFCC) with the first and second derivatives are used. The modelling and recognition of phonemes and words of the vocabulary are based on Hidden Markov Models (HMM). The acoustical models are realised as HMMs of the triphones with mixture Gaussian probability density functions. HMMs of triphones have 3 meaningful states (and 2 additional states intended for concatenation of the triphones in the word models).

The size of the vocabulary for the task of voice command recognition is quite small, and the baseline system of the speech recognition can be applied in this case [291]. Nevertheless several original approaches are being introduced in the automatic speech recognition system in order to apply the SIRIUS system for the task of Russian speech recognition, in particular, the morphemic level of the representation of the Russian speech and language. The Russian language is highly inflective and contains over 3 millions of different word-forms on the whole. Figure 7.6 presents the process of speech recognition in the SIRIUS system from feature extraction to formation of the recognition hypothesis. Morphemic representation of the recognition vocabulary reduces its size in several times, while

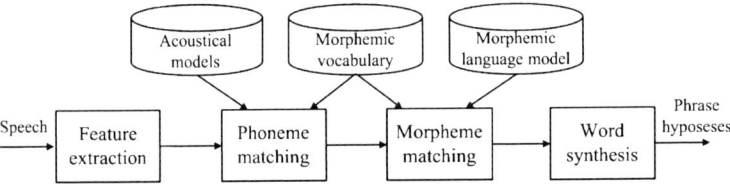

Fig. 7.6 The architecture of the morpheme-based automatic speech recognition system SIRIUS.

introduction of the morphemic analysis and synthesis increases the speed and accuracy of continuous speech recognition [289].

Now the operator has to use a special program such as MS Windows On-Screen Keyboard to enter any text in a computer working with ICANDO. This program is a virtual keyboard on the computer desktop like in a Personal Digital Assistant. Alternatively the Dasher text entry interface [292], that is a computer accessibility tool enabling users to enter text efficiently using a natural continuous pointing device rather than a keyboard, can be applied. However, in future the ICANDO multimodal interface is to be enhanced with the dictation system based on SIRIUS engine that will accelerate essentially the speed of text entering.

7.2.2.3 Head Tracking and Mouse Pointer Control

It is known that the head tracking can be performed with two different ways: hardware and software-based methods. When choosing a hardware technique a user has to wear some special devices on his/her head. Therefore ICANDO interface employs optical flow video processing for tracking natural operator's head motions instead of hand-controlling motions. A software method for tracking operator's head motions was elaborated for ICANDO. It is based on the software library Intel Open Source Computer Vision Library that realizes many algorithms for image and video processing.

An original approach providing video processing in the real-time mode was developed for the head tracking task. Just one USB web-camera is applied for the head tracking that provides the calculation of the head position coordinates in the 2D space. To control the mouse pointer it is enough to have the 2D model because the screen desktop is flat. But the usage of the second camera allows creating a 3D model of head motion. In some applications (for instance, games) the third coordinate could be used for zoom a picture on the screen coming the head near to the screen. The present head tracking method has two stages of functioning: calibration and tracking. In the starting stage a face position is defined in the video stream. It is realised by the software module, which uses the Haar-based object detector to find rectangular regions in the given image that can likely contain human's face [293]. Such region has to be larger than 250x250 points that allows the system to find only one a biggest face on an image accelerating the video processing. Then the second stage is started to track natural markers of the opera-

7.2 ICANDO Multimodal Interface

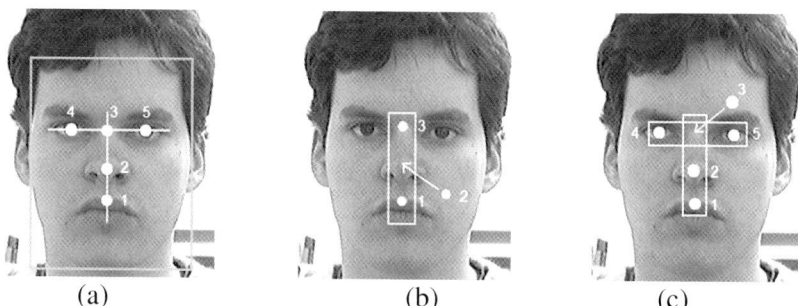

Fig. 7.7 The 2D model of facial points for the software tracking of operator's head.

tor's face. Some researches [294] said that the most suitable point on a face for tracking is the tip of nose. It is the center of the face and when we make any gestures with the head (turn to the right, left, up or down) the position of the tip of nose moves to that direction and thus it can indicate the position of the mouse cursor on a desktop. However in contrast to the Canadian hands-free mouse system Nouse [294], where the nose position is employed for tracking and controlling the mouse pointer only, the system of 5 natural points is applied in ICANDO to improve the robustness of the operator's head tracking. These points are: the center of the upper lip, the tip of the nose, the point between the eyebrows, the left eye (iris) and the right eye points (Fig. 7.7, a). These points form two perpendicular lines: vertical (points 1-3) and horizontal (points 3-5). When a face is found on the image the points of interest, displayed on the screen, are marked automatically by colour points applying the proportions of an average face. It is clear that proportions of faces for different people are different. Therefore the system has a possibility to tune manually the proportions of user's face at the system installation. The correction of proportions and points' initial positions can be made by changing the arguments of the command line. The user's aim at the calibration stage is to combine maximally the colour points (markers) on the image with his/her real facial objects. Then five points are captured and the tracking algorithm starts. This algorithm uses the iterative Lucas-Kanade algorithm for optical flow [295] that is an apparent motion of image brightness.

To increase the robustness the method for operator's face tracking was enhanced by the method for automatic restoring of lost tracking points. The latter uses rectangular areas containing vertical and horizontal lines of tracking points (Fig. 7.7, b). If the system looses a point then its proper position is restored taking into account the coordinates of two other tracking points in line. For instance, if the point 2 (Fig. 7.7, b) goes out of the rectangle formed by the points 1 and 3, this fact is detected and the coordinates of the point 2 are restored as the linear combination of the coordinates of the points 1 and 3 taking into account initial proportions between coordinates of all the tracking points. Point 3 is included in both of the rectangular areas, so it can be considered as the most reliable marker in the system of points (Fig. 7.7, c). The algorithm rarely looses more than one point at once, but sometimes it happens because of the change of light conditions or very fast motion of head or moving out of the zone of video capturing. In this case the automatic restoring of the points is not possible and the head tracker calibration should be started again. There is a special voice command "Calibration" to start the calibration process. Moreover it was found during the experiments that the iris points tracking is not robust for people with light (blue or light-gray) eyes due to the lack of brightness contrast so the tracking of two these points could be disabled from the command line.

For calculation of the current position of the mouse cursor on the screen the linear combinations of the coordinates of the points 1 3 (for X coordinate) and the coordinates of the points 3 5 (for Y coordinate) are applied. To calculate X coordinate of the mouse pointer the points 1-3 are used only, but the points 4-5 are not taken into account. The reason is that a Cartesian coordinate system is applied. Indeed the points 4 and 5 being far from the middle of the face, when the operator rotates his head on the left or on the right the motion of the eye points is not linear but rotary, consequently those points are not representative of the real movement of the operator's head in a Cartesian coordinate system. For Y coordinate the situation is the same and it is calculated based on the coordinates of the points 3-5 (without points 1-2). So the pointer is shifted proportionally to the shift of the points between the consecutive frames of the video stream multiplied by the actual speed of the mouse pointer.

The present method for mouse cursor control can adjust the mouse pointer speed. The question is the resolution of web-camera is limited to 640x480 points (pixels) and the user's head is located in some work area on the image that is not more than 200 points per each axis. However the typical resolution of a modern LCD monitor is 1280x1024 points at least. Here the problem arises that it is required to transform adequately the dynamics of head motion in the work zone of the camera into the correct dynamics of pointer motion on the screen. Thus it is needed to introduce multiplication factors for each coefficient to transform head motion inside the work zone into pointer motion at the screen.

However usage of the multiplication factors leads to the problem of non-smooth motion of the pointer and it is difficult to select a graphical object of a small size because of instability of the pointer. In order to solve this problem two different coefficients of mouse pointer speed depending on the speed of motion of the operator's head are applied in the module. If a user wants to shift the pointer to a big distance (for instance, from one screen corner to another one) he/she moves the head quite fast thus the bigger coefficient (more than 5.0 depending on the screen resolution) is applied. If a user needs to select any graphical object on the desktop then he/she makes minor movements by the head and the smaller multiplication coefficient (1.0 – 2.0 depending on the screen resolution) is employed.

The application of the proposed method for the mouse pointer speed adaptation in the ICANDO interface allows one to reach easily the screen corners by the pointer controlled by the head as well as selecting any small graphical objects on the desktop.

7.2.2.4 Modalities Synchronization and Information Fusion

An analysis of the head motion speed function and actions of the testers who worked with the interface has allowed us to discover the important matter: the coordinates of the mouse pointer for the information fusion should be taken at a moment before the user starts pronouncing a voice command instead of the moment of appearance of a hypothesis of speech recognition. It is connected with the problem that during phrase pronouncing a user involuntarily slightly moves his/her head and to the end of the speech command recognition the cursor can indicate to another graphical object. Moreover a voice command appears in the brain of a human in a short time before the beginning of phrase input.

Thus the synchronization of the information streams is activated by the automatic speech recognition module and performed by the following way: a mouse cursor position, which is calculated continuously by the head tracking system, is taken just before pronouncing of a voice command. The head tracking module processes video stream with 10–20 frames per second (depending on the computer configuration), but the speech is processed with 100 segments per second. So at the moment of finding an useful speech fragment in the audio signal the current mouse pointer position is stored that correspond to some short time earlier phrase pronouncing owing to some delay of the video frame processing. The delay between the current video and audio frames processing can vary up to 100 ms.

Figure 7.8 illustrates the process of modalities synchronization and information fusion in the system. This figure shows the process of fulfilment of one scenario for hands-free work with Internet Explorer for obtaining some information at a web-portal (the sequence of voice commands simultaneously with user's head motions: "Left", "Scroll down" and "Left"), selecting a fragment of a web-page (voice commands "Left down", "Left up" and head motions), copying this text into the memory buffer (command "Copy"), opening the MS Word (commands "Start" and "Left") and pasting the text from the buffer into the text editor (command "Paste").

A black circle on the Fig. 7.8 means that a recognised command is multimodal one (for instance, the command "Left down") and a white circle denotes that the command has unimodal nature (speech-only, for instance, the command "Paste"). The automatic speech recognition module works in the real-time mode ($< 0.1 \times RT$);

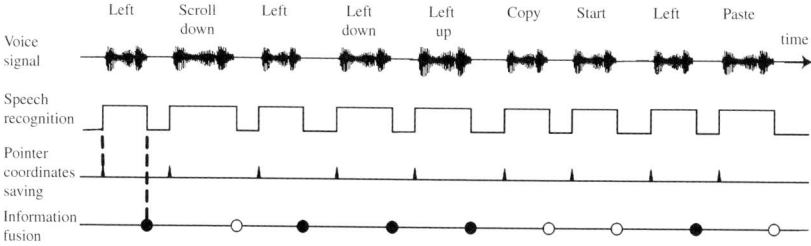

Fig. 7.8 The mechanism of synchronization and multimodal fusion.

since the vocabulary of voice commands is small one, there are minor delays between an utterance of a phrase and fulfilment of the recognised multimodal command and these delays may be excluded from the consideration.

For information fusion the frame method is employed when the fields of some structure are filled in by required data and on completion of the speech recognition process the action corresponding to the multimodal command is executed. The fields of this structure are: text of a speech command, X coordinate of the mouse cursor, Y coordinate and kind of the speech command (multimodal or unimodal). If a speech command has multimodal nature (mouse manipulation commands) then it has to be combined with saved coordinates of the mouse pointer and then the Windows message to the mouse device of the operating system is sent automatically. If a voice command is unimodal the coordinates are not taken into account and the message to the keyboard device is sent. The head movements without the speech modality cannot produce any active commands, but they can be used for drawing in a graphical editor.

7.3 Experimental Results

The testing and application of ICANDO was made in collaboration with St. Petersburg professional rehabilitation center (PRC) where people with special needs are studied and take courses of rehabilitation. The multimodal interface was installed on a laptop with the processor Intel Core 2 Duo 1.66 GHz, 1 Gb RAM and 17" wide display. The application of the dual-core processor is preferred to avoid processing delays because it recognizes two information streams in parallel.

The testing of the system was fulfilled by five inexperienced users, who had minor experience of work with a computer as well as by one man with no hands, who is PRC student. The system was tested on the task of GUI control for the operational system MS Windows. The test scenario in the experiments was connected with the work by Internet Explorer for finding the weather forecast at the web-portal http://www.rbc.ru, selecting, copying and saving this information in a MS Word file and printing this file. The task is divided into several elementary actions, which were accomplished by the multimodal way and by the standard way (mouse + keyboard). Table 7.1 shows the results of experiments and comparison of two ways of operation with a computer. Averaged values of speech recognition accuracy and the time required to fulfil the test scenario are presented.

Table 7.1 The comparison of the multimodal and standard ways of human-computer interaction

Command recognition rate, %	Time to perform test scenario	
	Multimodal way, s.	Standard way, s.
96.5	82	43

As the result it was experimentally determined that the proposed multimodal way of human-computer interaction is slower than the traditional way by a factor of 1.9. However, this decrease of interaction speed is acceptable since the developed system is intended mainly for motor-impaired operators.

The speech recognition system was trained on the user's voice personally. A speaker-dependent automatic speech recognition is more adequate for the present task than a speaker-independent one because it provides a lower word error rate. Taking into account the main purpose of the multimodal interface as well as low quality of an audio signal captured by the web-camera the speaker-dependent recognition was applied. The accuracy of voice command recognition was over 96.5% for each tester who used ICANDO interface.

The same operators worked with a computer by the multimodal interface for the Internet surfing during several days. The following analysis of the logs of the ICANDO has shown that among over 750 voice commands, given by all the users, some commands are more frequent than others, but some commands were not used at all. It was easy to predict that the most frequently command used is "Left" command to click the left button of the mouse (approximately 30%), because it replaces the button that is used very often. Internet surfing includes the necessity to type some text in search forms and all the users performed it by the On-Screen keyboard program giving the command "Left". All the other commands are distributed among the rest 70% being 7% at the maximum. It was found also that novices used the common command "Left" together with pointing gesture more often, than experienced users who preferred to give special voice commands. Totally above 55% of the speech commands were given in the multimodal way (simultaneously with head motion) and 45% of the commands were unimodal (speech-only) ones. It was discovered also that sometimes operators pronounced out-of-vocabulary (OOV) voice commands. The total amount of such commands was about 8% among all the commands. The most frequent out-of-vocabulary commands that were said are: "Click" instead of "Left" and "Back" as equivalent to "Previous" command. After considering such cases two these voice commands were added to the recognition vocabulary expanding it to 25 items.

The presented results of testing and exploitation allow us to say that the developed Intellectual Computer AssistaNt for Disabled Operators is successfully used by the persons with disabilities of their arms for contactless work with a personal computer. The prototype of the ICANDO multimodal interface was awarded with the grand prix at the Low Cost Multimodal Interfaces Software Contest.

7.4 MOWGLI Multimodal Interface

This section presents MOWGLI (Multimodal Oral With Gesture Large display Interface) which is a multimodal interface associating computer vision-based gesture recognition with automatic speech recognition.

7.4.1 Objectives

Although large displays could theoretically allow several users to work together and to move freely in a room, their associated interfaces are limited to contact devices that must generally be shared. Moreover when interacting together, persons use natural powerful ways of communication like voice and gestures, but when interacting with a computer, they are still limited to poor means of interaction like pushing buttons.

The goal of MOWGLI is to provide a system allowing two users to simultaneously interact with very large display, thus allowing the users to collaborate, being aware of what the other user is doing, sharing a common view [307], and interacting with the computer using the same speech and gesture modalities using more natural interaction provided by speech and gesture. The multi-person version of MOWGLI is SHIVA (Several-Human Interfaces with Video and Audio) [296].

7.4.2 System's Description

Within the multimodal MOWGLI system (Fig. 7.9), gesture recognition and speech recognition results are fused and interpreted by MOWGLI depending on the application context. Pointing and selection are involved in most applications, whereas commands and context are application dependant. Therefore speech vocabulary for addressing commands and context fusion has to be trimmed for each application. MOWGLI system is illustrated with a chess game application, but has also been used, after adaptation, for other applications such, virtual manipulation of 3D objects (Fig. 7.17) and navigation (Fig. 7.18).

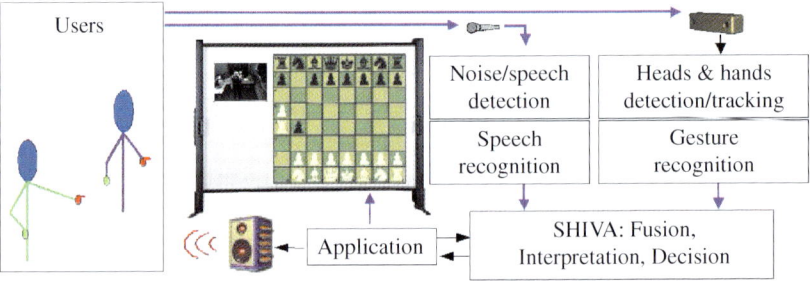

Fig. 7.9 The different speech and gesture components of MOWGLI and the communication with the application. The user is heard by a microphone and seen by a camera. The audio signal is processed by a noise/speech detector and the speech recognition depends on the application vocabulary. From the video signal, head and hands positions are computed then a pointing and selection gesture recognition is done. Results from gesture and speech recognition are fused and interpreted by MOWGLI depending on the application context. Closing the loop as in cite [276], the user perceives the audiovisual feedback resulting from its interaction

7.4.2.1 Automatic Speech Recognition

To recognize speech, signal is sampled at 8 kHz in 16 bits on a linear scale. MFCC (*Mel Frequency Cepstrum Coefficients*) coefficients are computed, each 16 ms, over 32 ms signal frames. Recognition system uses the frame energy, 8 cepstral coefficients and an estimation of the first and second order derivatives. Thus, observation vector has 27 dimensions.

The decoding system uses Hidden Markov Models. The sentences to be recognised are built using syntax rules to build categories (Table 7.2). The vocabulary used by the chess game remains below 50 words. Each word is obtained by concatenation of context dependant phonetic units (allophones). The system outputs the n-best results and is speaker independent.

A noise/speech detector component filters the input signal to provide the decoder only with speech signal surrounded by silence frames.

The decision of the beginning of detection is not causal: the detection component provides several frames which precede the speech detection decision. But as the decoder is faster than real time, it recovers from the non-causality of the detection component.

To detect end of speech, some consecutive silent frames must be observed. These frames are sent to the decoder. The best solution can be provided as soon as the last frame has been received. Computing the n-best solutions generates a negligible lag compared to the lag due to the trailing silence frames. The number of frames to detect the end of speech is a parameter of the noise/speech component set to 15. The lag between the end of speech and the result of the recognition is thus 240 ms.

Table 7.2 The 28 different individual word or speech items below can generate far many more sentences by combining the individual items according to the rules of association of the categories

Category	Content
Piece	"queen" + "rook"+ "pawn"+ "bishop" + "knight"+ "king";
Pronoun	"the" + "this"+ "that";
Colour	"black" + "white";
_Rel_Location_	"left" + "right" + "top"+ "bottom";
_Abs_Location_	"here" + "there";
Piece	(_Pronoun_+()).(_Rel_Location_+()).(_Colour_+())._Piece_;
_Imperative_Verb_	"move" + "take";
_Active_Verb_	"moves" + "takes";
_Passive_Verb_	"is moved" + "is taken by";
_Move_Verb_	_Imperative_Verb_+_Passive_Verb_+_Active_Verb_;
_Move_Action_	(_Piece_+())._Move_Verb_.(_Piece_+()).(_Abs_Location_);
_Other_Move_Action_	"castle";
_Other_Command_	"undo"+ "new game"+ "quit"+ "close";
Syntax	Move_Action_+ _Other_Move_Action__Other_Command_;

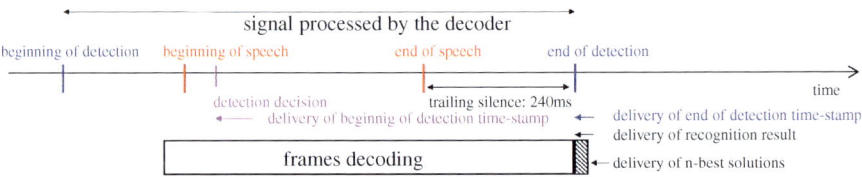

Fig. 7.10 Different times related to speech recognition. Speech recognition is only available after a delay following speech signal

The start and end detection times are sent to MOWGLI as soon as the start and end of speech decisions are available. These times include the before speech and after speech silent frames which differ from the times of start and end of speech. These former can be computed from the noise/speech parameters. All the times constraints are summarised in the Fig. 7.10.

The different moves of a chess game can be described using the written international chess notation. Thus each chess move could be described with {d2d4} type oral commands. But such a notation is used only by players with a sufficient level of expertise. For others players, it would require them to find the coordinates of both piece squares involved in a move (with the standard chess board labels from "a" to "h" on the bottom row and from "1" to "8" on the left column as in Fig. 7.20 to help the reader) and such command are not intuitive. From an automatic speech recognition point of view, {d2d4} type utterances are short and thus can lead to more errors than longer utterances.

Thus, unless the pieces or positions involved in a chess move are unique (for example "queen takes queen"), most moves will involved speech and gesture multimodal commands.

The ASR process is running continuously but results are taken into account only when a user points. In order to make the difference between an effective oral command and an oral comment to a third party similar to an oral command, only oral commands performed together with a pointing gesture anywhere on the chess board are taken into account, even for oral commands such as "Undo" or "New Game".

Oral commands are built using the syntax described in Table 7.2. Categories are defined by underscore signs whereas atomic speech items are defined by quote signs. The plus sign defines an "OR" relation and the dot sign a AND relation between words of words' classes whereas brackets indicate that an item is optional.

All the possible speech sentences to be recognised by the multiple speaker automatic speech recognition engine are automatically generated according to the given syntax.

7.4.2.2 Body Tracking and Mouse Pointer Control

A user has different means to point at objects located in space. With gaze [298] or head orientation [299], it is difficult to detect the user intention to point and interact. In [300], the authors find the pointing direction of a hand seen by a wearable

7.4 MOWGLI Multimodal Interface

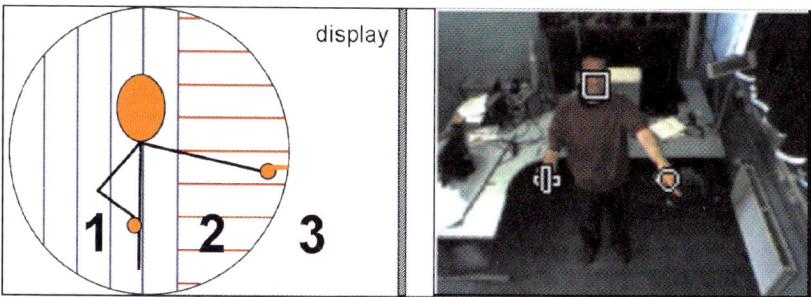

Fig. 7.11 Left: Schematic user related-space with 1-vertical stripes: user rest area, 2-horizontal stripes: user action area, 3-beyond sphere: hand non detection area. Right: a view of a user with both hands in the action area

computer. In [301, 302], the direction of a stretched arm is used, but such a pointing convention quickly tires a user. Pointing with only a forearm as in [299, 303], is less tiring than with the preceding stretched arm convention. In [303], the direction of the forearm is obtained from a 3D articulated model of the upper body. In [299], the pointing precision reached with the forearm is lesser than the precision obtained with a head-hand pointing, but when practically approximated by a head-raised hand convention, the pointing hand hides the feedback cursor on the display.

For the MOWGLI system, we have decided to artificially raise the true position of the tracked pointing hand. It allows to cope with this hidden cursor problem, and to let the user's hand in a lower and more comfortable position with the elbow against the torso. The pointed location is given by the head-raised hand convention observed with a feedback cursor.

Detection and tracking of body parts rely on skin-colour, disparity and movement information. Skin colour is obtained by a broad filter look-up table obtained on different users and under different lighting conditions. Obtained from the two images of a stereo camera, the disparity gives the 3D position of a pixel with respect to the camera coordinates. Disparity cannot be precisely determined in homogeneous colour zones or in areas with large depth difference. A filtered disparity image is computed for each user for which the observations too far away (> 1.3 m) from the head centre are discarded (once the head has been detected). The movement is obtained by background subtraction which is updated every image so that a still person quickly fades in the background. Once a head is detected, its 3D position is used to define the body space involve in the detection of hands and to detect the pointing purposiveness of the user.

Biometric constraints limit the search space to a centroid centred on the face. Furthermore, it is reasonable to admit that, when interacting with the display, the user moves its dominant hand towards the display and sufficiently away from its face (more than 30 cm). Thus the hand search space is restricted to a volume delimited by a sphere and a plane, a volume called the 'action area' (Fig. 7.11, left). Behind this plane, lies the rest area which allows the user not continuously interacting. The action area allows a user expressing its intentionality to interact through gesture or speech. Owing to morphological constraints, the hand non detection area cannot be reach by a hand. These areas are face referenced, so that their absolute positions change as a user moves but the system behaviour remains the same.

When the second hand is used to select a pointed object, for example grabbing a chess piece in Fig. 7.15, a similar use of the user related-space is made. When a second hand is used to control a 3D axis such as a zoom, a different description of the user related-space is used. The second hand is not detected as long as it stays close to the torso. It is detected as it moves towards the screen but is kept inactive as long as it remains in a low position, below the chest which corresponds to a rest area. When the second hand rises, it enters its action area, where, as it moves forwards and backwards, it can continuously control an axis (for example a zoom in Fig. 7.17). The first hand spontaneously brought forward by a user is labelled pointing hand whereas the second hand is labelled control hand and can be used for object selection or third axis control for 3D interaction. The system works both for right-handed or left-handed users without the need for left or right hand labels.

7.4.2.3 Body tracking

Upon detection, body parts are 3D tracked in real time with an EM algorithm, which uses a colour histogram and a 3D Gaussian for each of the body parts (Figures 7.12 and 7.13). Full details of the tracking algorithm, for one or two persons, can be found in [296, 297].

7.4.2.4 Mouse Pointer Control

The obtained raw position of the cursor is rather unstable, owing to the tracking algorithm of the body parts that does not aim at finding the most stable body part position, but the best position that can be estimated using current and past observations. This instability is demultiplied by the head-arm lever and distance to screen. In order to obtain a stable pointing, an adaptive filtering of the cursor position is performed.

$$Xc_t = Dc_t \hat{X}c_t - (1 - Dc_t)Xc_{t-1}, \quad Dc_t = \left| \hat{X}c_t - \hat{X}c_{t-1} \right|$$

with Xc_t and $\hat{X}c_t$ the normalised cursor position displayed on screen at time t, and its estimation from tracking cursor position, Dc_t the estimated from tracking cursor displacement.

For small pointing movement, this filtering takes into account the previous cursor position allowing to stabilize the cursor with a time lag cost. For large displacement, the current position is privileged, thus allowing for fast but less precise movement. This choice is coherent with Fitts law (Fitts 1954) learning where a movement can be described with two periods. During the first period needed to approximately reach

7.4 MOWGLI Multimodal Interface

a target, movement is fast, too fast for a user to be aware of cursor instability. During the second period, slow movements allow to precisely point at the target.

Two experiences relative to pointing precision and to pointing stability on a 2.21 × 1.66 meter screen are related in Fig. 7.14.

Fig. 7.12 Left: camera image (blue: face, red: pointing hand, green: second hand). Right: observations assigned to one of the four models depending on their probabilities (blue: face, red: hand1, green: hand2, light grey: discard, white: pixels ignored in EM). For one person tracking, observations behind the head of the person tracked are ignored

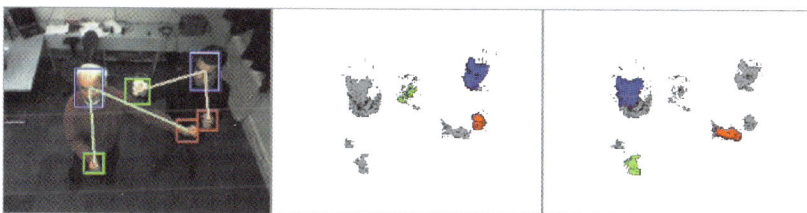

Fig. 7.13 Left: camera image (blue: face, red: pointing hand, green: second hand). Middle and Right: observations assigned to one of the four models depending on their probabilities (blue: face, red: hand1, green: hand2, light grey: discard, white: pixels ignored in EM) for the first and second detected person. Body part pixels of a person correspond to discard pixels for the other person. For two persons tracking, only observations behind their heads are ignored

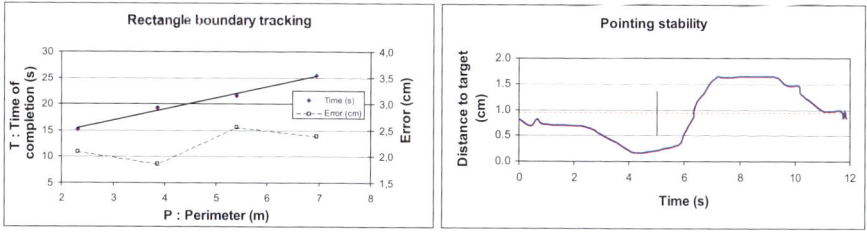

Fig. 7.14 Left: Rectangular boundary drawing task. Completion time task and pointing error (mean distance to boundary) versus perimeter of rectangles drawn on a 2.21x1.66 meter (1280x1024 pixels) screen. Right: Pointing stability: The stability of the pointing when a user is instructed to keep pointing a cross located at the center of the screen for a period of 10 s. Mean is 0.94 cm and standard deviation is 0.43 cm, as shown with dotted lines

7.4.2.5 Mouse Selection

The mouse selection can be performed using speech through "select" and "unselect" (or "drag" and "drop") oral commands, using a gesture of the non pointing hand (Fig. 7.15) or by recognizing pointing and grasping hand postures with a camera located above the user (Fig. 7.16).

7.4.2.6 Modalities Synchronization

Pointing gesture is a permanent process (20 Hz). Gesture selection event is a discrete event occurring on one frame, whereas speech is an event with a beginning and an end. Synchronization is triggered by speech event or gesture selection event. In the case of action triggered by speech, we make the hypothesis that speech and gesture are temporally aligned and we consider the mean pointing position that occurs during speech. As [275], the mean pointing position between time $[t_b, t_e]$, with $t_b = t_{bs} - 240\ ms$, and $t_e = t_{es} + 240\ ms$, with t_{bs} and t_{es} the beginning of speech and end of speech times. The 240 ms is the delay used to isolate speech from silence.

Indeed it has been experimentally observed that users first stabilize their pointing process before uttering and that they usually wait for the answer of the ASR before beginning pointing to another location. But with long oral utterance or as user become confident that ASR will succeeds, users tend to anticipate and move to the next pointing positions which leads to erroneous mean pointing.

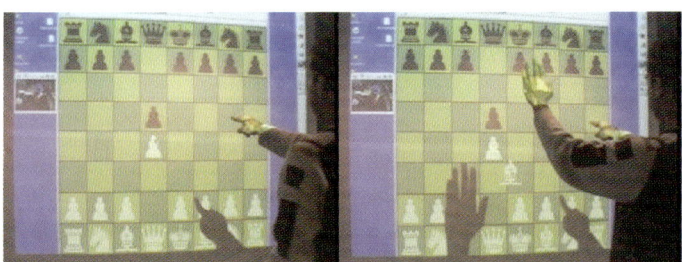

Fig. 7.15 Interaction with a 2D object Left: dominant hand pointing is used to indicate a chess piece. Right: the other hand is moved towards the screen to select the pointed piece, which can then be moved with the dominant hand as long as the other hand is kept towards the screen

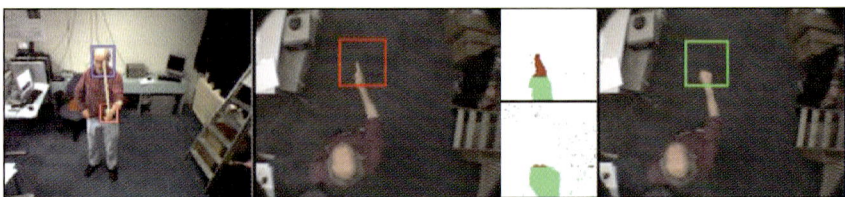

Fig. 7.16 Far left: 3D body parts tracking camera view. Hand posture classification top camera. Left: view of user pointing. Middle: Background segmented hand, finger-wrist classification (top: user pointing, bottom: user grasping). Right: view of user grasping [304]

Fig. 7.17 Interaction with a 2D panoramic image and fusion between gestures. Left: pointing is obtained with the dominant hand and the other hand is moved towards the screen to zoom into the image. Right: the user moves non pointing hand away from the screen to zoom out of the image

The advantage of a mean position over a single discrete pointing position (whether synchronised with begin/end or result time) is that it allows users not having to be aware of synchronization occurring at a specific time or being aware of when exactly speech begins or ends (Fig. 7.10).

$$X_m = \frac{1}{N}\sum_{t_b}^{t_e} \alpha_t X_t \quad \alpha_t = 1 - \sqrt{\frac{V_t - V_{min}}{V_{max} - V_{min}}} \quad V_t = Xc_t - Xc_{t-1} \quad (7.1)$$

Since when a user wishes to interact we observe that the cursor position is stabilised, the speed of the cursor is used to weight the mean cursor position. Thus X_m is the weighted cursor position is given by the above equation, with N the number of pointing sample during time interval $T = [t_b, t_e]$, Xc_t the cursor position at time t and α_t the weight at time t. V_t is the instant cursor X_{Ct} speed, and V_{min} and V_{max} are the minimum and maximum speed during time interval T.

Gesture or speech synchronization allows to establish a reference for the cursor (and thus the object selected behind the cursor) but also for hands reference position for subsequent bi-manual gesture. Thus, fusion occurs between gestures (Fig. 7.17) or between speech and gesture (Fig. 7.18 and Fig. 7.19).

As said earlier, oral commands which do not need to be complemented with a gesture information are taken into account only if performed together with a pointing gesture anywhere on the screen in order to make the difference between an effective oral command addressed to the machine and an oral comment, addressed to a third party, that would be similar to an oral command.

7.4.2.7 Information Fusion with Application Context

When gesture occurs, information can be retrieved from an application, for instance the nature and the properties of the pointed object. In a chess game, it allows to know which moves are concerned by the pointed chessboard square, whether empty or occupied by a piece. Thus, fusion occurs between speech, gesture and application context (Fig. 7.20) [306]

Fig. 7.18 User rotating a 3D object and fusion between speech and gesture. Left: the user puts his hand in a reference position and utters "Rotation". Right: Subsequent hands' movements are interpreted as a 3D rotation for the object until the user utters "Stop"

Fig. 7.19 User resizing a 3D object and fusion between speech and gesture. Left: the user puts his hand in a reference position and utters "Resize". Right: Subsequent hands' movements are interpreted as a 3D resize for the object until the user utters "Stop" [305]

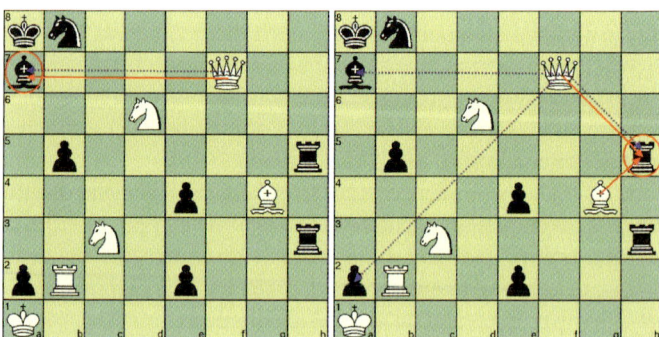

Fig. 7.20 White to play. Pointing on a black piece implies that white move involves taking a black piece. Red circles indicate the pointed square. Red arrows are for moves corresponding to pointing information. Hyphened blue arrows show moves corresponding to speech information. Left: Speech and gesture are both redundant and explicit. Pointing occurs on the Black bishop. Context indicates only one possible move as well as speech result "Take bishop". Merging speech, gesture and context multimodal information leads to a unique command "queen takes bishop". Right: Speech and pointing are complementary and explicit. Pointing occurs on top black rook. Context indicates that both the white queen and bishop can take the black rook. Speech information "queen takes", leads to three possible moves. Fusion and multimodal information leads to a unique command "queen takes rook"

7.4 MOWGLI Multimodal Interface

Fusion with context involving one pointing location, allows for more natural and faster interaction time (Table 7.3), than interaction involving two pointing locations.

Computer vision-based contactless device can handle a large interaction space; large enough to handle simultaneous pointing by two users (Fig. 7.21). Currently, simultaneous multi-user gesture interactions, for example with DiamondTouch table, are limited to the area covered by a hand displacement [307].

Table 7.3 Comparison of typical chess move interaction time for different modalities

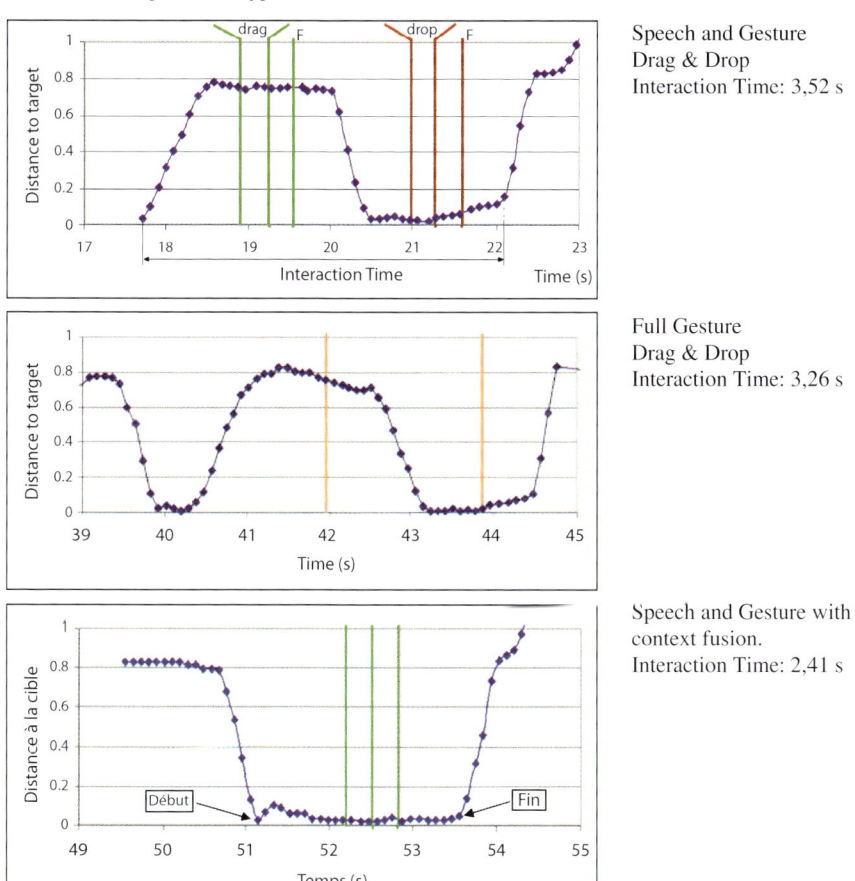

Speech and Gesture
Drag & Drop
Interaction Time: 3,52 s

Full Gesture
Drag & Drop
Interaction Time: 3,26 s

Speech and Gesture with context fusion.
Interaction Time: 2,41 s

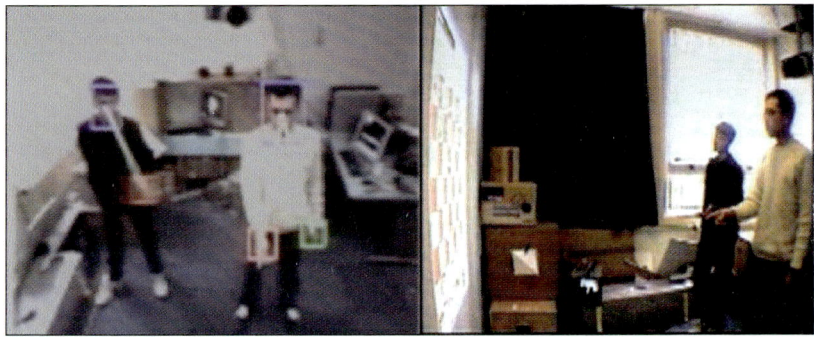

Fig. 7.21 Left: The head and hands 3D positions of the two users are detected and tracked with the help of a stereo camera. The person on the right of the image is pointing and can choose to used gesture or speech to select a chess piece. The other player is waiting for his turn. Right: Both users share the large display located in front of them, with which both can interact

7.5 Interfaces Comparison and Evaluation

This section proposes the results of qualitative comparison and quantitative evaluation of the ICANDO and MOWGLI multimodal interfaces.

Table 7.4 shows the correspondence of 12 different parameters for the two above described interfaces. It can be seen from the table that the systems are different in many parameters however the purpose of both systems is the same – contactless human-computer interaction.

Table 7.4 A qualitative comparison of the multimodal and standard ways of human-computer interaction

Parameter meaning	ICANDO	MOWGLI
Target group of users	hand-disabled persons	ordinary users
Number of users	single user	one or two users
Kind of interaction	Human-computer interact.	HCI, human-human interact.
Hardware equipment	low cost web camera	high cost stereo camera
Speech recognition	small vocabulary	small vocabulary
Languages	English, French, Russian	French
Gesture recognition	2D vision-based	3D vision-based
Recognition objects	head and face	hands and head
Operating area	short distance interaction	long distance interaction
Synchronization kind	by speech modality	by speech or gesture
Context awareness	no support	in chess game
Main applications	assistive systems for hands-free HCI, games	edutainment, cooperative HCI, games, virtual reality

7.5 Interfaces Comparison and Evaluation

To evaluate contactless multimodal devices, we have chosen the ISO 9241-9 standard methodology [308], which is based on Fitts' law [309] and on the work carried out by Soukoreff and Mackenzie [310]. Users are instructed to point and to select the 16 different targets, with a circular layout so hand movements are carried out in different directions. When selection occurs, the former target is shadowed and the next target is displayed (Fig. 7.22). The index of difficulty (ID), measured in bit, is given, according to the Shannon formulation, by $ID = \log_2 \left(D/W + 1 \right)$, where D is the distance between targets and W the target's width. Fitts' law states that the movement time MT between two targets is a linear function of ID.

The three tested contactless devices are ICANDO interface, MOWGLI two hands gesture (Fig. 7.22), and MOWGLI speech and gesture device (Fig. 7.23).

The methodology and the evaluation of the performances of the two hands gesture (Fig. 7.22) and the speech and gesture device (Fig. 7.23) are fully described in [311], along with comparison with other contactless vision-based device long distance interaction [277] and short distance interaction [272, 273].

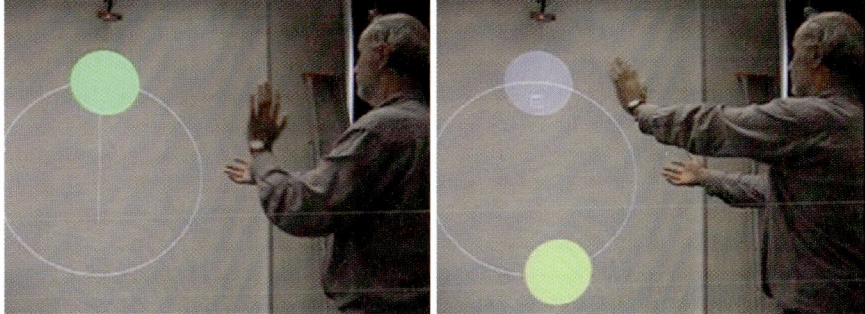

Fig. 7.22 Using the two hands gesture device, a right-handed user pointing with his dominant hand. The initial target before being selected (left) and just after being selected as the user moves his left hand forward (right)

Fig. 7.23 Using the speech and gesture speech and gesture device. Pointing has move from the top towards the bottom target when speech occurs (left). A user moves his hand in a direction opposite the bottom target as the next target appears only after a 240 ms lag owing to speech recognition (right)

Figure 7.24 shows the movement time versus index of difficulty for MOWGLI and ICANDO contactless devices.

Although ICANDO and MOWGLI interfaces function in very different conditions and rely on different computer vision and speech recognition techniques, they have very similar performances, suggesting that performances may be limited on one hand by the low frequency rate of cameras involved in computer vision processes and on the other hand by time scale involved in human speech.

Also such standard input devices as mouse, touchpad, trackball and touch screen were tested in order to evaluate the performance of proposed contactless devices. Table 7.5 shows averaged values of *ID* (index of difficulty), *MT* (movement time) and *TP* (throughput) parameters as well as path length ratio for all devices tested.

The best results were shown by contact-based touch screen and mouse device and Fitts' law results obtained by MOWGLI and ICANDO interfaces are quite close to trackball device. However the key advantage of the devices developed is that they provide natural contactless human-computer interface that is similar to human-human communication.

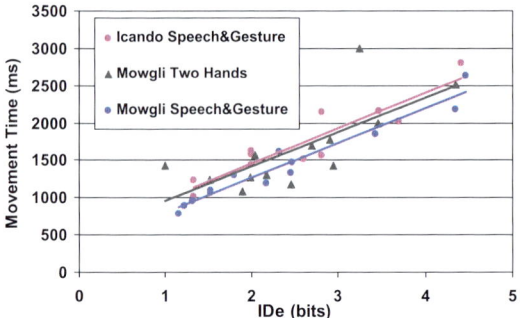

Fig. 7.24 Comparison of *MT* vs. *IDe* for MOWGLI and ICANDO interfaces

Table 7.5 Fitt's law comparison between ICANDO, MOWGLI Speech and Gesture (SG) and Two Hands (2H) interfaces and contact-based pointing devices

Input device	ID (bits)	MT (s)	Path length ratio	TP (bits/s)
ICANDO	1.32-4.41	1.74	1.28	1.45
MOWGLI SG	1.32-4.64	1.43	1.16	1.84
MOWGLI 2H	1.32-4.64	1.66	1.23	1.48
Mouse	1.32-4.41	0.65	1.08	3.84
TouchPad	1.32-4.41	1.21	1.06	2.08
Trackball	1.32-4.41	1.29	1.19	1.95
Touchscreen	1.32-4.41	0.68	1.02	3.94

7.6 Main Future Challenges

To develop a more effective interface we believe that the following scientific challenges have to be addressed in the interface design:

- *Improvement of the unimodal recognition:* The state-of-the-art continuous speech recognition and gesture recognition systems are still very far from the human's abilities. They can work quite reliable in "ideal" laboratory conditions, but when one tries to apply them for real exploitation they essentially lose in quality. The key challenges for speech recognition are robust voice activity detection, noise suppression and speaker localization by microphone array in the case when many persons are in an operating room, garbage speech modelling in order to discard non-informative or accidental phrases, accurate spontaneous speech recognition. The problems for gesture recognition are change of the illumination conditions, dynamic background and presence of several persons in the image, etc. All these topics must be comprehensively studied in order to improve the quality of speech & gestures interfaces.
- *Error handling in multimodal interfaces:* Error handling currently remains one of the main interface problems for recognition-based technologies like speech and gestures. However if there is some redundancy in speech and gesture inputs it is possible to apply methods of mutual disambiguation between the signals. Mutual disambiguation involves recovery from unimodal recognition errors within a multimodal architecture, because semantic information from each input mode supplies partial disambiguation of the other mode, thereby leading to more stable and robust overall system performance [269].
- *Inclusion of additional modalities:* Other natural modalities could help increasing the accuracy robustness of the interfaces involving gestures and speech. For instance, eye gaze can be considered as a complementary pointer to a spatial object on the screen. The interfaces combing speech, gesture and gaze are temporally cascaded and here it is possible to process advance information from the gaze to constrain accurate interpretation of the following gesture and speech. Moreover usage of facial expressions (for instance, lip motions) could enhance the automatic speech recognition especially in acoustical noisy environments. Human's speech is always related to the corresponding movements of lips and this passive modality can give new informative features for speech processing. An interface able to process both modalities in parallel is known as audio-visual speech recognition.
- *Adaptation to a user and a context:* Efficient multimodal interfaces should be able to take into account user's requirements and needs. For instance, hand impaired people cannot use manual gestures, however in this case head gestures can replace the missing modality. Fast automatic adaptation to user's voice parameters, height, skin colour, clothes is a very important property of prospective speech and gestures multimodal systems. An ability of an interface to recognize current context, to change dialogue model in correspondence with this information, as well as to process out-of-vocabulary voice commands, semanti-

cally rich gestures and to add dynamically new items in the recognition vocabulary should be the object of the further studies too. Such technology will help avoiding many errors of recognition due to restriction of commands acceptable in a given situation.

- *Elaboration of the methods for evaluation and usability of the interfaces:* Currently used ISO methods for evaluation pointing devices (for instance, Fitts' law) are quite adequate for pointing gestures, but not very suitable for evaluation of speech & gestures interfaces having richer semantics of the verbal and manual control commands. Some ISO 9241 standards are in the process of being elaborated and concern instance tests for evaluation command dialogues, direct manipulation dialogues, haptic and tactile interactions in multimodal environments. Additional researches are needed to create suitable tests for comparison between contactless multimodal interfaces.

Successful solution of all these issues should lead to a new generation of speech and gestures multimodal interfaces that will be more intelligent, natural, accurate and robust.

7.7 Acknowledgements

A. Karpov and A. Ronzhin would like to thank A. Cadiou for his contribution in development of the head tracker and F. Sudarchikov for demonstration and user feedback of ICANDO interface. S. Carbini and J.E Viallet would like to thank O. Bernier and L. Delphin-Poulat, for their respective contributions in computer vision and automatic speech recognition. This work has been supported by the EU-funded SIMILAR Network of Excellence in multimodal interfaces research.

Chapter 8
Multimodal User Interfaces in Ubiquitous Environments

Fabio Paternò, Carmen Santoro
ISTI-CNR, Pisa, Italy

8.1 Introduction

Ubiquitous technology is continually spreading computing and communication power in our life. Recent technological advances (including increasing availability of various types of interactive devices, sensors and communication technology) are more and more enabling novel interactive software to support users in several contexts for different objectives, by profiting from the resources available in smart environments. Such environments are expected to be able to capture and interpret what the user is doing, possibly anticipating what the user will do or wish to do, so being not only reactive but also proactive, by assisting the user through recording and analyzing what the user is doing for later use. This trend is expected to occur more and more in different situations of our life, from work settings, to home environment, to recreation environments. Pervasive environments are characterised by seamless connectivity and various types of devices, through which a user can interact with applications. This creates challenges in implementing the means for interaction, and several interaction modalities e.g. graphical, vocal, gesture, touch, and tactile can potentially be used. Also, different devices have different operating platforms, and user interfaces must be described in different ways: for example some devices support browsers and the user interface can be implemented with XML-based mark-up languages e.g. X/HTML, VoiceXML, X+V etc., while in other devices access to system level software is required to enable the functionality for supporting specific modalities.

Model-based approaches can provide useful support to address the challenges raised by pervasive computing. Models can provide a structured description of the relevant information, and then intelligent environments can render the associated elements taking into account the actual context of use. The goal is to achieve pervasive usability: the ability of still supporting usability even when interactions can be performed under the 'anytime–anyhow–anywhere' paradigm. This means users are able to seamlessly access information and services regardless of the device they are using or their position, even when the system and/or the environment change dynamically. The combination of ubiquitous environments with multimodal interfaces can provide innovative solutions able to support natural interaction, where users interact with systems in a way similar to how they communicate among them.

By offering to users usable ubiquitous access to any service from any available interactive device (TV, computer screen, mobile, PDA, etc.) requires that users not to have to learn diverse procedures to access different services. Consistent interfaces should be provided in all devices, with similar interaction procedures but still adapting the user interface to the available interaction resources, thus enabling more natural and personalised interaction obtained by exploiting different modalities supporting the mobile user.

One important aspect of ubiquitous environments is to provide users with the possibility to freely move about and naturally continue the interaction with the available applications through a variety of interactive devices (i.e. cell phones, PDAs, desktop computers, digital television sets, intelligent watches, and so on). Indeed, in such environments one big potential source of frustration is that people have to start their session over again from the beginning at each interaction device change. Migratory interactive services can overcome this limitation and support continuous task performance. This implies that interactive applications be able to follow users and adapt to the changing context of use while preserving their state. General solutions for migratory interactive services can be obtained by means of addressing three aspects: adapt and preserve the state of the software application parts dedicated to interacting with end users; support mechanisms for application logic reconfiguration; and define suitably flexible mechanisms from the underlying network layers.

In this paper we fist discuss the state of art in the area of ubiquitous user interfaces, we then introduce migratory user interfaces and discuss the relevant design dimensions. Then, we briefly describe an example architecture developed in our laboratory supporting migratory interfaces and show a couple of example applications. Lastly, some conclusions are drawn along with indications for future evolutions in this area.

8.2 Related Work

Model-based user interface design is a research area that was started several years ago. However, in the early work (see for example, UIDE [312], Mastermind [313], Humanoid [314]) the type of models developed were mainly oriented to represent abstractions of graphical user interfaces, without a clear distinction between models independent of the interaction platform and models dependent on this information, which is important when addressing multi-device environments. In recent years, such interest has been accompanied by the use of XML-based languages to represent such logical descriptions (examples are XIML [315], UIML [316], TERESA-XML [317], USIXML [318]). However, such approaches have usually focused on providing support in the design phase in order to help designers to

8.2 Related Work

efficiently obtain different versions for the various possible target platforms rather than to support adaptation at runtime so that the user interface is dynamically transformed while users interact with it because of a device change. Various techniques for redesigning Web applications for small devices have been proposed see for example [319, 320], both orientated to obtaining a thumb-nailed view of the original page. PUC [321] dynamically generates user interfaces able to control a domestic appliance starting with its logical description, but it does not address the possibility of supporting users in continuing task performance when moving dynamically through different devices. Supple [322] is an application and device independent system that enumerates all possible ways of laying out the interface and chooses the one which minimizes the user's expected cost of interaction. Supple addresses mainly generation of graphical interfaces, (even if its authors have built a module that allows Supple-generated GUIs to be controlled with speech), and it does not support interface migration. ICrafter [323] is a solution to generate adaptive interfaces for accessing services in interactive spaces. It generates interfaces that adapt to different devices starting with XML-based descriptions of the service that must be supported. However, ICrafter is limited to creating support for controlling interactive workspaces by generating UI for services obtained by dynamic composition of elementary ones and does not provide support for continuity of task performance across different devices. Aura [324] provides support for migration but it is obtained by changing the application depending on the resources available in the device in question, while there is a need for solutions able to generate interfaces of the same application that adapt to the interaction resources available. Bharat and Cardelli [325] addressed the migration of entire applications (which is problematic with limited-resource devices and different CPU architectures or operating systems), while we focus on the migration of the UI part of a software application. Kozuch and Satyanarayanan [326] identified a solution for migration based on the encapsulation of all volatile execution state of a virtual machine. However, their solution mainly supports migration of applications among desktop or laptop systems by copying the application with the current state in a virtual machine and then copy the virtual machine in the target device. This solution does not address the support of different interaction platforms supporting different interaction resources and modalities, with the consequent ability to adapt to them. Chung and Dewan [327] proposed a specific solution for migration of applications shared among several users. When migration is triggered the environment starts a new copy of the application process in the target system and replays the saved sequence of input events to the copy to ensure that the process will get the state where it left off. This solution does not consider migration across platforms and consequently does not support run-time generation of a new version of the UI for a different platform. A discussion of some high-level requirements for software architectures in multi-device environments is proposed in [328] without presenting a detailed software architecture and implementation indicating a concrete solutions at these issues. In this paper, we also introduce a specific architectural solution, based on a migration/proxy server, able to support migration of user interfaces associated with applications hosted by different content servers. While in our previous work [329] we found a solution based on the use of pre-computed interfaces for different platforms that are dynamically activated, in this paper we also report on a solution, along with an engineered prototype, supporting also migration through different modalities (such as voice) and briefly show a couple of example applications.

8.3 Migratory User Interfaces

Migratory user interfaces are user interfaces able to support device changes and still allow the user to continue the task at hand. They imply two main features: adaptation to the changing context of use and interface state persistence. Indeed, in order to support task continuity, one key aspect in the migration process is the ability to capture the state of the migrating interface, which is the result of the history of user interactions with the application.

Migration involves one or multiple source devices and one or multiple target devices. In order to determine the *target device(s)* support for device discovery and context management is necessary. Then, a migration environment uses all this information in order to calculate the migration transformation and preserve the interface state. This processing identifies two main aspects: the interface state, which is composed of all the user interface data resulting from the user interactions; and the interaction resources and modalities associated with the target device(s). In this process, a number of migration mappings are defined in order to determine which activities will be supported by which devices as a result of the migration process.

Then, the migration engine identifies how to adapt the user interface and its state to the target devices taking into account various aspects, such as user preferences. The result of this phase is the specification of the user interface adapted to the new context of use. Lastly, run-time support provides the resulting user interfaces in the target devices with the state updated with the state of the source user interface when migration was triggered.

There are many applications that can benefit from migratory interfaces. In general, applications that require time to be completed (such as games, business applications) or applications that have some rigid deadline and thus need to be completed wherever the user is (e. g.: online auctions). Other applications that can benefit from this flexible reconfiguration support are those that have to provide users with continuous support during the whole day through different devices (for example, in the assisted living domain).

8.4 The Dimensions of Migration

Migratory interfaces are interfaces that can transfer among different devices/ contexts, and thus allow the users to continue their tasks from where they left off.

8.4 The Dimensions of Migration

Especially in heterogeneous environments (namely, environments in which different types of devices exist), the concept of migratory user interfaces raises a number of design issues that should be appropriately analysed and addressed in the attempt to identifying an effective migration architecture/process. A suitable framework for migration should consider at least the dimensions described hereafter. Such dimensions are:

Activation Type: How the migration is triggered. The simplest case is on demand, in which the user actively selects when and how to migrate. Otherwise, in automatic migration, it is the system that activates the device change (depending on e. g. mobile device battery consumption level, device proximity, etc.).

Type of Migration: This dimension analyses the 'extent' of migration, as there are cases in which only a portion of the interactive application should be migrated. A number of migration types can be identified:

- Total migration allows basically the user to change the device used to interact with the application.
- Partial migration is the ability to migrate a portion of the UI (the remaining portion remains in the source device).
- In the distributing migration the user interface is totally distributed over two or more devices after migration.
- The aggregating migration performs the inverse process: the interface of multiple source devices are grouped in the user interface of a single target device.
- The multiple migrations occur when both the source and the target of the migration process are multiple devices.

Number/Combinations of Migration Modalities: This dimension analyses the modalities involved in the migration process. Mono-modality means that the devices involved in the migration adopt the same modality interaction. Trans-modality, means that the user can migrate changing the interface modality. An example of migration from graphical interface to vocal interface is the case of users navigating the Web through a PDA or Desktop PC and afterwards migrate the application to a mobile phone supporting only vocal interaction. Lastly, with multi-modality the migratory interface contemporaneously supports two or more interaction modalities at least in one device involved in the migration. Work in this area often has mainly focused on graphical interfaces, investigating how to change the graphical representations depending on the size of the screens available.

Type of Interface Activated: This dimension specifies how the user interface is generated in order to be rendered on the target device(s). With precomputed user interfaces the UI has been produced in advance for each type of device. Thus, at runtime there is no need for further adaptation to the device but only for adaptation of the state of the UI. On the contrary, if a runtime generation of user interfaces is considered, the migration engine generates the UI according to the features of the target device when migration occurs. In an intermediate approach the migration engine adapts dynamically to the different devices using some 'templates' previously created.

Granularity of Adaptation: The adaptation process can be affected at various levels: the entire application can be changed depending on the new context or the UI components (presentation, navigation, content).

Migration time: The device changes that characterize migration can occur in different temporal evolutions:

- Continuous: The user interface migration occurs immediately after its triggering.
- Postponed: in this case the request of migration can be triggered at any time but it actually occurs after some time. This can happen when the target device is not immediately available (for example when migrating from the desktop in the office to the vocal device in the car, which has to be first turned on by the user when he enters the car).

How the UI is Adapted: Several strategies can be identified regarding how to adapt user interfaces after a migration process occurs:

- Conservation: this strategy maintains the arrangement and the presentation of each object of the user interface: one possible example is the simple scaling of the user interface to different screen sizes.
- Rearrangement: in this case all the UI objects are kept during the migration but they are rearranged according to some techniques (e.g.: using different layout strategies).
- Increase: when the UI migrates from one device with limited resources to one offering more capabilities, the UI might be improved accordingly, by providing users with more features.
- Reduction: this technique is the opposite of increase and it can be applied when the UI migrates from desktop to mobile device because some activities that can be performed on the desktop might result unsuitable on a mobile device.
- Simplification: in this case all the user interface objects are kept during the migration but their representation is simplified, for example, different resolutions are used for figures or figures are substituted with textual descriptions.
- Enhancement: this technique represents the opposite of simplification (e.g.: a textual description might be substituted with multimedia information).

The Impact of Migration on Tasks: The impact of migration on tasks depends on how the user interface is adapted because reduction and increase can produce some change on the range of tasks supported by each device. Differently, conservation and rearrangement do not produce any effect on the set of tasks supported. Then some possible cases are: 1) after a partial or distributing migration some tasks can be performed on two or more devices in the same manner (task redundancy), which means, for instance, that the decomposition of the different tasks

8.4 The Dimensions of Migration

into subtasks is unchanged, as well as the temporal relationships (sequencing, concurrency, etc.) occurring among them; 2) after a partial or distributing migration a part of a task can be supported on one device and the other part/s is/are available on different devices (task complementarity). Additional cases are when the number of task supported 3) increases (task increase) or 4) decreases (task decrease) after migration. Obviously, a final case might be identified when the migration has no impact on tasks, as they remain substantially unchanged.

Context Model. During adaptation of the UI the migration process can consider the context in terms of description of device, user and environment. Generally, the context dimension that is most taken into account with the aim of producing usable UI is the device and its properties, together with the surrounding environment.

Context Reasoning. A context modelling process begins by identifying the context of interest. This depends on the specific task which should be associated with the context of interest. The context of interest can be a primitive context, which can directly be captured by employing a sensor; or, it can be a higher-level context, which is a result of manipulation of several primitive contexts. If a context of interest is a higher-level context, a reasoning scheme is inadvertently required, which can be either a logic-based reasoning scheme or a probabilistic reasoning scheme. A logic-based reasoning scheme considers a primitive context as a factual data while a probabilistic reasoning scheme does not. Depending on the nature of the sensed data available and the way the data are manipulated, ignorance can be classified as follows:

- Incompleteness: refers to the fact that some vital data about the real-world situation to be reasoned about is missing; the available data, however, are considered to be accurate.
- Imprecision: refers to inexact data from sensors. Inexact data arises due, partly, to the physical limitations of the sensing elements employed. Different sensors have different resolution, accuracy, and sensing range. Besides, the performance of physical sensors can be influenced by external factors such as surrounding noise or temperature.
- Uncertainty: refers to the absence of knowledge about the reliability of the data sources – this knowledge might be information about the parameters listed above to determine (characterize) the degree of imprecision incorporated in sensory data.

Implementation Environment. The migration process can involve different types of applications. Probably due to their diffusion, the most recurrently considered applications are web-based systems (static/dynamic pages), but also other applications (Java, Microsoft.NET etc.) can be considered.

Architecture. With regard to the architecture of the migration support environment there are different strategies, for example: proxy-based, in which there is an intelligent unit managing all migration requests and sending all data to target devices; and peer to peer, where the devices directly communicate and negotiate the migration parameters.

8.5 An Architecture for a Migration Platform

In this section, we briefly describe an example of architecture supporting migratory interfaces designed and implemented at the HIIS Laboratory of ISTI-CNR. The architecture is based on a migration/proxy server, which receives the access and migration requests and performs interface adaptation and interface runtime state management. Users who want to access the service have to load the migration client software onto their devices. This is a light weight piece of software used to communicate with the server to send migration requests, indicate when the device is available for migration and additional information related to the device, which can be useful in the migration process. It also permits gathering runtime information concerning the state of the interface running on the device, which results from the history of user interactions. However, the application that should migrate does not need to be manipulated for this purpose except an automatic inclusion of a JavaScript used to gather the application state by the migration/proxy server, neither there is any particular transformation that takes place in the client device or in the application server.

The migration/proxy server can be used for many purposes:

Simple proxy: The user accesses the Web application through a device that belongs to the same platform type for which the pages were created for (e. g.: the desktop). The migration/proxy server retrieves the required page from the Web server and passes it on to the client.

Interface redesign: The user accesses the Web application through a platform different from that for which the pages were created, for example a vocal platform client accessing the pages designed for the desktop. First, the migration/proxy server retrieves the required page from the Web server, then the page is redesigned to be suitable for vocal interaction and the resulting UI is activated on the vocal device.

Migration from Desktop to non-Desktop platform: The user accesses the Web application through a device whose platform is the same for which the pages were created and at a certain point migration towards a different platform is required. Let us consider for example a desktop to PDA migration. First, the target device of the migration has to be identified. Thus, the page that was accessed through the source device is stored and automatically redesigned for the PDA. This is enough to support platform adaptation, but there is still one step to be performed in order to preserve task performance continuity. The runtime state of

8.5 An Architecture for a Migration Platform

the migrating page must be adapted to the redesigned page. In addition, the server has to identify the page to activate in the target PDA; this page is the one supporting the last task performed by the user, before migration. At the end of the process, the adapted selected page is sent to the PDA from which the user can continue the activity.

Migration from non-Desktop platform to a different non-Desktop platform: The user accesses the Web application through a device whose platform is different from the one for which the pages were created, and at a certain point migration towards another different platform is required. This is the most complex and interesting case and it differs from the previous one because both the source and the target device of migration do not match the platform of the original Web pages (the desktop one), and consequently both of them require that the associated interfaces be dynamically generated starting from the desktop version taking into account their different interaction capabilities. Indeed, the environment retrieves the original pages for the desktop from the Web and analyses them, thereby obtaining logical descriptions at different levels and used to generate a redesigned page for the platform accessing the page or for the platform onto which the page is migrating. This double transformation (from the desktop platform to the source platform, when the application is accessed, and, then when the application is migrated, from the desktop platform to the target platform) is necessary because it is not taken for granted that a direct transformation exists from every user interface object of the source platform to every logical object in the target platform. Instead, the desktop version, which is the most complete platform from the point of view of interaction capabilities, can be considered the 'reference' platform because for each interaction object considered a mapping exists from this platform and to this platform. It is worth pointing out that having the desktop version available is not a limiting assumption because very often this is the only version available. This solution based on two transformations also allows the environment to associate the state of the source application to the target application through a double mapping: the elements of the logical description of the source version are first mapped onto those for the desktop version and then to those for the target version.

As an example we can consider an application developed for the desktop, accessed through a PDA and migrating to a mobile vocal phone. The steps to perform this migration example shown in Fig. 8.1 are:

1. The user loads a Web page from a PDA
2. The server asks the content server for the original desktop page
3. The page is retrieved and stored in the migration/proxy server
4. The desktop page is redesigned for PDA going through the generation of the necessary logical descriptions
5. The redesigned page is sent to the PDA; The user interacts with the application
6. The user sends a migration request toward a vocal platform

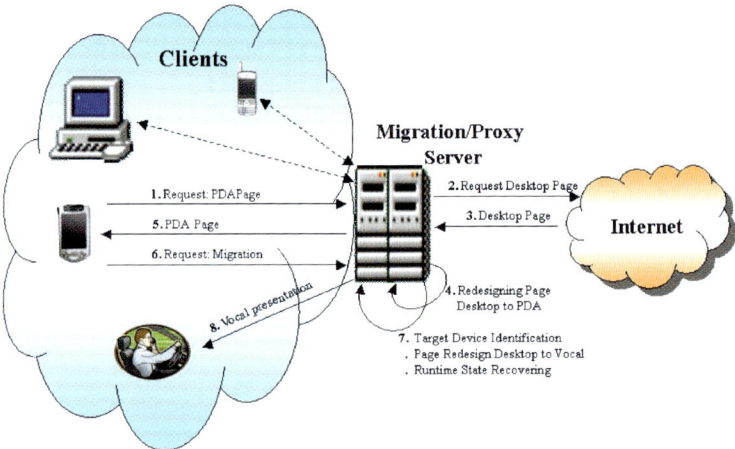

Fig. 8.1 The Migration Service Architecture

7. The server recognizes and identifies the target platform and redesigns the interface for the vocal platform. Thus, also the runtime state of the migrating page is adapted to the newly generated user interface.
8. The vocal interface containing the adapted state is sent to the target vocal platform.

In order to make such scenarios possible our environment supports various phases (see Fig. 8.2):

- Device Discovery, a dynamic discovery of the devices that are available for migration and their main features;
- Migration Trigger and Target Identification, the activation of the migration process along with an identification of the target device, both of them can be either user or system-generated;
- Reverse Engineering, the dynamic construction of the logical specification of the source user interface;
- Semantic Redesign, transforming the results of reverse engineering to obtain a logical description of the target user interface taking into account the interaction resources of the target device;
- State Extraction and Adaptation to the Target Version, in order to provide support for task continuity it is necessary to extract the state resulting from the input entered by the user in the source device and associate such values to the corresponding elements in the target user interface;
- Run-time Interface Generation and Activation, generation of the UI in an implementation language supported by the target device with the state updated to make it consistent with that of the source UI and its activation at the same point in which the user left off on the source device.

8.5 An Architecture for a Migration Platform

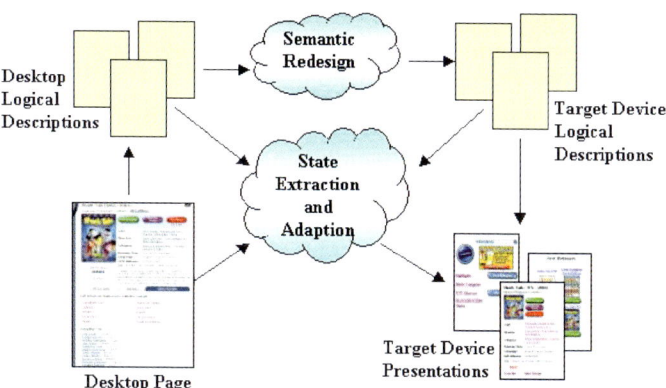

Fig. 8.2 Processing performed on the server side for supporting migration

More precisely, when a request for migration arrives, no matter which agent (the user or the system) starts the process, the Migration Manager, which acts as the main server module, retrieves the original version of the Web page that has to be migrated by invoking the HTTP Proxy service and retrieving the Web page(s) the user has accessed. Once the Web page(s) are retrieved, the Migration Manager builds the corresponding logical descriptions, at a different abstraction level, by invoking the Reverse Engineering service of our system. The result of the reverse engineering process is then used as input for the Semantic Redesigner service, in order to perform a redesign of the user interface for the target platform, taking into account the logical descriptions it received from the Migration Manager. Once the target presentations have been generated, the Presentation Mapper service is used to identify the page that should be first uploaded to the target device (which is the one supporting the last basic task performed on the source device). In order to support task continuity throughout the migration process, the state resulting from the user interactions with the source device (filled data fields, selected items, cookies …) is gathered through a dynamic access to the DOM of the pages in the source device. Then, such information is associated to the corresponding elements of the newly generated pages and adapted to the interaction features of the target device, in a totally user-transparent way. Regarding the migration trigger, the migration environment activates a dialogue, which asks the users whether they agree to accept the incoming application before activating the application in the target device. In addition, it can be activated only from the source device (either by a user action or a system event), and the source interface is closed after migration to avoid that someone else picks up the source device and corrupt the input.

8.6 Example Applications

In this section we describe two applications supported by the migration environment introduced beforehand.

8.6.1 Domotic Scenario (mobile vocal+graphical->graphical desktop)

The example application considered in this case is a domotic application and we introduce it through a scenario. Francesca uses her Smartphone in the car to set some parameters of home scenario while she is going at office. She uses her Smartphone in a multimodal way, being able to trigger UI actions via touch-screen and/or vocal commands, while the Smartphone provides graphical and vocal output.

The domotic application is organised in such a way to support a number of predefined scenarios (see Fig. 8.3), which capture frequently occurring user needs and are associated with specific settings of a number of domestic appliances. Figure 8.4 shows some settings associated with the Leave Home scenario.

Indeed this morning she was in a bit of a hurry and she was unable to set the adequate scenario before leaving the house: putting up the rolling shutters of rooms where there are plants, shut down all the lights and activate the house alarm. She initially planned to have lunch at home. However, afterwards she decides to have lunch in the city center with a friend: while waiting for the meeting time, she accesses the home application through the mobile device but since there is a PC available nearby she decides to migrate the user interface to the desktop system in order to better modify some scenario parameters: configure the thermostat so that she will have a comfortable temperature in the home when she will come back to home, and program the video recorder so that it will record her fa-

Fig. 8.3 The multimodal mobile control for setting the home scenarios

8.6 Example Applications

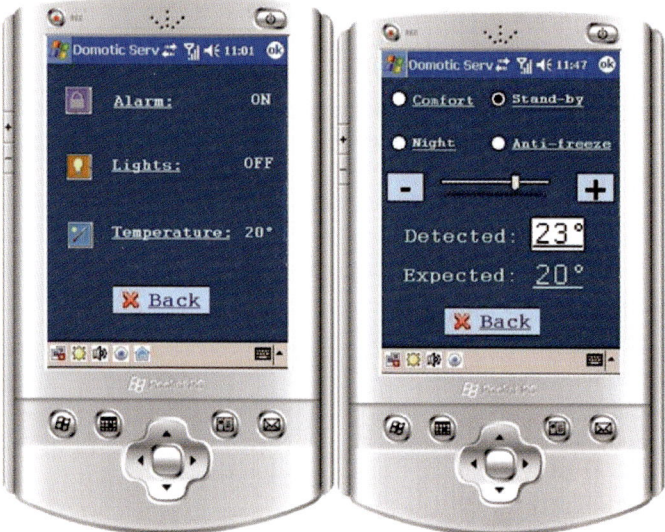

Fig. 8.4 Some settings associated with the Leave Home scenario

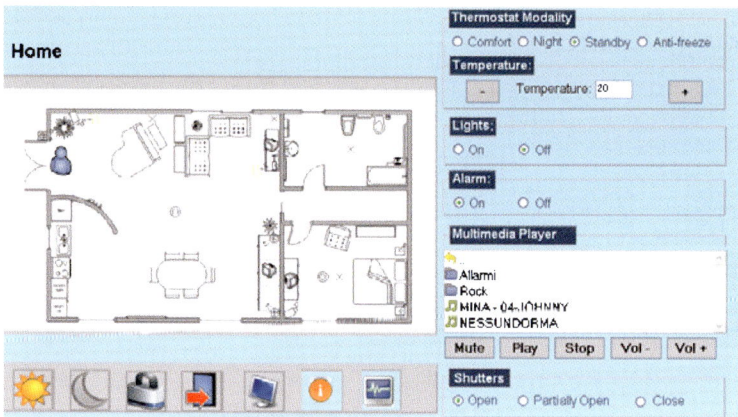

Fig. 8.5 The multimodal mobile control for setting the home scenarios

vourite TV movie and change to close the state of the shutters because of the bad weather conditions (see Fig. 8.5).

8.6.2 Museum Scenario (pda->digital tv)

The example application considered in this case is a museum game application. Visitors in a museum can use their PDAs to access to a number of games that are associated to the artworks shown in the museum. While they are in the museum, they start to play the game: more specifically, they access the service, identify themselves and start to access the different games.

When they arrive at home they may want to continue to play the games, in order to test the knowledge they gained during the visit. It may occur that while in the museum they started to solve some questions associated to an artwork, but without finishing it. When they return back home after the visit, they may want to continue to play the game. Therefore, they require migrating the user interface onto the large screen of the digital TV available, without having to re-enter the previously specified options.

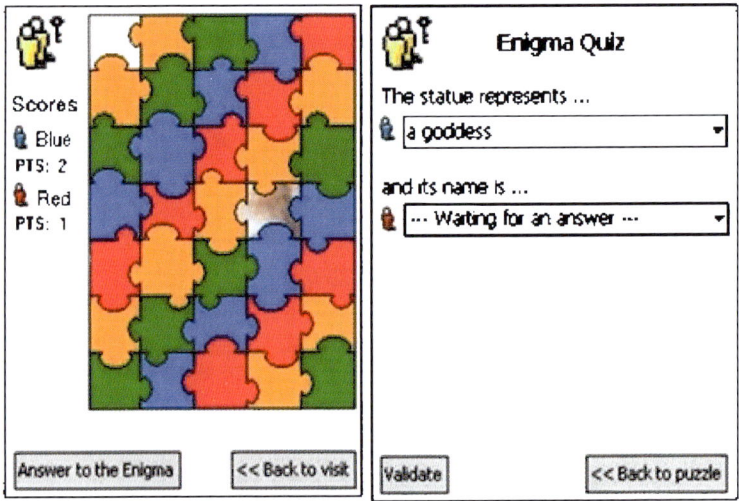

Fig. 8.6 The PDA interface of the museum games

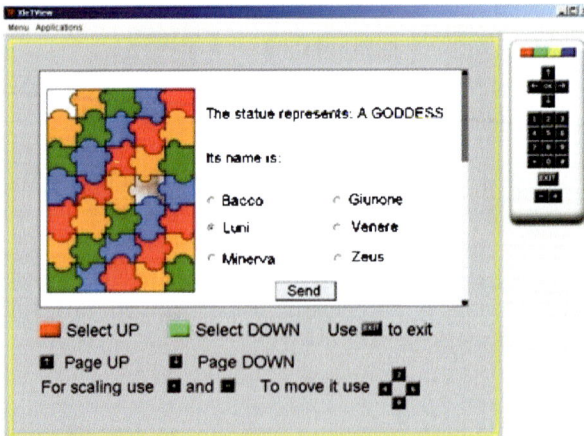

Fig. 8.7 The digital TV interface for the museum games

After the interface migration, they can still find the option they specified before (see Fig. 8.7) and continue to play the game by solving the remaining question.

Figure 8.6 shows the user interface of an enigma as it is visualised onto the PDA. Due to the small capability of the PDA, the user interface is presented into two separate screens: one dedicated to the visualization of the image associated to the enigma, the other one showing the questions to be answered. Figure 8.7 shows the user interface after the user require the migration on the digital TV. As you can see the choice the user has already carried out on the PDA (what the statue represents) is visualised as soon as the user interface is loaded on the user interface of the digital TV. Then, the user can continue his/her interaction by selecting the name of the represented artwork. As you can see, thanks to the larger screen of the digital TV, the user interface shows in a single presentation both the image of the enigma and the associated questions.

8.7 Conclusions

There are still many issues to solve in order to achieve pervasive usability in ubiquitous environments. The steadily increasing availability of new interactive device types in the mass market calls for support for migratory interactive services in order to better exploit task continuity in multi-device environments. Multimodal interfaces can further improve the user experience in such dynamic contexts of use, if designed carefully, providing more natural interaction techniques.

In general, in next years further work needs to be dedicated in order to obtain environments able to further empower end users. Web 2.0 and End-User Development [330] are interesting developments for this purpose by providing innovating mechanisms to allows users to easily generate content for interactive applications, and modify their appearance and behaviour even when such users are non professional software developers.

Chapter 9
Software Engineering for Multimodal Interactive Systems

Laurence Nigay[1], Jullien Bouchet[1], David Juras[1], Benoit Mansoux[1], Michael Ortega[1], Marcos Serrano[1], Jean-Yves Lionel Lawson[2]

[1] Université Joseph Fourier, Grenoble Informatics Laboratory (LIG), 385 rue de la Bibliothèque, BP 53, 38041 Grenoble cedex 9, France
{firstname.lastname}@imag.fr
[2] Université Catholique de Louvain, TELE Lab, Place du Levant, 2, B-1348 Louvain-la-Neuve, Belgium
jean-yves.lawson@uclouvain.be

9.1 Introduction

The area of multimodal interaction has expanded rapidly and since the seminal "Put that there" demonstrator [62] that combines speech, gesture and eye tracking, significant achievements have been made in terms of both modalities and real multimodal systems. The goal of multimodal interfaces is to extend the sensory-motor capabilities of computer systems to better match the natural communication means of human beings. The purpose is to enhance interaction between the user and the computer by using appropriate modalities to improve:

- the information bandwidth between the human and the computer; that is the amount of information being communicated,
- the signal-to-noise ratio of conveyed information; that is the rate of information useful for the task being performed [332].

Parallel to the development of the Graphical User Interface technology, natural language processing, computer vision, 3-D sound, and gesture recognition have made significant progress [333]. In addition recent interaction paradigms such as perceptual User Interface (UI) [85], tangible UI [85] and embodied UI [335] open a vast world of possibilities for interaction modalities including modalities based on the manipulation of physical objects such as a bottle and modalities based on the manipulation of a PDA and so on.

We distinguish two types of modalities: the active and passive modalities. For inputs, active modalities are used by the user to issue a command to the computer (e. g., a voice command). Passive modalities are used to capture relevant information for enhancing the realization of the task, information that is not explicitly expressed by the user to the computer such as eye tracking in the "Put that there" demonstrator [62] or location tracking for a mobile user.

In addition to many modalities that are more and more robust, conceptual and empirical work on the usage of multiple modalities (CARE properties [336], TYCOON design space [337], etc.) are now available for guiding the design of efficient and usable multimodal interfaces. These aspects have been studied in Chapter 4 (Theoretical Approach and Principles for Plastic Multimodal Human Computer Interaction).

Due to this conceptual and predictive progress and the availability of numerous modalities, real multimodal systems are now built in various domains including medical [62, 338] and military ones. One of our application domains is military: In [339] we study multimodal commands in the cockpit of French military planes. For example while flying, the pilot can mark a point on the ground by issuing the voice command "mark" (active modality) and looking at a particular point (passive modality). Another example that we will reuse in this chapter is a multimodal driving simulator that is based on multimodal driver's focus of attention detection as well as driver's fatigue state detection and prediction [62]. Moreover multimodal interfaces are now playing a crucial role for mobile systems since multimodality offers the required flexibility for variable usage contexts, as shown in our empirical study of multimodality on PDA [340].

Although several real multimodal systems have been built, their development still remains a difficult task. Tools dedicated to multimodal interaction are currently few and limited in scope. The power and versatility of multimodal interfaces that is highlighted in Chapter 4, results in an increased complexity that current software design methods and tools do not address appropriately. As observed by B. Myers, "user interface design and implementation are inherently difficult tasks" [341]. Myers's assertion is even more relevant when considering the constraints imposed by multimodal interaction. In particular, multimodal interaction requires (see chapter 4):

- the fusion (CARE properties – Chapter 4) of different types of data originating from distinct interaction techniques as exemplified by the "put that there" paradigm,
- the management of multiple processes including support for synchronization and race conditions between distinct interaction techniques.

Thus, multimodal interfaces make necessary the development of software tools that satisfy new requirements. Such tools are currently few and limited in scope. Either they address a very specific technical problem such as media synchronization [342], or they are dedicated to very specific modalities. For example, the Artkit toolkit is designed to support direct manipulation augmented with gesture only [343]. In this chapter, we focus on this problem of software development of multimodal interfaces. The structure of the chapter is as follows: first we present our PAC-Amodeus software architectural model as a generic conceptual solution for addressing the challenges of data fusion and concurrent processing of data of multimodal interfaces. We then focus on software tools for developing multimodal interfaces. After presenting several existing tools, we then describe and illustrate the ICARE and OpenInterface platforms.

9.2 PAC-Amodeus: a Conceptual Architectural Solution

Our conceptual solution for supporting the data fusion and in particular the CARE properties (see Chapter 4) as well as the concurrent processing of data draws upon our software architecture model: PAC-Amodeus [344, 345]. The solution is conceptual since it is independent of any development technique and tool that will be studied afterwards in section 3.

As shown in Fig. 9.1, the PAC-Amodeus model reuses the components of Arch [62] but refines the dialogue controller in terms of PAC agents [346]. This refinement has multiple advantages including an explicit support for concurrency.

The Functional Core (FC) implements domain specific concepts in a presentation independent way. The Interface with the Functional Core (IFC) maps Domain objects from the Functional Core onto Conceptual objects from the Dialogue Controller and vice versa. The Dialogue Controller (DC) is the keystone of the model. It has the responsibility for task-level sequencing. Each task of the user corresponds to a dialogue thread. This observation suggests a multi-agent decomposition where an agent, or a collection of agents, can be associated with each thread. The Presentation Techniques Component (PTC) defines two multi-valued mapping functions that link Presentation and Interaction objects. The PTC describes the presentation (i. e., input and output interfaces). It implements the perceivable behaviour of the application for output and input commands in a device independent way. The PTC is device independent but language dependent.

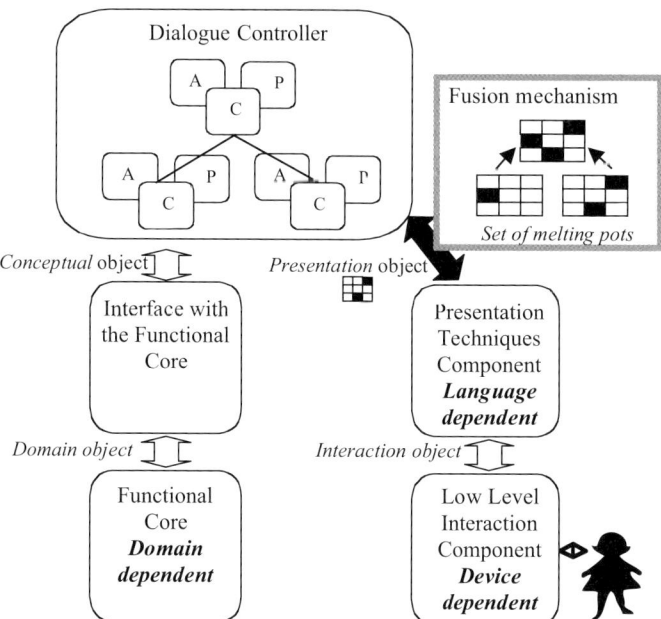

Fig. 9.1 The software components of the PAC-Amodeus model and the interfaces between them

The Low Level Interaction Component (LLIC) denotes the underlying software and hardware platform. It supports the physical interaction with the user. The LLIC is therefore device dependent. It manages user's events from different media (time-stamps and queues) and has the responsibility for their lexical analysis. Some of the low-level events are not transmitted to the Presentation Technique Component. Indeed, lexical tasks such as window resize, are locally performed by the LLIC. In addition, in the case of spoken-utterances, this component can include mechanisms for confirmation allowing the user to intercept a wrong recognition.

As a conclusion, two components are modality dependent, the Low Level Interaction Component (LLIC) and the Presentation Techniques Component (PTC). The three remaining components (DC, INF and NF) are modality independent. We need now to show how concurrent processing and data fusion are performed within our architectural framework.

9.2.1 Concurrent Processing of Data

Concurrent processing of data is achieved at different levels of abstraction. Raw data is captured in the LLIC component by event handlers. There is one event handler per input device. Event handlers process in parallel. Concurrency is also supported in the PTC which receives low level events (Interaction objects) from the LLIC and transforms them into more abstract interaction techniques. For example, a mouse click is transformed into the Select interaction technique. There is one abstracting process per supported language (modality). A modality, for instance a gestural language, may be supported by different physical devices (and so different events handlers), such as a mouse and a data glove combined with computer vision. Finally, the multi-agent architecture of the Dialogue Controller offers an interesting conceptual framework to support concurrency. Agents can process data in parallel.

9.2.2 Data Fusion

The complementary and redundancy of the CARE properties imply fusion of data from different modelling techniques. Each technique is associated with a modality. As discussed in [344], fusion occurs at every level of the PAC-Amodeus components. For example, within the LLIC, typing the option key along with another key is combined into one single event. More generally, fusion at the signal level, for example combining the flow of two video streams, is performed in LLIC. In this chapter focusing on multimodal interaction, we are concerned with data fusion to obtain an elementary task sent to the Dialogue Controller. Fusion is performed between the PTC and the DC, resulting in an elementary task that is modality independent since the DC is modality independent.

9.2 PAC-Amodeus: a Conceptual Architectural Solution

Fusion is performed on the presentation objects received from the PTC. These objects obey a uniform format: the melting pot. As shown in Fig. 9.2, a melting pot is a 2-D structure. On the vertical axis, the "structural parts" model the composition of the task objects that the Dialogue Controller is able to handle. For example request slots can be the name of a command and two parameters corresponding to the structural parts of an elementary command (e.g. <put that there>). Events generated by user's actions are abstracted through the LLIC and PTC and mapped onto the structural parts of the melting pots. In addition, LLIC events are time-stamped. An event mapped with the structural parts of a melting pot defines a new column along the temporal axis. The structural decomposition of a melting pot is described in a declarative way outside the engine. By so doing, the fusion mechanism is domain independent: structures that rely on the domain are not "code-wired". They are used as parameters for the fusion engine.

Figure 9.3 illustrates the effect of a fusion on three melting pots: at time t_i, the user selected one icon with the mouse while s/he has uttered the sentence "put that there" at time t_i+1. The melting pot on the bottom left of Fig. 9.3 is generated by the mouse selection action. The speech act triggers the creation of the bottom middle melting pot: the slot "name" is filled in with the value "Put". The fusion engine combines the two melting pots into a new one where the name and a pa-

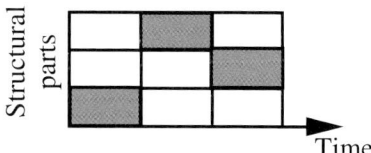

Fig. 9.2 Melting pot: a common representation for data fusion

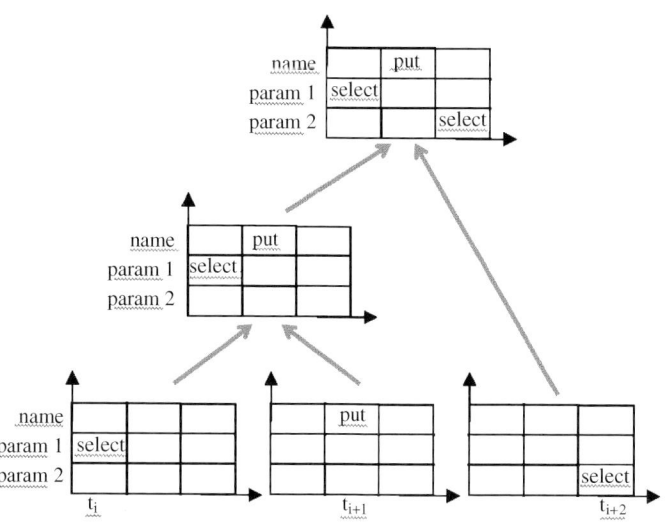

Fig. 9.3 Fusion of three melting pots. Elementary task or command:<put that there>

rameter are both specified. At time ti+2, the user selected a second icon: the bottom right melting pot of Fig. 9.3 is generated by this selection action. The fusion engine finally combines the two melting pots into a new one, resulting in the melting pot at the top of Fig. 9.3 that corresponds to a complete command.

The two main criteria for triggering fusion are the complementarity/redundancy of melting pots and the time. When triggered, the engine attempts three types of fusion in the following order: microtemporal fusion, macrotemporal fusion, and contextual fusion. The three types of fusion are fully described in [345].

- Microtemporal fusion is used to combine related informational units produced in parallel or in a pseudo-parallel manner. It is performed when the structural parts of input melting pots are complementary and when these melting pots are close in time: their time interval overlaps.
- Macrotemporal fusion is used to combine related informational units produced sequentially or possibly in parallel by the user but processed sequentially by the system, or even delayed by the system due to the lack of processing resources (e. g., processing speech input requires more computing resources than interpreting mouse clicks). Macrotemporal fusion is performed when the structural parts of input melting pots are complementary and when the time intervals of these melting pots do not overlap but belong to the same temporal window.
- Contextual fusion is used to combine related informational units produced without attention to temporal constraints.

For implementing the fusion mechanism, a set of rules [345] are defined. For example one rule is given priority to parallelism at the user's level and corresponds to the microtemporal fusion. To perform such fusion, the mechanism is based on a set of metrics associated with each melting pot. Figure 9.4 portrays the metrics that describe a melting pot m_i. A melting pot encapsulates a set of structural parts $p_1, p_2 \ldots p_n$. The content of a structural part is a piece of information that is time-stamped. Time stamps are defined by the LLIC when processing

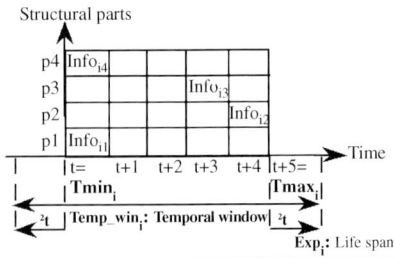

Fig. 9.4 Metrics used to define a melting pot m_i. From [345]

user's events. The engine computes the temporal boundaries (T_{max} and T_{min}) of a melting pot from the time stamps of its pieces of information: So for $m_i = (p_1, p_2, \ldots p_n)$, $Tmax_i=Max(Tinfo_{ij})$ and $Tmin_i=Min(Tinfo_{ij})$. The temporal window of a melting pot defines the temporal proximity (+/- Δt) of two adjacent melting pots: for $m_i=(p1, p2, \ldots pn)$, $Temp\_win_i= Tmin_i-\Delta t, Tmax_i+\Delta t$. Temporal windows are used to trigger macrotemporal fusion. The last metrics used to manage a melting pot is the notion of life span, Expi: $Expi=Tmax_i+\Delta t=Max(Tinfo_{ij})+\Delta t$. This notion is useful for removing a melting pot from the set of candidates for fusion.

The software architecture model, PAC-Amodeus, augmented with the fusion mechanism support the software design of multimodal systems. PAC-Amodeus along with the fusion mechanism fulfils specific requirements of multimodal systems such as data fusion and parallel processing. The fusion mechanism is responsible for combining data specified by the user through different modalities (i.e., a combination of devices and interaction languages). Based on criteria such as time and structural complementarity, the mechanism is generic and reusable. Each melting pot processed may have any number of structural parts (e.g., lines) that can be filled independently. Consequently, the PAC-Amodeus model, along with the fusion mechanism, define a conceptual solution for implementing multimodal systems. In light of this conceptual solution, we now study the existing tools.

9.3 Software Tools for Multimodality

As defined in [347], in the general context of user interface software tools, tools for multimodal interfaces must aim to have a low threshold (easy to use) while providing a high ceiling (how much can be done with the tool).

Moreover for the case of multimodal interaction, software tools should support the data fusion and concurrent processing studied at a conceptual level in the previous paragraph. The tools should also be extendable. Indeed in order to take account of the ever-widening world of modalities, the tools must be easily extendable, an extensibility that we address in our ICARE and OpenInterface platforms by considering a component-based approach.

9.3.1 Existing Tools

The existing frameworks dedicated to multimodal interaction are currently few and limited in scope.

Existing tools mainly focus on input multimodality, either by addressing a specific technical problem or by being dedicated to specific modalities. Tools that address a specific technical problem focus on one of the components of the PAC-Amodeus model (Fig. 9.1). Such tools include:

- the implementation in C language [345] of the fusion mechanism described in section 2 or the one presented in [81],
- the composition of several devices [81] (LLIC in Fig. 9.1),
- mutual disambiguation [81, 349], (LLIC in Fig. 9.1).

Other tools are dedicated to specific modalities such as gesture recognition [350] speech recognition [351] or the combined usage of speech and gesture [352]. Going one step further than providing a particular modality or generic reusable mechanisms (i. e. fusion and mutual disambiguation mechanisms), Quickset [353] defines an overall implementation architecture as well as the Open Agent Architecture (OAA) [354]. Quickset mainly focuses on input multimodality based on speech and gesture and has been applied to the development of map-based systems.

For outputs, several studies have been performed in the context of the conversational paradigm, also called intelligent multimedia presentation in which seminal work is presented in [62]. The system is designed here as a partner for the user (computer-as-partner [355]): an output communicative act as part of a natural dialogue between the user and the system is made perceivable by a multimodal presentation. Moreover the main focus of such existing output multimodal frameworks is to automatically generate the output presentation, also called presentation planning systems, based on a speech act, a context such as the current available interaction resources and a user's profile. For example in the Embassi demonstrator [356], the architecture is based on OAA and includes a dedicated agent to achieve the combination of output modalities. Based on a speech act, the current context and the user's profile, Embassi generates multimodal presentations that are rendered by a dynamic set of distributed agents.

Focusing on the direct manipulation paradigm (computer-as-tools [355]), very few tools are dedicated to the design and development of output multimodal interfaces. MOST (Multimodal Output Specification Tool) [357] is a recent framework for multimodal output interaction which focuses on automatic generation of multimodal presentation based on the interaction context defined as the triplet <user, system, environment>. MOST includes a rule-based selection mechanism for generating the multimodal presentation. MOST therefore defines a reusable framework for developing adaptive multimodal systems and its focus is not on the design of multimodality but more on adaptability by providing an editor for specifying the adaptation rules.

Tools for both multimodal inputs and outputs include the two platforms described in the following two paragraphs, namely ICARE and OpenInterface. A related tool to these two platforms is CrossWeaver [85]: it is a prototyping tool dedicated to non-programmer designers. The created prototypes may involve several input modalities and two output modalities: visual display and text-to-speech synthesis. CrossWeaver divides the design process into three steps. First, the designer makes various sketches to form a storyboard. The right part of Fig. 9.5 shows an example of a sketch, a map. She/he also decides which combinations of input and output modalities will be available for each sketch and for transitions

9.3 Software Tools for Multimodality

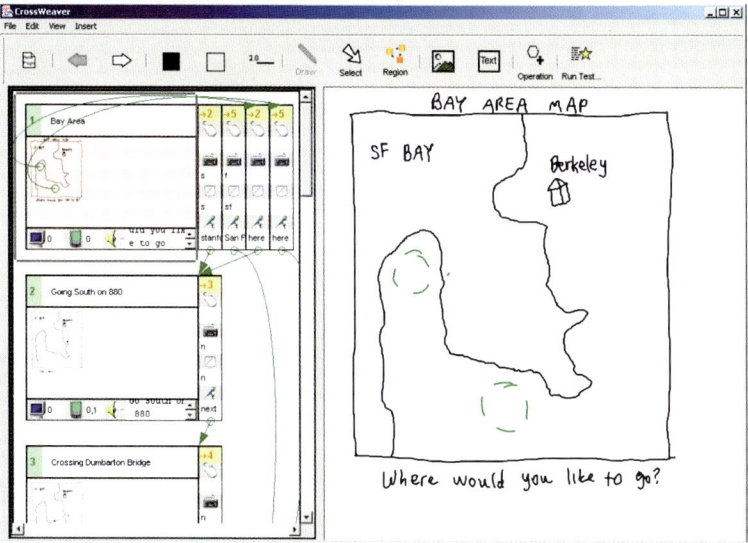

Fig. 9.5 Crossweaver prototyping tool. (guir.berkeley.edu/projects/crossweaver/)

between sketches. The left part of Fig. 9.5 shows an example of several sketches linked by modalities: the input modalities include the mouse, the keyboard, gestures and speech commands. Then, the user tests the prototype with the available input and output modalities. Finally, thanks to a log of the user's actions in the previous step, the designer can analyze how multimodality is handled by the user, and can quickly change the combination of modalities to adapt the interaction. Implementation of CrossWeaver is also based on OAA. As opposed to CrossWeaver that is a prototyping tool, the two following platforms ICARE and OpenInterface are development tools.

9.3.2 ICARE Platform

ICARE stands for Interaction-CARE (Complementarity Assignment Redundancy Equivalence). The ICARE platform enables the designer to graphically manipulate and assemble ICARE software components in order to specify the multimodal interaction dedicated to a given task of the interactive system under development. From this specification, the code is automatically generated. To fully understand the scope of the ICARE platform we show in Fig. 9.6 where the ICARE components are located within the complete code of the interactive system structured along the PAC-Amodeus software architectural model of Fig. 9.1.

In order to present the ICARE platform, we first explain the underlying ICARE conceptual model by presenting the types of components manipulated by the plat-

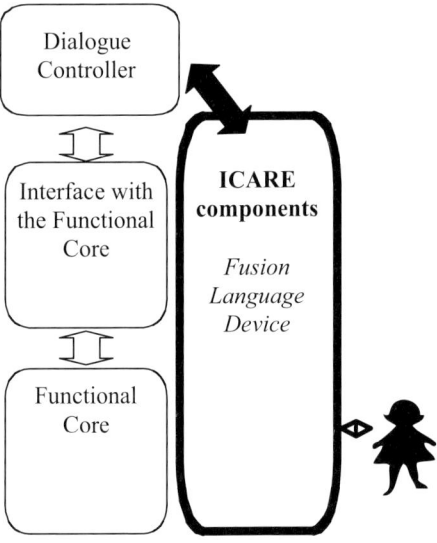

Fig. 9.6 ICARE components within the PAC-Amodeus software architecture of Fig. 9.1

form. We then present the graphical editor that enables the design and the code generation of multimodal input interaction.

9.3.3 ICARE Conceptual Model

The ICARE conceptual model includes elementary and composition components. They are fully described in [70] while their developments are described in [359].

Elementary components are building blocks useful for defining a modality. Two types of ICARE elementary components are defined: Device components and Interaction Language components. We reuse our definition of a modality presented in Chapter 4.

Composition components describe combined usages of modalities and therefore enable us to define new composed modalities. As opposed to ICARE elementary components, Composition components are generic in the sense that they are not dependent on a particular modality. The ICARE composition components are defined based on the four CARE properties (Chapter 4): the Complementarity, Assignment, Redundancy. Because the CARE properties have been shown to be useful concepts for the design and evaluation of multimodal interaction (Chapter 4), we decided to reuse those concepts to make them explicit during the software development. We define three Composition components in our ICARE conceptual model: the Complementarity one, the Redundancy one, and the Redundancy/Equivalence one. Assignment and Equivalence are not modelled as components in our ICARE model. Indeed, an assignment is represented by a single link between two components. An ICARE component A linked to a single component

B implies that A is assigned to B. As for Assignment, Equivalence is not modelled as a component. When several components (2 to n components) are linked to the same component, they are equivalent.

9.3.4 ICARE Graphical Editor

Figure 9.7 presents the user interface of the ICARE platform: it contains a palette of components, an editing zone for assembling the selected components and a customization panel for setting the parameters of the components. The palette of components is organised along the three types of ICARE components: Composition, Interaction Language and Device components. The user of the ICARE platform selects the modalities (Device and Language components) and specifies the combination of modalities by selecting a Composition component, all by graphically assembling software components without knowing the details of the code of the components. After graphically specifying the multimodal interaction for a given task, the user of the ICARE platform can generate the code of the corresponding multimodal interaction.

Using the ICARE platform, we developed several multimodal systems:

- A Multimodal Identification system (MID) combining speech, keyboard and mouse modalities for identification.
- A mobile system MEMO: MEMO allows users to annotate physical locations with digital notes which have a physical location and are then read/removed by

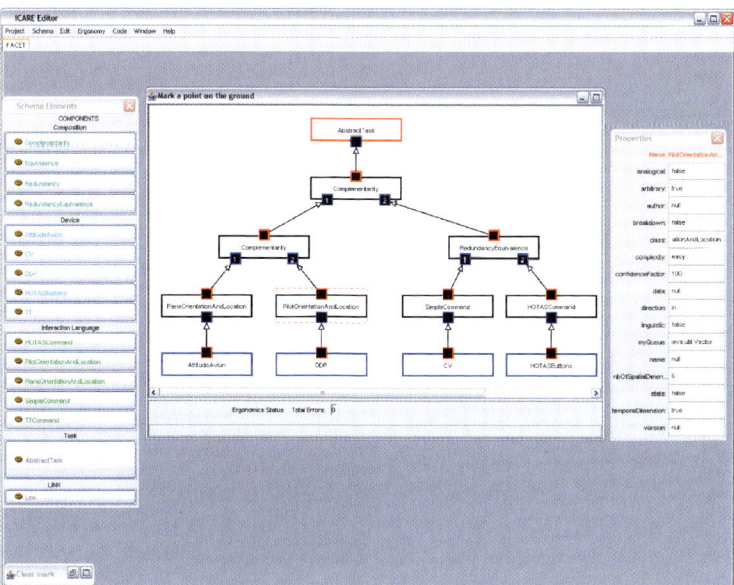

Fig. 9.7 ICARE graphical platform

other mobile users. Two versions are currently running, one using a PDA and another one using a Head-Mounted Display. The development of MEMO based on ICARE is described in [70].
- A Yellow Pages system, allowing the user to specify queries using speech, keyboard and mouse modalities. In addition the user can specify synergistic commands combining speech and direct manipulation. For example the user can issue the voice command "Zoom here" while selecting a point on the displayed map using a stylus.
- A larger system, IAS, a flight simulator of a military fighter: the multimodal system is a real-time complete flight simulator of a French military plane, for studying future interaction techniques that will be embodied in the cockpit. Several input modalities are included in IAS, making IAS a good candidate for studying rich multimodal interaction in various contexts. The input modalities include the HOTAS (Hands On Throttle And Stick), a helmet visor, speech inputs and a tactile surface. Surface. The development of IAS based on ICARE is fully described in [339, 359]. In Fig. 9.8, we give a simple example of an ICARE diagram for the task of switching equipment on/off. The ICARE specification of Fig. 9.8 is quite simple since only two modalities can be used for switching equipment on/off: the pilot selects a button displayed on the tactile surface and/or (redundancy/equivalence composition) uses speech.
- An augmented reality puzzle: The goal is to complete a physical puzzle and the multimodal system helps the player to correctly orient the puzzle pieces [338]. For providing the guidance information, the output modalities are graphics displayed on a localised mini-screen fixed to the puzzle piece as well as voice messages.
- An augmented surgery system, PERM, a computer assisted kidney puncture described in [338]. Our goal is to be able to quickly explore several design alternatives with the surgeon using ICARE, focusing on the multimodal way to

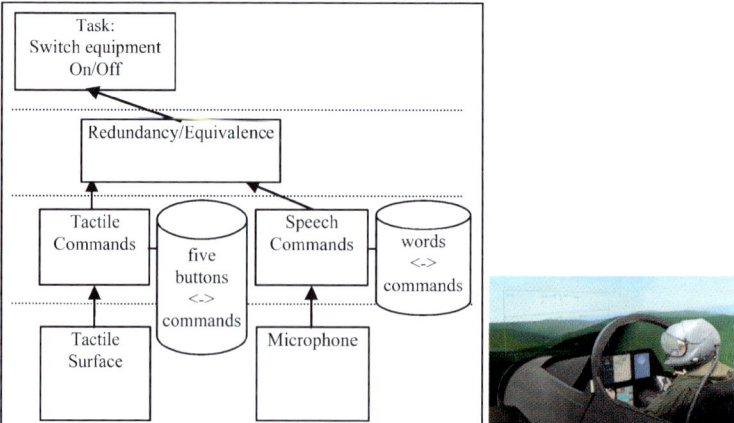

Fig. 9.8 ICARE specification of a redundant/equivalent usage of two modalities for the task <Switch equipment on/off> in the IAS system. From [339]

present the guidance information to the surgeon according to the different phases of the surgery.

9.3.5 OpenInterface Platform

OpenInterface is a component-based platform developed in C++ that handles distributed heterogeneous components. The connection between heterogeneous components is mainly based on existing technologies. OpenInterface supports the efficient and quick definition of a new OpenInterface component from an XML description of a program. Although the platform is generic, in the context of the SIMILAR[1], the OpenInterface platform is dedicated to multimodal applications. In SIMILAR, we define a multimodal application as an application that includes multimodal data processing and/or offers multimodal input/output interaction to its users. In this chapter we then focus on the OpenInterface platform for multimodal interaction and multimodal data processing.

Figure 9.9 gives an overview of the platform. The OpenInterface platform is an open source project, free to download at www.openinterface.org. Each component

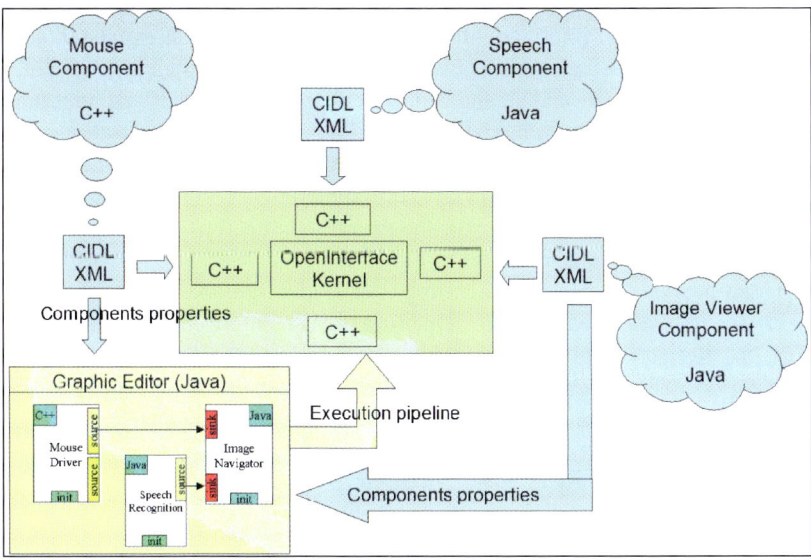

Fig. 9.9 OpenInterface platform

---

[1] SIMILAR: FP6 European Network of Excellence (FP6-507609). WP2-OpenInterface platform. www.similar.cc.

public interface is described using the Component Interface Description Language (CIDL) and then registered in OpenInterface platform. The registered component properties are retrieved by the Graphic Editor (developed in Java). Using the editor the user can edit the component properties and compose the execution pipeline (by connecting the components) of the multimodal application. The pipeline is then executed by the OpenInterface kernel (C/C++) that instantiates components and performs the connection between components.

OpenInterface is designed to serve three levels of users: programmers, application designers (AD) and end-users. Programmers are responsible for the development and integration of new components into the platform. The application designers focus on end-user's needs and are aware of the resources provided by the platform. The AD will use the graphical editor to assemble components in order to develop a multimodal application. End-users interact with the final application whose components are executed within the platform.

We illustrate the OpenInterface platform by considering a multimodal driving simulator. The simulator developed using the platform is fully described in [62]. OpenInterface components of the multimodal driving simulator have been manually assembled: we did not use the graphical editor of Fig. 9.9. It is based on multimodal driver's focus of attention detection as well as driver's fatigue state detection and prediction. Capturing and interpreting the driver's focus of attention and fatigue state is based on video data (e.g., facial expression, head movement, eye tracking). While the input multimodal interface relies on passive modalities only (also called attentive user interface), the output multimodal user interface includes several active output modalities for presenting alert messages including graphics and text on a mini-screen and in the windshield, sounds, speech and vibration (vibration wheel). Active input modalities are added in the meta-User Interface to let the user dynamically select the output modalities.

For the development, we reuse the GPL driving simulator TORCS[2] that we extend to be multimodal. Figure 9.10 show the overall architecture of the multimodal driver along the PAC-Amodeus model of Fig. 9.1.

The video analysis system for capturing user's state (passive input modalities in Fig. 9.10) is composed of two OpenInterface components as shown in Fig. 9.11: the Retina component that enhances the input data and extracts different information. The second component computes the user face analysis and outputs different indicators related to the user's state. These outputs are used by the component presenting the alarms. Three outputs are alarms related to the estimation of the driver fatigue level. An alarm is sent when the user closes his eyes for more than a specified duration (we experimentally fix it to 200ms). Another is sent when the driver yawns. Also, an alarm is generated when the user moves his head longer than a specified period (we experimentally fix it to 300ms).

---

[2] TORCS Driver Simulator: torcs.sourceforge.net

9.3 Software Tools for Multimodality

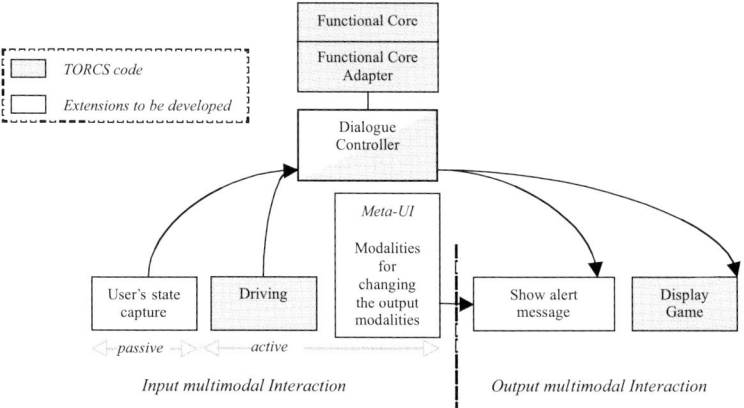

Fig. 9.10 TORCS code and extensions to be developed within our architecture. From [62]

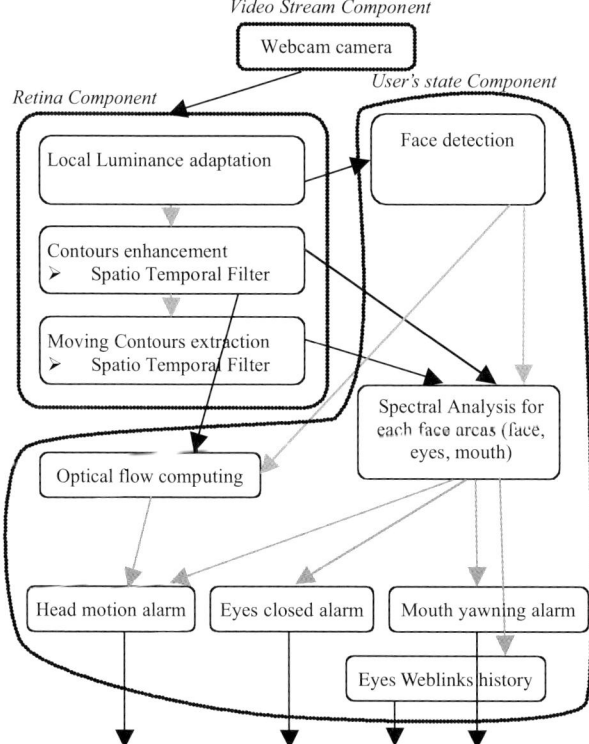

Fig. 9.11 Implemented OpenInterface components for capturing the user's state. Starting from the input provided by the Video Stream component, two OpenInterface components, namely Retina component and User's State component, have been implemented for providing four outputs: Focus of attention (head motion), duration of eyes closed, yawning and eye blinking history from [62].

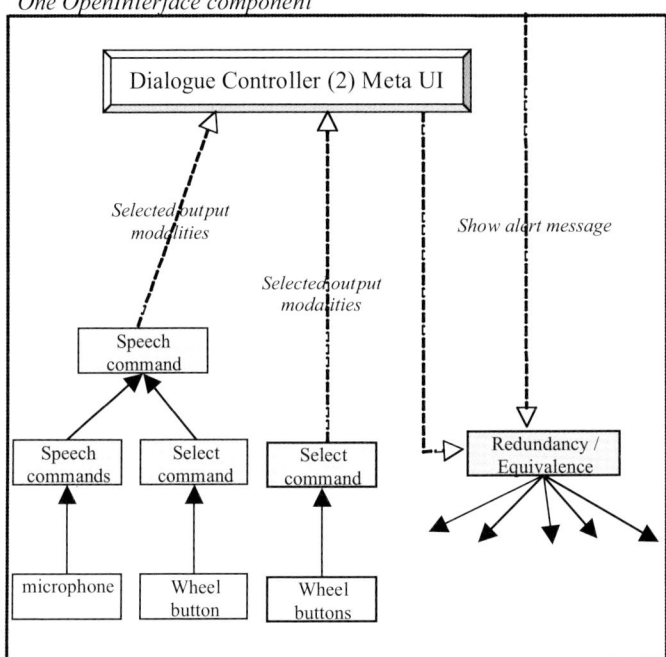

Fig. 9.12 Implemented OpenInterface component for presenting the alarm in a multimodal way. Several ICARE components are encapsulated in one OpenInterface component. From [62]

For output interaction, presenting the alarm in a multimodal way, one OpenInterface component has been developed. To do so we reused a complete pipeline defined using ICARE. As a consequence, the OpenInterface component includes several ICARE components developed in JavaBeans. Figure 9.12 shows the OpenInterface component and its encapsulated ICARE components.

9.4 Conclusion

It is impossible to discuss the software engineering of multimodal Human-Computer Interaction and, in particular, software architecture models and tools without first focusing on the user interface changes that will require new results in Engineering for Human-Computer Interaction (EHCI). Indeed, innovation in EHCI follows innovation in UI design (see Chapter 4); developments in EHCI make sense only when you know the types of interfaces for which you are building models and tools. Therefore in this chapter we started by identifying key issues for multimodal user interfaces: data fusion, concurrent processing of data and a huge variability in available and used interaction modalities that can be pure or combined, passive or active. We then presented a conceptual solution independent of

9.4 Conclusion

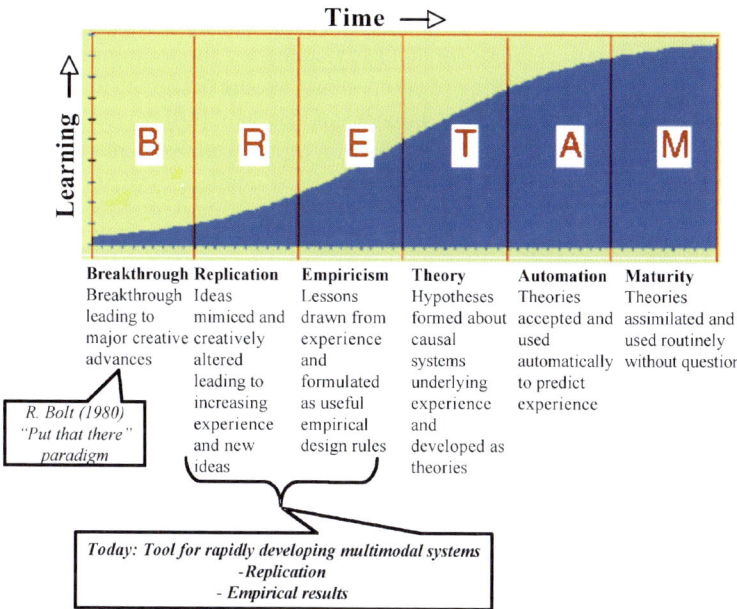

Fig. 9.13 B. Gaines's model on how science technology develops over time, applied to multimodal interaction

implementation issues before studying existing software tools. Such tools dedicated to multimodal interfaces are crucial for moving away from WIMP (Windows, Icons, Menus, and Pointing Devices). Indeed as pointed out by B. Gaines' model [360] on how science technology develops over time, it is now timely to make a step change in the domain of multimodal interaction. As shown in Fig 9.13, after the initial breakthrough phase [361] we are now at the stage of replication.

As explained in this chapter, the existing tools dedicated to multimodal interaction are currently few and limited in scope. Either they address a specific technical problem including the fusion mechanism, the composition of several devices and mutual disambiguation, or they are dedicated to specific modalities such as gesture recognition, speech recognition or the combined usage of speech and gesture. Two platforms ICARE and OpenInterface are more general and are able to handle various modalities. ICARE is explicitly based on the concepts of devices and languages as well as the CARE properties for data fusion. The various multimodal interfaces developed using ICARE highlight the problem of the reusability of the language components that are in general too specific to be used for another device. Moreover ICARE is dependent on a particular technology, namely JavaBeans, as opposed to the OpenInterface platform. Indeed the first issue addressed by the OpenInterface platform is to be able to handle distributed heterogeneous components. By using a graphical editor, multimodal interaction is developed by graphically assembling components, but the editor does not support a model dedicated to multimodal interaction as in ICARE.

As on-going work in the OpenInterface project (FP6-35182, www.oi-project.org), starting from the OpenInterface platform that is extended for supporting more features including the dynamic assembling of components, we are designing and developing an OpenInterface Interaction Development Environment (OIDE) for building changing and monitoring the multimodal interface in order to analyze, interpret and evaluate interaction data. The OIDE [362] addresses these challenges via its component repository and its construction, debugging and logging tools explicitly based on multimodal concepts. The OIDE when fully designed and developed should allow multimodal user interfaces to be created more quickly and to be able to be integrated cost-effectively into new ubiquitous and mobile environments. The OIDE should enable rapid prototyping and therefore more iterations as part of an iterative design method for achieving usable multimodal user interfaces [347] as well as providing a means of enabling end-users and other stakeholders to become directly involved in the development and configuration of systems over which they can take genuine ownership and control.

9.5 Acknowledgments

The work on the ICARE platform is funded by DGA (French Army Research Dept.) under contract INTUITION #00.70.624.00.470.75.96, while the work on the OpenInterface platform described in this paper is supported by the European FP6 SIMILAR Network of Excellence (www.similar.cc).

Chapter 10
Gestural Interfaces for Hearing-Impaired Communication

Oya Aran[1], Thomas Burger[2], Lale Akarun[1], Alice Caplier[2]

[1] Dep. of Computer Engineering, Bogazici University 34342 Istanbul, Turkey, aranoya@boun.edu.tr, akarun@boun.edu.tr
[2] GIPSA-lab/DIS, 46 avenue Félix Viallet, 38031 Grenoble cedex 1, France, thomas.burger@lis.inpg.fr, alice.caplier@lis.inpg.fr

10.1 Introduction

Recent research in Human-Computer Interaction (HCI) has focused on equipping machines with means of communication that are used between humans, such as speech and accompanying gestures. For the hearing impaired, the visual components of speech, such as lip movements, or gestural languages such as sign language are available means of communication. This has led researchers to focus on lip reading, sign language recognition, finger spelling, and synthesis. Gestural interfaces for translating sign languages, cued speech translators, finger spelling interfaces, gesture controlled applications, and tools for learning sign language have been developed in this area of HCI for the hearing impaired.

Gestural interfaces developed for hearing impaired communication are naturally multimodal. Instead of using audio and visual signals, hearing impaired people use multiple vision based modalities such as hand movements, shapes, position, head movements, facial movements and expressions, and body movements in parallel to convey their message.

The primary means of communication between hearing-impaired people are sign languages. Almost all countries and sometimes regions within countries have unique sign languages that are not necessarily related with the spoken language of the region. Each sign language has its own grammar and rules [363]. Instead of audio signals, sign languages use hand movements, shapes, orientation, position, head movements, facial expressions, and body movements both in sequential and parallel order [364]. Research on automatic sign language communication has progressed in recent years. Several survey papers are published that show the significant progress in the field ([365, 366]). Interfaces are developed that handle isolated [367] and continuous sign language recognition ([368, 369]). Interactive educational tools have also been developed for teaching sign language [370].

Fingerspelling is a way to code the words with a manual alphabet which is a system of representing all the letters of an alphabet, using only the hands. Fin-

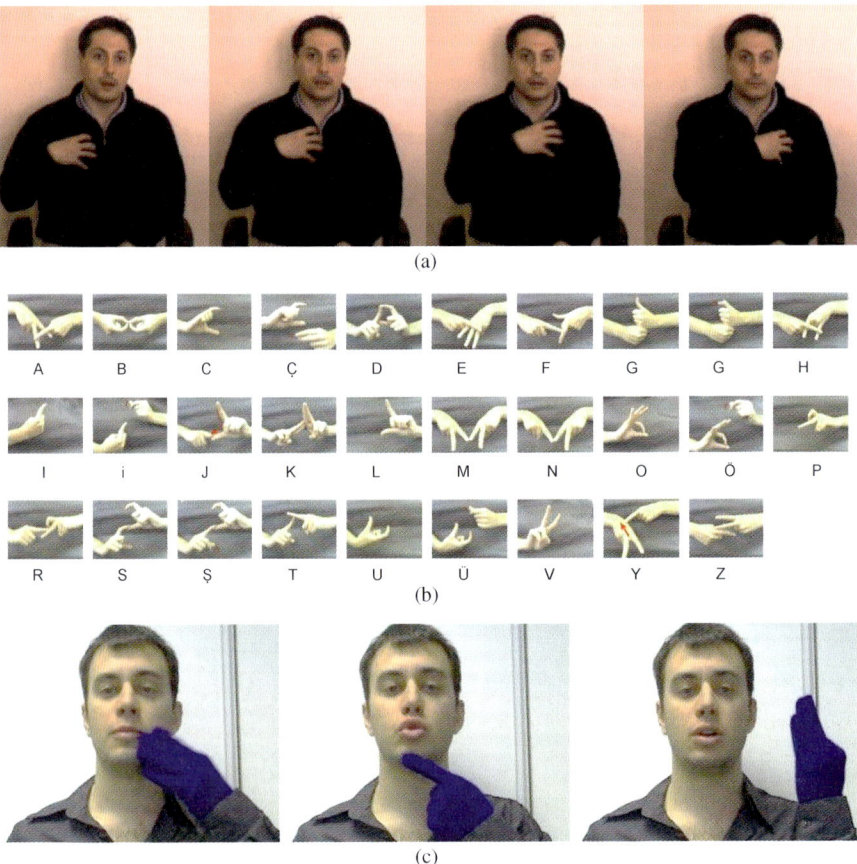

Fig. 10.1 (a) An example sign "anne (mother)" from Turkish Sign Language (TSL), (b) Fingerspelling alphabet of TSL (Dynamics are indicated by red arrows), and (c) an example of French cued speech "bonjour (good morning)"

gerspelling is a part of sign languages and is used for different purposes. It may be used to represent words which have no sign equivalent, or for emphasis, clarification, or when teaching or learning a sign language ([371, 372]).

Cued Speech (CS) is a more recent and totally different means of communication, whose importance is growing in the hearing-impaired community. It was developed by Dr. Cornett in 1967 [373]. Its purpose is to make the natural oral language accessible to hearing-impaired people, by the extensive use of lip-reading. But lip-reading is ambiguous: for example, /p/ and /b/ are different phonemes with identical lip shape. Cornett suggests replacing invisible articulators (such as vocal cords) that participate to the production of the sound by hand gestures. Basically, it means completing the lip-reading with various manual gestures. Then, considering both lip shapes and hand gestures, each phoneme has a specific

10.2 Modality Processing and Analysis

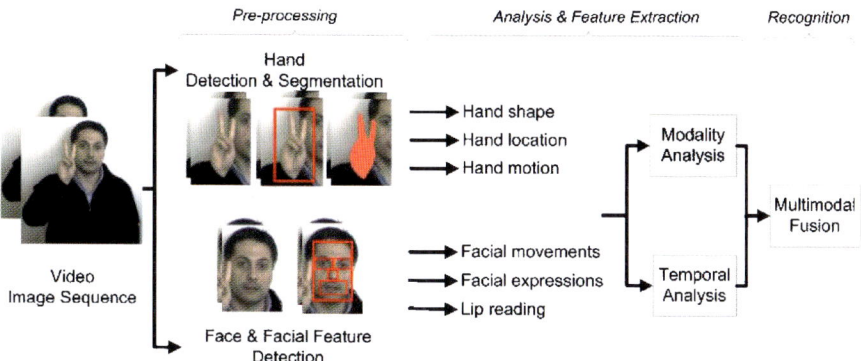

Fig. 10.2 Multimodal gestural interfaces for the hearing impaired

visual aspect. There are three modalities in CS: lip motion, hand shape and hand location with respect to the face.

Figure 10.2 shows the overall architecture of the multimodal gesture based interfaces for the hearing impaired communication. In the next section, we discuss and review analysis techniques for the modalities that are used in hearing impaired communication. We concentrate on the individual modalities: hand, face, lips, expression and treat their detection, segmentation, and feature extraction. In the *Temporal analysis* section, we focus on the temporal analysis of the modalities, specifically in sign languages and in CS. The following section presents temporal modelling and belief-based multimodal fusion techniques. In the last section, we give two example applications: a sign language tutoring tool and a cued speech interpreter.

10.2 Modality Processing and Analysis

The modalities involved in gestured languages can be discussed from several points of view:

- *The part of the body that is involved*: Hands, head, facial features, shoulders, general standing, etc. For example, sign languages use the whole upper body, hands, head, facial features, and body/shoulder motion, whereas in cued speech, only a single hand and lips are in action.
- *Whether the modality conveys the main message or a paralinguistic message*: The hand shapes, locations and the lips in a CS phrase jointly convey the main message. On the other hand, in sign languages, paralinguistic elements can be added to the phrase via the non-manual elements or the variations of the manual elements. In sign languages, the main message is contained jointly in the manual (hand motion, shape, orientation and position) and non-manual (facial

features, head and body motion) modalities where the non-manual elements are mainly used to complement, support or negate the manual meaning.
- *Whether the modality has a meaning by itself or not*: In CS, both modalities contain an ambiguity if they are used independently. The hand shapes code the consonants and the hand locations code the vowels. A hand shape-location pair codes several phonemes that are differentiated by the lip shape. In sign languages, the manual component has a meaning by itself for most of the signs. For a small number of signs, the non-manual component is needed for full comprehension.

In this section, we present analysis and classification methods for each of the modalities independently. The synchronization, correlation, and the fusion of modalities are discussed in the next sections.

10.2.1 Preprocessing

Vision based systems for gestural interfaces provide a natural environment in contrast to the cumbersome instrumented gloves with several sensors and trackers that provide accurate data for motion capture. However, vision based capture methodology introduces its own challenges, such as the accurate detection and segmentation of the face and body parts, hand and finger configuration, or handling occlusion. Many of these challenges can be overcome by restricting the environment and clothing or by using several markers such as differently coloured gloves on each hand or coloured markers on each finger and body part. In communication of the hearing impaired, the main part of the message is conveyed through the hands and the face. Thus the detection and segmentation of hands and face in a vision based system is a very important issue.

10.2.1.1 Hand Detection

Hand detection and segmentation can be done with or without markers. Several markers are used in the literature such as single coloured gloves on each hand, or gloves with different colours on each finger or joint. With or without a marker, descriptors of colour, motion and shape information, separately or together, can be used to detect hands in images [369, 374, 375]. Similar techniques are used to detect skin coloured pixels or the pixels of the glove colour. Colour classification can be done either parametrically or non-parametrically. In parametric methods, a distribution is fitted for the colour of interest, and the biggest connected region is extracted from the image (see Fig. 10.3 from [376]).

A popular non-parametric method is histogram-based modelling of the colour [377]. In this approach, a normalised histogram is formed using the training pixels and the thresholds are determined. The similarity colour map of the image is ex-

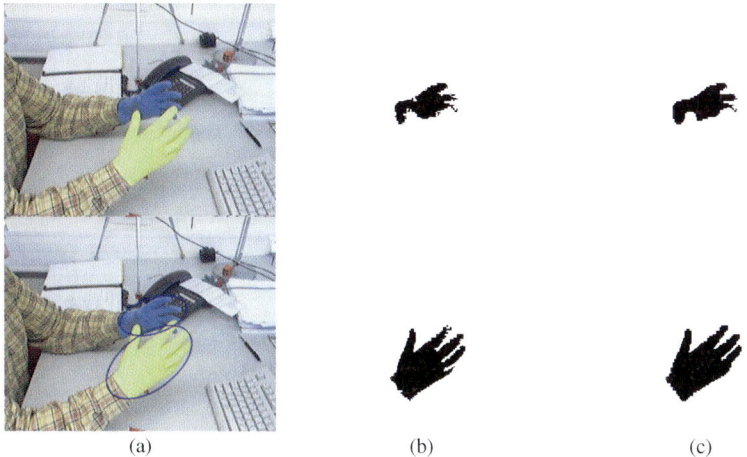

Fig. 10.3 Hand segmentation by automatically defined thresholds. (a) Original image and the detected hands, (b) thresholding & connected components labelling, (c) region growing

tracted using the histogram bins. Similar steps, thresholding, connected components labelling and region growing, are applied to obtain the segmented hand [370].

The main advantage of using a marker is that it makes tracking easier and helps to resolve occlusions. In a markerless environment, hand tracking presents a bigger challenge. In sign languages, the signing takes place around the upper body and very frequently near or in front of the face. Moreover the two hands are frequently in contact and often occlude each other. Another problem is to decide which of these two hand regions correspond to the right and left hands. Similarly in CS, the frequent face/hand contacts are difficult to deal with. Thus, the tracking and segmentation algorithm should be accurate enough to resolve these conditions and provide the two segmented hands.

10.2.1.2 Face Detection

The user face conveys both verbal and non-verbal communication information. The first step is to localize the user's face during the gesture analysis process.

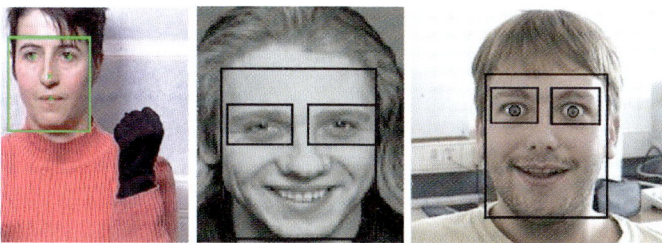

Fig. 10.4 Face localization

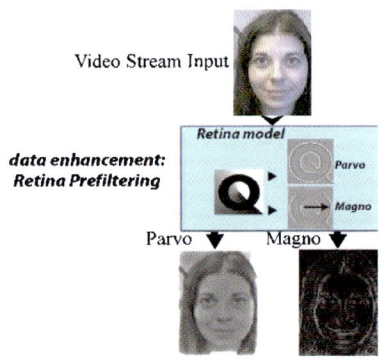

Fig. 10.5 Retina filtering effect

Face localization have been widely studied ([378, 379]). The most popular face detector is the detector developed by [380] whose code is freely available (MPT). Independent of the technique employed, the output of the face detector is a bounding box around the face and the position of some facial features, as shown in Fig. 10.4.

10.2.1.3 Retinal Pre-processing

In the human retina, some low level processing is done on video data. This processing is very efficient in order to condition the data for high level processing.

In the human retina [381], two steps of filtering (OPL and IPL filtering) are done so that two information channels are extracted: the Parvo (parvocellular) channel dedicated to detail analysis (details: static contours enhancement) and the Magno (magnocellular) channel dedicated to motion analysis (moving contours enhancement). For a more detailed description of the retina modelling and properties see [382, 383].

In the sequel of this chapter, we provide several examples where the properties of the retina are used to condition video data before high level processing.

10.2.2 Hand Shape

Hand shape is one of the main modalities of the gestured languages. Apart from sign language, CS or finger spelling, hand shape modality is widely used in gesture controlled computer systems where predefined hand shapes are used to give specific commands to the operating system or a program. Analysis of the hand shape is a very challenging task as a result of the very high degree of freedom of the hand. For systems that use limited number of simple hand shapes, such as hand gesture controlled systems (hand shapes are determined manually by the system designer) or in CS (the French, Spanish, English and American CSs are based on

10.2 Modality Processing and Analysis

Fig. 10.6 French cued speech hand shapes

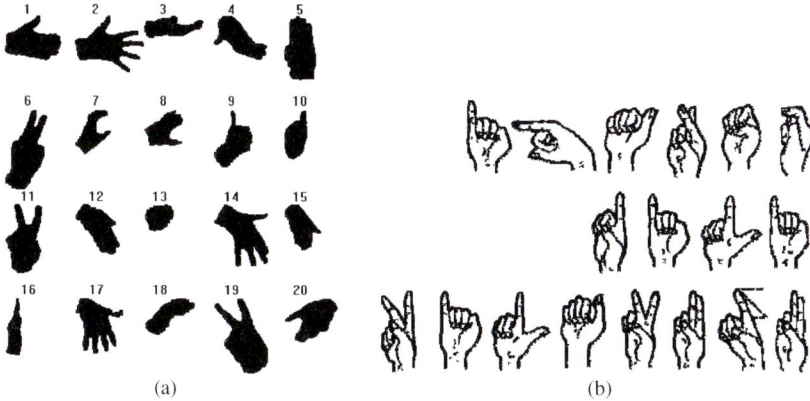

Fig. 10.7 Example hand shapes from (a) ASL and (b) TSL

eight predefined hand shapes), the problem is easier. However, for sign languages, the unlimited number and the complexity of the hand shapes make discrimination a very challenging task, especially with 2D vision based capture systems.

In CSs, there are eight hand shapes that mainly differ by open and closed fingers. The CSs coding is ideally performed in 2D. Thus, the hand is supposed to be frontal, and all the hand rotations are supposed to be planar, although it is not the case in practical situations (which is the source of one of the main difficulties). French cued speech (FCS) hand shapes are presented in Fig. 10.6.

In sign languages, the number of hand shapes is much higher. For example, without considering finger spelling, American Sign Language (ASL) has around 150 hand shapes, and in British Sign Language there are 57 hand shapes. These hand shapes can be further grouped into around 20 phonemically distinct subgroups. Example hand shapes from ASL and TSL are given in Fig. 10.7.

10.2.2.1 Inertial Study of the Hand

It is possible to compute the global direction of the hand shape using principal axis analysis. Then, a hand rotation in order to work on a vertical shape is considered in order to make the whole study easier.

The Distance Transform of the binary image of an object associates to each pixel of the object its distance to the closest pixel of the background, and associates the value 0 to all the pixels of the background.

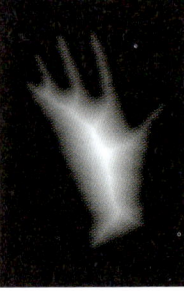 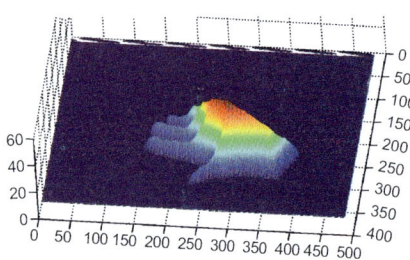

Fig. 10.8 Illustration of the distance transform in gray level (black pixels belong to the background and the lighter the pixels the higher its value after the distance transform) and in 3D

Fig. 10.9 (a) Palm delimitation, (b) once the "V" is removed, the shape is instable, (c) after the "V" is filled with a disc whose radius linearly varies between the two sides of the "V"

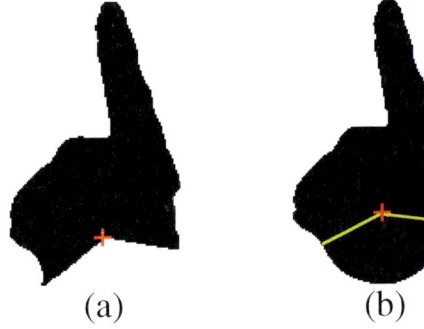

Obviously, the centre of palm is one of the points of the hand which is the furthest from the contour of the hand (Fig. 10.8). As a consequence, the palm of the hand can be approximated by a circle whose radius is related to the maximum value given by the distance transform of the binary hand image.

Once the palm is approximated by a circle, the wrist (or eventually the forearm) is removed as illustrated in Fig. 10.9.

The next step is to detect particular fingers. The main application is the study of a pointing (or deixis) gesture. The deixis gesture may be performed by the whole arm, and sometimes by the gaze of the eye. However, we consider hand shape for the pointing gestures:

- The general hand shape orientation is used to indicate a direction. In such a case, it is straightforward to deal with as the first principal axis of the bounding box corresponds to the deixis direction.
- The longest unfolded finger is used to materialize a pointing zone (for instance, a cursor gesture for HMI).
- The position of the extremity of a particular finger is considered depending on the hand shape.
- The precise deixis gesture with a single finger is replaced by a deixis gesture where the pointing element does not belong to the hand, but to its convex hull

10.2 Modality Processing and Analysis

(linear or polynomial). This case is practically very likely in human gestures, including CSs, for which the pointing rules are supposed to be really strict.

In the case of CSs, the deixis gesture is of prime importance, as it is required to determine the location of the hand with respect to the face.

The location is determined by the position of the pointing finger with respect to the face. It is theoretically the longest one, but, in practice, (1) parallax errors, (2) wrist flexion, and (3) the use of the convex hull of the hand shape modify the problem. Then a more robust algorithm, using fusion of information from hand shape and respective positions, must be used [384].

In [386], we consider the use of a thumb presence indicator, which returns a non-zero positive value if the thumb is unfolded and 0 otherwise. This is useful when (1) the thumb-up gesture is used, or when (2) the thumb presence has a particular meaning. The approach uses the polar parametric representation of the binary hand shape. The peaks of this representation correspond to potential fingers of the hand shape. Thresholds, derived from statistics on the morphology of the hand [387], are defined in order to materialize the region of the thumb extremity when it is unfolded. If a finger is detected within the thumb area, then, it is the thumb. The corresponding peak height is measured with respect to the lowest point between the thumb and the index. This value provides a numerical indicator of the presence of the thumb.

10.2.2.2 Hand Shape Descriptors

To analyze the hand shape, appearance or 3D-model based features can be used [388]. Appearance based features are preferred due to their simplicity and low computation times, especially for real time applications. Region based descriptors (image moments, image eigenvectors, Zernike moments, Hu invariants, or grid descriptors) and edge based descriptors (contour representations, Fourier descriptors, or Curvature Scale Space descriptors) can be used for this purpose. A survey on binary image descriptors can be found in [388].

A combination of several appearance based features is used as hand shape features for recognizing ASL signs [370]. Half of the features are based on the best fitting ellipse (in least-squares sense) to a binary image, as seen in Fig. 10.10a. The rest are calculated using area filters as seen in Fig. 10.10b. The bounding box

Fig. 10.10 (a) Best fitting ellipse, (b) Area filters. Green and white pixels indicate the areas with and without and hand pixels, respectively

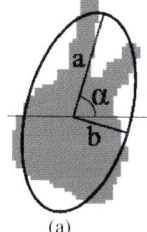
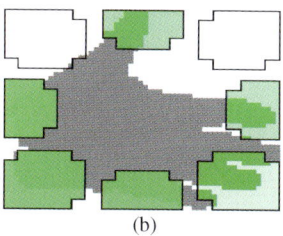

of the hand is divided into eight areas, in which percentage of on and off hand pixels are calculated.

Hu invariants are successful in representing hand shapes [386, 390, 391]. Their purpose is to describe the binary shape region via several moments of various orders, on which specific transforms ensure invariance properties.

The centred scale invariant inertial moments of order $p+q$ are calculated as follows:

$$n_{pq} = \frac{m_{pq}}{m_{00}^{\frac{p+q}{2}+1}} \quad \text{with} \quad m_{pq} = \iint_{xy}(x-\bar{x})^p (y-\bar{y})^q \delta(x,y)\, dx\, dy \qquad (10.1)$$

where $\bar{x}$ and $\bar{y}$ are the coordinates of the center of gravity of the shape and $\delta(x,y) = 1$ if the pixel belongs to the hand shape and 0 otherwise. Then, the seven Hu invariants are calculated:

$$\begin{aligned}
S_1 &= n_{20} + n_{02} \\
S_2 &= (n_{20} + n_{02})^2 + 4 \cdot n_{11}^2 \\
S_3 &= (n_{30} - 3 \cdot n_{12})^2 + (n_{03} - 3 \cdot n_{21})^2 \\
S_4 &= (n_{30} + n_{12})^2 + (n_{03} + n_{21})^2 \\
S_5 &= (n_{30} - 3 \cdot n_{12}) \cdot (n_{30} + n_{12}) \cdot \left((n_{30} + n_{12})^2 - 3 \cdot (n_{03} + n_{21})^2\right) - (n_{03} - 3 \cdot n_{21}) \cdot (n_{03} + n_{21}) \cdot \left(3 \cdot (n_{30} + n_{12})^2 - (n_{03} + n_{21})^2\right) \\
S_6 &= (n_{20} + n_{02}) \cdot \left((n_{30} + n_{12})^2 - (n_{03} + n_{21})^2\right) + 4 \cdot n_{11}^2 \cdot (n_{30} + n_{12}) \cdot (n_{03} + n_{21}) \\
S_7 &= (3 \cdot n_{21} - n_{03})(n_{30} + n_{12}) \cdot \left((n_{30} + n_{12})^2 - 3 \cdot (n_{03} + n_{21})^2\right) - (n_{30} - 3 \cdot n_{12}) \cdot (n_{03} + n_{21}) \cdot \left(3 \cdot (n_{30} + n_{12})^2 + (n_{03} + n_{21})^2\right)
\end{aligned}$$
(10.2)

We have used Hu invariants as descriptors of CS hand shapes [386]. The experiments show that Hu invariants have an acceptable performance which can be improved by the addition of the thumb information presence.

The Fourier-Mellin Descriptors (FMD) are an interesting alternative [392]. The Fourier-Mellin Transform (FMT) of a function f corresponds to its Mellin transform result represented in terms of Fourier coefficients. The FMT is defined for all real positive function $f(r,\theta)$ in polar coordinates (the shape to describe) so that the Mellin transform is 2π-periodic:

$$M_f(q,s) = \int_{r=0}^{\infty} \int_{\theta=0}^{2\pi} r^{s-1} e^{-iq\theta} f(r,\theta)\, dr\, d\theta \quad \text{with} \quad q \in \mathbb{Z}, s = \sigma + iv \in \mathbb{C}, \text{ and } i = \sqrt{-1}$$
(10.3)

Then the application of the delay theorem and the extraction of the module of the FMT lead to a set of descriptors indexed by q and s. They are rotation invariant, and normalization by $M_f(0, \sigma)$ makes them scale invariant. The translation invariance is derived from the choice of the centre of development (the origin of (r,θ) coordinates).

In case of digital images, it is necessary to digitalize the FMT and to convert the sampled Cartesian space into a polar space. In practice, $M_f(0, \sigma)$ is approximated by:

10.2 Modality Processing and Analysis

$$M_f(q, \sigma + iv) \approx \sum_{\substack{k,l \\ 0 \leq (k^2+l^2) \leq r^2_{max}}} h_{p,q}(k,l) \cdot f(\bar{k}-k, \bar{l}-l)$$

with
$\begin{cases} (\bar{k}, \bar{l}) \text{ centre of development of } \textit{the} \text{ FMT (here, the gravity centre of the hand)} \\ r_{max} \text{ superior bound for } r \\ h_{p,q}(k,l) = \dfrac{1}{(k^2+l^2)^{1-\frac{\sigma}{2}}} \cdot \exp\left(i \cdot \left(\dfrac{p}{2}\ln(k^2+l^2) - q \cdot \arctan\left(\dfrac{l}{k}\right)\right)\right) \end{cases}$

(10.4)

These descriptors are particularly efficient to discriminate hand shapes, even in cases of (1) multi-coder (when the morphologic variability is introduced), (2) unknown coder, (3) imprecise classifier tuning [384].

10.2.3 Hand Location

The location of the hand must be analysed with respect to the context. It is important to determine the reference point on the space and on the hand. In sign languages, where both the relative location and the global motion of the hand are important (see Fig. 10.11), the continuous coordinates and the location of the hand with respect to body parts should be analysed. This analysis can be done by using the center of mass of the hand. On the other hand, for pointing signs, using center of mass is not appropriate and the coordinates of the pointing finger and the pointing direction should be used. Since signs are generally performed in 3D space,

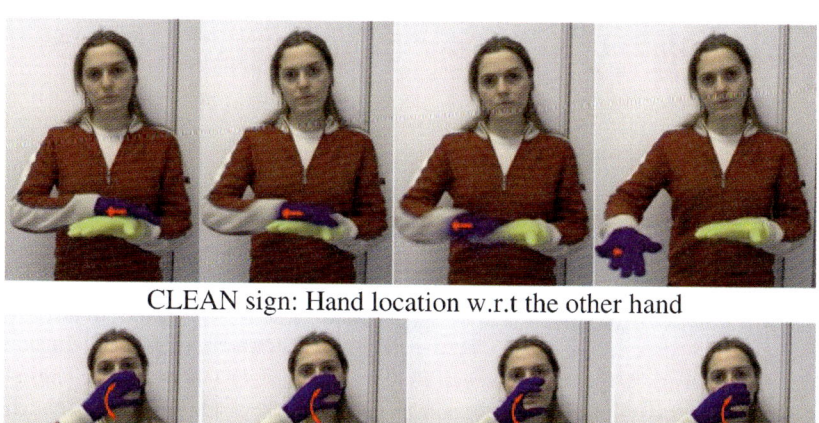

CLEAN sign: Hand location w.r.t the other hand

DRINK sign: Hand location w.r.t the mouth

Fig. 10.11 Possible reference points on the signing space. Examples from ASL

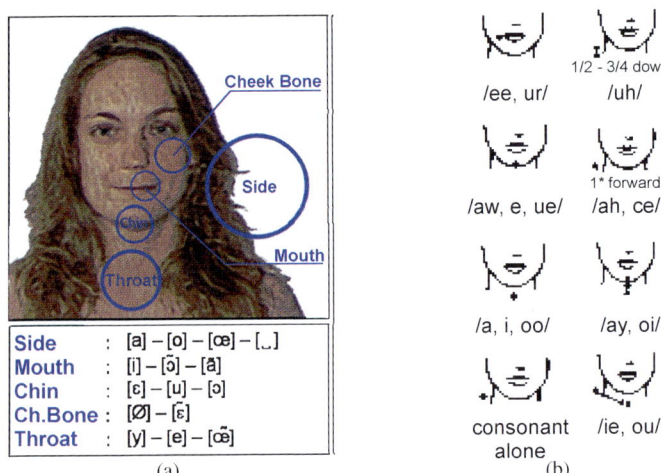

Fig. 10.12 (a) French and (b) American cued speech hand locations, and their phonemic meaning

location analysis should be done in 3D if possible. Stereo cameras can be used to reconstruct 3D coordinates in vision based systems.

The hand locations in CS are determined by the pointing location of the hand with respect to the coder's face. For example, in French CS, "mouth", "side", "throat", "chin", and "cheek bone" are used as five different locations on the face (see Fig. 10.12a). Once the pointing finger and the face features are located of the image, determining the location component of the gesture is rather simple.

10.2.4 Hand Motion

In gestured communication, it is important to determine whether the performed hand motion conveys a meaning by itself.

From a linguistic point of view, FCS is the complete visual counterpart of oral French. Hence, it has a comparable prosody and the same dynamic aspect. From a gesture recognition point of view, the interpretation is completely different: each FCS gesture {Hand shape + Location} is a static gesture (named a *target gesture*) as it does not contain any motion and can be represented in a single picture or a drawing. Then, a coder is supposed to perform a succession of targets. In real coding, the hand nevertheless moves from target to target (as the hand cannot simply appear and disappear) and *transition gestures* are produced. We consider that FCS is inherently static: target images are sufficient to decode the continuous sentence: as a consequence, complete transition analysis is most of the time useless to be processed [384, 393].

In sign languages, the hand motion, together with the hand shape and location, is one of the primary modalities that form the sign. Depending on the sign, the

10.2 Modality Processing and Analysis

characteristic of the hand trajectory can change, requiring different levels of analysis. For example, some signs are purely static and there is no need for trajectory analysis. The motion of the dynamic signs can be examined as either of two types:

1. Signs with global hand motion: In these signs, the hand center of mass translates in the signing space.
2. Signs with local hand motion: This includes signs where the hand rotates without any translation, or where the finger configuration of the hand changes.

10.2.4.1 Trajectory Analysis

Trajectory analysis is needed for signs with global hand motion. For signs with local motion, the change of the hand shape over time should be analysed in detail, since even small changes of the hand shape convey information content.

The first step of hand trajectory analysis is tracking the center of mass of each segmented hand. Hand trajectories are generally noisy due to segmentation errors resulting from bad illumination or occlusion. Thus a filtering and tracking algorithm is needed to smooth the trajectories and to estimate the hand location when necessary. Moreover, since hand detection is a costly operation, hand detection and segmentation can be applied not in every frame but less frequently, provided that a reliable estimation algorithm exists. For this purpose, algorithms such as Kalman filters and particle filters can be used. Kalman filters are linear systems with Gaussian noise assumption and the motion of each hand is approximated by a constant velocity or a constant acceleration motion model. Particle filtering, also known as the condensation algorithm [385], is an alternative with non-linear and non-Gaussian assumptions. The main disadvantage is its computational cost which prevents its usage in real time systems.

Based on the context and the sign, hand coordinates should be normalised with respect to the reference point of the sign, as discussed in the previous section. In addition to the coordinates, the velocity and the acceleration can be used as hand motion features.

Several methods are proposed and applied in the literature for modelling the dynamics of signs or hand gestures. These include Hidden Markov Models (HMM) and its variants, Dynamic Time Warping (DTW), Time Delay Neural Networks (TDNN), Finite State Machines (FSM), and temporal templates. Some of these techniques are only suitable for simple hand gestures and cannot be applied to complex dynamic systems. Among dynamic systems, HMM and its variants are popular in sign language recognition, and hand gesture recognition in general.

10.2.4.2 Static Gestures in Dynamic Context

In order to take advantage of the static nature of some gestures, let us assume that it is possible to extract target gestures from the surrounding transition motions using low-level kinetic information that can be extracted before the complete recognition process.

This hypothesis is motivated by the analysis of FCS sequences, and can be generalised directly to other static gestural languages. It shows that the hand is slowing down each time the hand is reaching a phonemic target. As a conse-

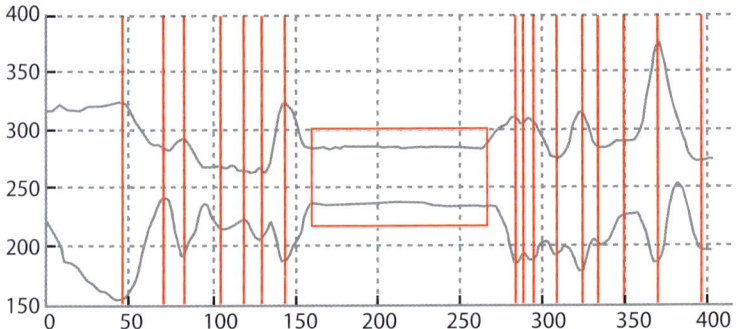

Fig. 10.13 Representation of the coordinates (vertical above and horizontal below) of the gravity centre of the hand shape during a CS sequence. The vertical lines correspond to images of minimal motion that are target images of hand location

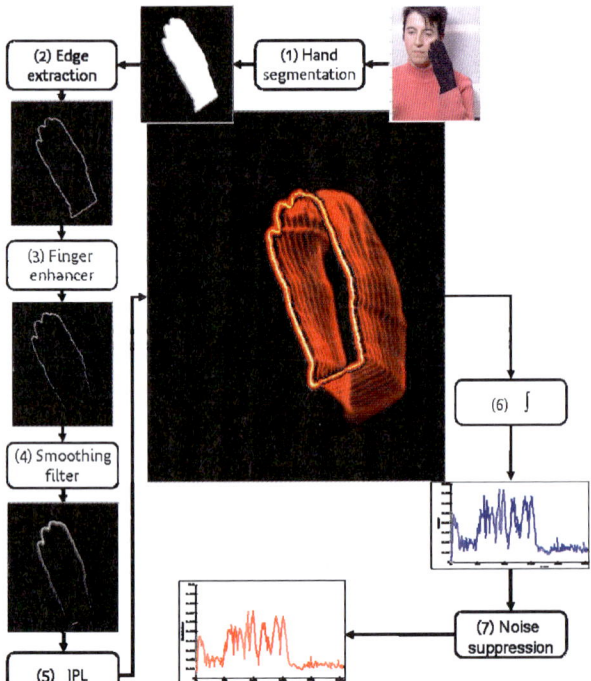

Fig. 10.14 Flowchart of the dedicated retina filter

quence, target gestures have slower hand motion than transition gestures. It nonetheless appears that there is almost always some residual motion during the realization of the target (because of the co-articulation).

In case the motion in which the static gesture is hidden is a global translation motion (i.e. the change of location in CS or any deixis gesture), any study of the rigid movement is likely to stress the variations of speed and the images on which the motion is small enough to be potentially considered as a target gesture. Figure 10.13 illustrates the trajectory of the hand gravity centre during a CS video sequence. It appears that each image for which the two components of the trajectory are stable (which corresponds to local minima in the speed evolution) corresponds to some location being reached.

In case of non-rigid motion, such as the deformation of the hand contour when its shape is considered, it is more difficult to define a cinematic clue that indicates when a target is reached or when the image represents a transitive motion. In order to do so, an approach based on the properties of the retina (and specially the IPL filter) has been proposed in [394]. A dedicated retina filter [386] has been defined to evaluate the amount of deformation of the hand contour along the sequence. It is made of several elements which are chained together (Fig. 10.14). As established in [384, 386] this method is particularly efficient.

10.2.5 Facial Movements

Thanks to the retina model, it is possible to efficiently detect some facial movements. The analysis of the temporal evolution of the energy of the Magno output related to moving contours has been used in [395] in order to develop a motion events detector. Indeed, in case of motion, the Magno output energy is high and on the contrary, if no motion occurs, the Magno output energy is minimum or even null. In Fig. 10.15, the case of an eye blink sequence is illustrated: the motion event detector generates a signal $\alpha(t)$ which reaches 1 each time a blink is detected (high level of energy on the Magno channel, frames 27, 59 and 115) and which is 0 if no blinks are present (the energy of the Magno channel is null).

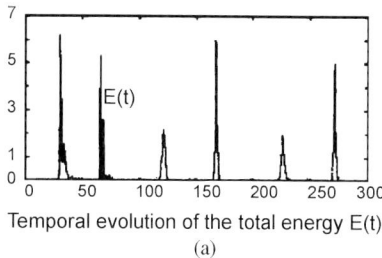
Temporal evolution of the total energy E(t)
(a)

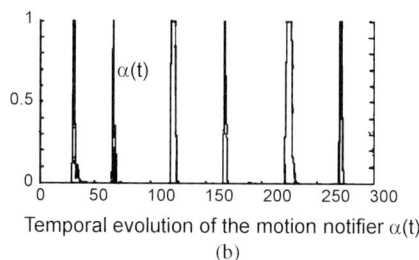
Temporal evolution of the motion notifier $\alpha(t)$
(b)

Fig. 10.15 (a) Temporal evolution of the Magno output in case of a blink video sequence; (b) temporal evolution of the motion events detector

10.2.6 Lip Reading

The main difference between SL and CS is that the CS message is partly based on lip reading: Although signers also use lip movements while they are signing, it is not a part of the sign language. However, for CS, it is as difficult to read on the lip without any CS hand gesture, than to understand the hand gestures without any vision of the mouth. The link between lip and oral message is included in the shape and the motion of the lips.

An important step for lip reading is lip contours extraction. Significant research has been carried out to accurately obtain the outer lip contour. One of the most popular approaches is using snakes [396], which have the ability to take smoothing and elasticity constraints into account [397, 398]. Another popular approach is using active shape models and appearance shape models. In [399] a statistical active model for both shape (AMS) and appearance (AAM) is presented. Shape and grey-level appearance of an object are learned from a training set of annotated images. Then, a Principal Component Analysis (PCA) is performed to obtain the main modes of variation. Models are iteratively matched to reduce the difference between the model and the real contour by using a cost function. Another approach is presented in [400], where a parametric model associated with a "jumping snake" for the initialization phase is proposed.

Relatively few studies deal with the problem of inner lip segmentation. The main reason is that inner contour extraction from front views of the lips without any artifice is much more difficult than outer contour extraction. Indeed, we can find different mouth shapes and non-linear appearance variations during a conversation. Especially, inside the mouth, there are different areas which have similar colour, texture or luminance than lips (gums and tongue). We can see very bright zones (teeth) as well as very dark zones (oral cavity). Every area could continuously appear and disappear when people are talking. Among the few existing approaches for inner lip contour extraction, lip shape is represented by a parametric deformable model composed of a set of curves. In [401], authors use deformable templates for outer and inner lip segmentation. The chosen templates are three or four parabolas, depending on whether the mouth is closed or open. The first step is the estimation of candidates for the parabolas by analyzing luminance information. Next, the right model is chosen according to the number of candidates. Finally, luminance and colour information is used to match the template. This method gives results, which are not accurate enough for lip reading applications, due to the simplicity and the assumed symmetry of the model. In [402], authors use internal and external active contours for lip segmentation as a first step. The second step recovers a 3D-face model in order to extract more precise parameters to adjust the first step. A k-means classification algorithm based on a non-linear hue gives three classes: lip, face and background. From this classification, a mouth boundary box is extracted and the points of the external active contour are initialised on two cubic curves computed from the box. The forces used for external snake convergence are, in particular, a combination of non-linear

10.2 Modality Processing and Analysis

hue and luminance information. Next, an inner snake is initialised on the outer contour, and then shrunk by a non isotropic scaling with regard to the mouth center and taking into account the actual thickness of the lips. The main problem is that the snake has to be initialised close to the contour because it will converge to the closest gradient minimum. Particularly for the inner lip contour, different gradient minima are generated by the presence of teeth or tongue and can cause a bad convergence. The 3D-face model is used to correct this problem, but the clone does not give accurate results for lip reading.

In [403], an AMS is build and in [404], an AMS and an AAM are built to inner and outer lip detection. The main interest of these models is that the segmentation gives realistic results, but the training data have to deal with many cases of possible mouth shapes.

Once the mouth contours have been extracted, lip shape parameters for lip reading have to be extracted. Front views of the lips are phonetically characterised with lip width, lip aperture and lip area. These lip parameters are derived from the inner and outer contours. In an automatic recognition task of lip-reading process, it is thus pertinent to consider these parameters.

10.2.7 Facial Expressions

A summary of the significant amount of research carried out in facial expression classification can be found in [405, 406]. One of the main approaches is optical flow analysis from facial actions [407–410]: These methods focus on the analysis of facial actions where optical flow is used to either model muscle activities or to estimate the displacements of feature points. A second approach is using model-based approaches [411–414]: Some of these methods apply an image warping process to map face images into a geometrical model. Others realize a local analysis where spatially localised kernels are employed to filter the extracted facial features. Once the model of each facial expression is defined, the classification consists in classifying the new expression to the nearest model using a suitable metric. A third group is fiducial points based approaches [415–418]: Recent years have seen the increasing use of geometrical features analysis to represent facial information. In these approaches, facial movements are quantified by measuring the geometrical displacement of facial feature points between the current frame and a reference frame.

We are going to illustrate the approach described in detail in [419]. In this work, the classification process is based on the Transferable Belief Model (TBM) [420] framework (see section on belief functions). Facial expressions are related to the six universal emotions, namely *Joy, Surprise, Disgust, Sadness, Anger, Fear*, as well as *Neutral*. The proposed classifier relies on data coming from a contour segmentation technique, which extracts an expression skeleton of facial features (mouth, eyes and eyebrows) and derives simple distance coefficients from every face image of a video sequence (see Fig. 10.16).

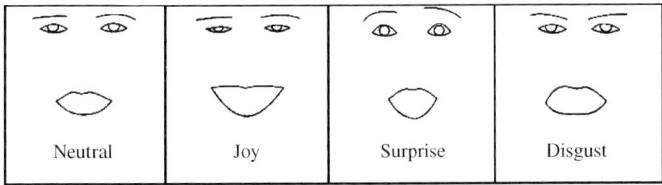

Fig. 10.16 Facial expression skeletons

The characteristic distances are fed to a rule-based decision system that relies on the TBM and data fusion in order to assign a facial expression to every face image. This rule-based method is well adapted to the problem of facial expression classification because it deals with confusing expressions (*Joy* or *Disgust*, *Surprise* or *Fear*, etc) and recognizes an *Unknown* expression instead of forcing the recognition of a wrong expression. Indeed, in the presence of doubt, it is sometimes preferable to consider that both expressions are possible rather than to choose one of them.

10.3 Temporal Analysis

In a multimodal interface, the correlation and synchronization of modalities must be clearly analysed. This is a necessary step prior to multimodal fusion.

10.3.1 Sign Language

The temporal organization of sign languages can be analysed in two: (1) The temporal organization within manual components (manual simultaneity), (2) the temporal organization between manual and non-manual components (manual/non-manual simultaneity).

The manual simultaneity is due to the usage of two independent moving elements: The two hands can perform different signs at the same time. We can classify the signs in a language as one or two-handed signs. In two-handed signs, the two hands are synchronised and perform a single sign. Whereas in one-handed signs, the performing hand is called the dominant hand and the other hand is idle in the isolated case. In continuous signing, as a result of the speed, while one hand is performing one sign, the other hand may perform the next sign, at the same time. From the recognition point of view, this property enforces the independent modelling of the two hands, while keeping their relation in case of two-handed signs.

The simultaneity of manual/non-manual components depends on the linguistic property of the performed sign. For example, non-manual signs for grammatical operators, such as negation and question, are performed over a phrase which generally includes more than one sign. On the other hand, the modifications on the meaning of a sign are performed via non-manual signs and they only affect the sign in focus. Of course, if these modifications affect a phrase, then the non-manual signs co-occur with one or more manual sign.

10.3.2 Cued Speech

In this section, we describe the temporal organization of the three modalities (hand shape, location, lips) of French Cued Speech. This description is based on the observation of numerous video sequences featuring a professional coder (hearing able translators) as well as hearing impaired people. A first study [421] has been published by Attina on the desynchronization between the labial motion and the manual one, but the desynchronization of the two modalities of the manual motion (the hand shape movement and the location movement) is not in its scope. Here, we summarize the principal results of [421] and we complete them with observations about hand shape/ location temporal organization.

The main point of [421] is a temporal scheme which synthesizes the structure of the code along time from a hand/lip delay point of view.

From this work it is possible to extract two remarks: The first is that the hand is in advance with respect to the lips, and apparently, the labial motion disambiguates the manual motion, and not the contrary. The second is that the variability of desynchronization is much too important to be directly used in a recognition system which automatically balances the desynchronization. Nevertheless, this scheme contains a lot of information which can be used to set the parameters of an inference system which purpose is to find a best matching between the modalities.

In general, the hand shape target is reached before the location target. This is easily explained by mechanic and morphologic arguments: in case of finger/face contact, the pointing finger must be completely unfolded before the beginning of the contact. As a consequence, hand shapes are in advance with respect to the locations. However, for some other hand shape/ location pairs, this observation is not valid [384]. As a consequence, it is really difficult to establish a precise enough model to forecast the desynchronization pattern. Nonetheless, the desynchronization are most of the time of intermediate amplitude (except at the beginning and the end of a sentence) so that computing a matching among the modalities in order minimize the desynchronization does not seem intractable.

10.4 Multimodal Fusion

There are two major difficulties in integrating modalities of gesture based communication: joint temporal modelling and multiplexing information of heterogeneous nature.

10.4.1 Temporal Modelling

In gesture based communication of the hearing impaired, multiple modalities are used in parallel to form the linguistic units such as signs/words in sign languages or phonemes in CS. The temporal relation between these modalities must be carefully investigated to obtain a good temporal model that will result in high accuracies in a recognition task.

10.4.1.1 Hidden Markov Models

Among the temporal modelling techniques for hand gestures HMMs draw much attention [422]. Their success comes from their ability to cope with the temporal variability among different instances of the same sign.

HMMs are generative probability models that provide an efficient way of dealing with variable length sequences and missing data. Among different kinds of HMM architectures, left-to-right HMMs (Fig. 10.17) with either discrete or continuous observations are preferred for their simplicity and suitability to the hand gesture and sign language recognition problems.

An HMM consists of a fixed number of states. Given a data sequence, the probabilities to determine the start state and transition probabilities, one can construct a state sequence. Each state generates an output (an observation) based on a probability distribution. This observation is the features observed at each frame of the data sequence.

For a sequence classification problem, one is interested in evaluating the probability of any given observation sequence, $\{O_1 O_2 \ldots O_T\}$, given a HMM model Θ.

In isolated sign language recognition, an HMM model is trained for each sign in the dictionary. The simplest case is to put the features of all the concurrent modalities in a single feature vector. The likelihood of each model is calculated and the sequence is classified in to the class of the model that produces the highest likelihood. Instead of concatenating the features into a single feature vector, a process can be dedicated for each modality with established links between the states of different processes. In [423], Coupled HMMs are proposed for coupling and training HMMs that represent different processes (see Fig. 10.18a).

When the synchronization of the modalities is weak, then it is not a good idea to process all the modalities in a single temporal model. Several models for each of the modalities can be used independently and integration can be done afterwards. An example is the Parallel HMM, as illustrated in Fig. 10.18b [424]. Belief based methods can also be used to fuse different models to handle the ambiguity in between, as we describe in the following sections.

An alternative is to use Input Output HMMs (IOHMM) (see Fig. 10.18c) which model sequential interactions among processes via hidden input variables to the states of the HMM [425].

10.4 Multimodal Fusion

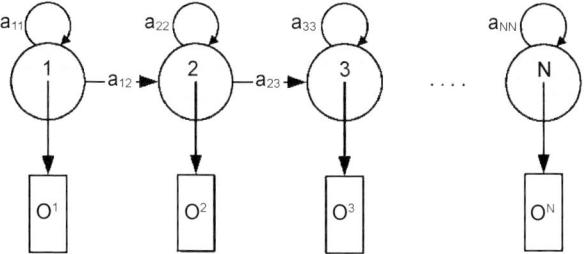

Fig. 10.17 Left-to-right HMM architecture

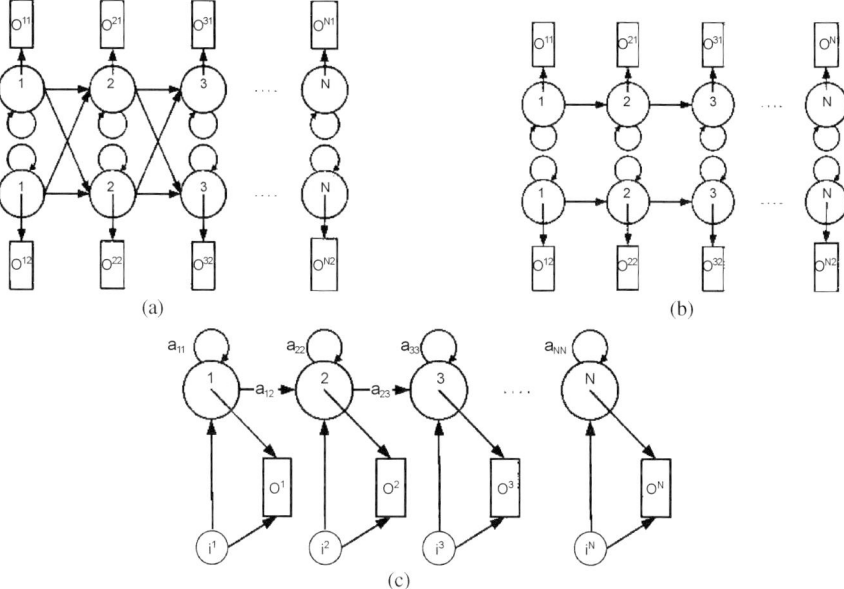

Fig. 10.18 (a) Coupled HMM, (b) Parallel HMM, (c) Input-Output HMM

10.4.1.2 Co-articulation

In continuous gestural language, the consequent signs affect the beginning and end of each other. This co-articulation phenomenon can also be seen in spoken languages. When an HMM for each sign is trained to recognize the signs, the performance will drop down since each sign in the continuous signing will be slightly different than their isolated equivalents. Many of the methods proposed for solving the co-articulation affect, rely on modelling the co-articulation by using pairs or triples of signs instead of a single one.

In case the modalities to be fused have a static nature which is classically hidden in a dynamic context because of a co-articulation phenomenon, we propose an alternative solution [384]. The main idea is to associate to any static gesture of the

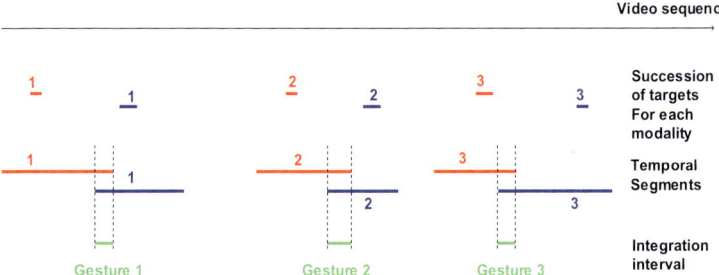

Fig. 10.19 Illustration of the definition of the temporal segments. Their overlapping deal with the dynamical aspect of the fusion of the modalities

modalities a temporal segment of the video sequence which is centred on the target image. This segment is supposed to represent the time interval in which it is not possible to get another static gesture for a minimum time interval is necessary to produce the transition movements which are required to reach and to leave the target of the gesture. Then, whatever the recognition process output within this segment, it is reasonable to assume that a single gesture has been produced during this time interval. As a consequence, even if the target gestures of each modality are not produced at the same time, it is possible to balance this lack of synchronization by matching the segments which overlap (Fig. 10.19).

Of course, such a process only allows balancing small desynchronization. If the desynchronization is larger than the segments associated to the target images, it is impossible to easily warp the modalities. On the other hand, this hypothesis of small desynchronization is not that an important restriction. In HCI systems, it is rather common to assume that the "gesturer" (coder/signer) produces an academic motion, which means, he/she is concentrated on limiting the desynchronization between the various component of his/her gestures.

In the general case, if multiple overlaps and/or empty intersection remain too numerous to allow a simple matching, then, the use of DTW methods or graph matching algorithm can be successfully applied to finalize the temporal matching of the modalities.

10.4.2 Heterogenic Multiplexing

The purpose of fusing the various gestural modalities is to provide a context in which taking a decision is less hazardous as the whole information is taken into account. Most of the time, such a strategy is more efficient than making a decision on each modality and grouping the independent decision afterward. In order to do so, the classical method is to associate probability scores to each possible decision for each modality and use them as input vectors in a probabilistic inference system

10.4 Multimodal Fusion

which fuses the pieces of knowledge under some rules expressed as conditional dependencies. Most of the time, such a framework is efficient as it corresponds to an excellent trade-off between complexity and accuracy. Nonetheless it suffers from several drawbacks. Here are few of them:

- The likelihood associated to a hypothesis is most of the time derived from a training algorithm. This guaranties a good generalization power in cases where the training data is representative.
- This likelihood is definitely derived from an objectivist point of view on probabilities, as statistical analysis of the training data are used, but probabilistic inference is deeply subjective.
- In the particular case of gesture interpretation, there is a lot of conflictive, contradictory, incomplete and uncertain knowledge, and there are other formalisms which are more adapted to this kind of situations.

Amongst all these formalisms, the one of belief function is really powerful. Moreover, it is close enough to the probabilistic formalism to keep some of its advantages and to allow an intermediate modelling where some interesting properties of both probabilities and belief functions can be used in common.

10.4.2.1 Belief Functions

Originally, this theory was introduced by [426] throughout the study of lower and upper bound of a set of probabilities, and it was formalised by Shafer in *A Mathematical Theory of Evidence* [427].

In this section, we recall the main aspects of belief functions from [427]. Let $X=\{x_1,\ldots, x_M\}$ be a set of M variables and Ω_X be the set of N exhaustive and exclusive multivariate hypotheses $\{h_1, \ldots, h_N\}$ that can be associated to X. Ω_X is the **frame of discernment** (of **frame** for short) for X. Let 2^{Ω_X} be the set of all the subsets A of Ω_X, including the empty set:

$$2^{\Omega_X} = \{A / A \subseteq \Omega_X\} \quad (10.5)$$

2^{Ω_X} is called the **powerset** of Ω_X. Let m a **belief mass function** (or BF for short) over 2^{Ω_X} that represents our belief on the hypotheses of Ω_X:

$$m : \begin{cases} 2^{\Omega} \to [0,1] \\ A \mapsto m(A) \end{cases} \text{ with } \begin{cases} \sum (m(A) | A \subseteq \Omega) = 1 \\ m(\emptyset) = 0 \end{cases} \quad (10.6)$$

$m(A)$ represents the belief that is associated exactly to A, and to nothing wider or smaller. A **focal set** is an element of 2^{Ω_X} (or a subset of Ω_X) with a non-zero belief. A **consonant BF** is a BF with nested focal sets with respect to the inclusion operator ($\subseteq$). The **cardinal** of a focal set is the number of elements of the frame it contains.

Let m be a BF over Ω_X and X and Y two sets of variables so that $X \subseteq Y$. The **vacuous extension** of m to Y, noted $m^{\uparrow Y}$ is defined so that:

$$m^{\uparrow Y}(A \times \Omega_{Y \setminus X}) = m(A) \qquad \forall A \subseteq 2^{\Omega_X} \tag{10.7}$$

Basically, it means that the vacuous extension of a BF is obtained by extending each of its focal sets by adding all the elements of Ω_Y which are not in Ω_X.

The combination of N BFs from independent sources is computed using the **Dempster's rule of combination**. It is a N-ary associative and symmetric operator, defined as follows:

$$(\cap): \overbrace{\mathcal{B}^{\Omega_{X_1}} \times \mathcal{B}^{\Omega_{X_2}} \times ... \times \mathcal{B}^{\Omega_{X_N}}}^{N} \to \mathcal{B}^{\Omega_X} \tag{10.8}$$
$$m_1 \; (\cap) \; m_2 \; (\cap)...(\cap) \; m_N \mapsto m_{(\cap)}$$

with $\mathcal{B}^{\Omega_{X_i}}$ being the set of BFs defined on Ω_{X_i} and with Ω_X being the **cylinder product** of the Ω_{X_i}:

$$\Omega_X = \Omega_{X_1} \times \left[\Omega_{X_2} \setminus (\Omega_{X_1} \cap \Omega_{X_2})\right] \times ... \times \left[\Omega_{X_{N-1}} \setminus \left(\bigcap_{i=1}^{N-1} \Omega_{X_i}\right)\right] \tag{10.9}$$

and with

$$m_{(\cap)}(A) = \frac{1}{1-\mathcal{K}} \cdot \sum \left(\prod_{n=1}^{N} m_n^{\uparrow X}(A_n) \middle| \bigcap_{n=1}^{N} A_n = A \right) \qquad \forall A \subseteq 2^{\Omega_X} \tag{10.10}$$

where the vertical bar indicating on its right the condition that A should fulfil in order to be taken account in the summation (we use this notation when the condition would be difficult to read on subscript under the summation sign). The normalizing constant

$$\mathcal{K} = \sum \left(\prod_{n=1}^{N} m_n^{\uparrow X}(A_n) \middle| \bigcap_{n=1}^{N} A_n = \emptyset \right) \tag{10.11}$$

quantifies the amount of incoherence among the BFs to fuse.

The **refinement** operation permits to express the knowledge in a more refined manner, by using a more precise frame than the one on which the original BF is defined. It is defined as follow: let two frames Ω_1 and Ω_2, and R an application from the powerset of Ω_1 to the powerset of Ω_2, so that:

- the set $\{R(\{h\}), h \in \Omega_1\} \subseteq 2^{\Omega_2}$
- the set $\{R(\{h\}), h \in \Omega_1\}$ is a partition of Ω_2 \qquad (10.12)
- $\forall A_1 \subset \Omega_1, R(\{A_1\}) = \bigcup (R(\{h\}) | h \in A_1)$

10.4 Multimodal Fusion

BFs are also widely connected to fuzzy set theory. It appears that membership functions on Ω are included in $\mathcal{B}^\Omega$. Consequently, fuzzy sets are BFs and moreover, they are particularly easy to manipulate and to combine with the Dempster's rule [426]. In that fashion, the link between the subjective part of the probabilities and the confidence measure in the fuzzy set theory is perfectly supported in the BF framework.

10.4.2.2 Derivation of new belief-based Tools for Multimodal Fusion

Evidential Combination of SVM. An efficient method to solve multi-classification problems is to use a bank of binary classifiers, such as SVM) and to fuse their partial results into a single decision. We propose to do so in the BF framework. As it is proved in [386, 394], the BF formalism is particularly adapted as it allows an optimal fusion of the various pieces of information from the binary classifiers. Thanks to the margin defined in SVMs, it is possible to implement the notion of hesitation easily and thus, to benefit from the richness of the BF modelling. In order to associate a BF to the SVM output, we rely on the strong connection between fuzzy sets and BFs, as explained in Fig. 10.20.

In order to make sure that the BF associated to each SVM are combinable via the Dempster's rule, it is necessary to apply a refinement from the frame of each SVM (made of two classes), to the frame of the entire set of classes, but then, it provides more accurate results than classical methods.

Evidential Combination of Heterogeneous Classifiers. In case the binary classifiers involved in the process are not SVMs, then our method is not applicable anymore. As no margins are defined altogether with the separation between the classes, there is no trivial support for the hesitation distribution. An alternative is to use one of the numerous classifiers which directly provide a BF, such as CrFM [428], Evidential K-NN, Expert systems, and Evidential NN [429–431].

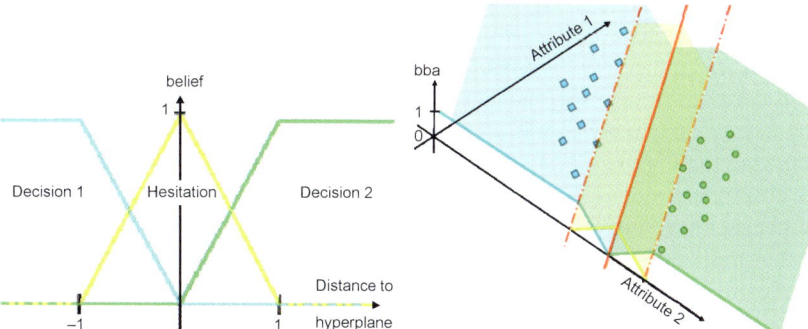

Fig. 10.20 Superposition of a membership function on the feature description where the SVM algorithm works

Another alternative is to use classical classifiers (no margins, no BF outputs), but to consider W the width of the support for the hesitation as an unknown value to determine by a learning or a cross validation.

The main interest of this evidential combination is to permit the simultaneous use of heterogeneous classifiers. As long as a classifier provides a BF, this latter can be fused with other BFs from other classifiers thanks to the conjunctive combination. This is particularly interesting when it is necessary to consider very wide sets of features which cannot be expressed in the same formalism.

Evidential Combination of Unary Classifiers. It is also possible to use a similar scheme (the definition of the support of the hesitation pattern via cross-validation) in order to extend the Evidential Combination of classifiers to the case of unary classifiers. In such a case, the point is to associate a generative model (without any discrimination power) to each class, to let them in competition. Each unary classifier provides a likelihood score between the generative model and the item to classify.

Then, it is possible to consider that the whole system provide an array of scores, each score being a likelihood value for each item to classify. If we assume that the highest the score, the more creditable the corresponding class (it corresponds to the first of the Cox-Jaynes axiom for the definition of subjective probabilities [432], then, it is possible to infer an evidential output with all the advantages it brings.

By considering the result of the algebraic comparison of the scores of each of the couple of classes, on obtains a series of values which are very similar to the precursors of the EBFs: they actually indicates the comparative membership of the item for each class of the two considered classes, in a equivalent way to a bank of SVM. The only difficulty remains to determine the values which separate the certitude of a class with respect to another one, or on the contrary, the doubt. Here again, we propose the use of the cross-validation.

10.4.2.3 Decision Making: Partial Pignistic Transform

When a decision is to be made with respect to a peculiar problem, there are two kinds of behaviour: to wait for the proofs of the trueness of one of the hypotheses, or to bet on one of them, with respect to its interest and risk. These two behaviours are considered as antagonist and it appears that no mathematical model allows making a decision which is a mix of these two stances. Consequently, we propose to generalize the Pignistic Transform, a popular method to convert BF into probabilities [420], in order to fill this lack [433].

Let γ be an **uncertainty threshold** and S^γ be the set of all the sets of the frame for which the cardinal is between 0 and γ (It is a truncation of the powerset to the elements of cardinal smaller than or equal to γ). We call S^γ the γth **frame of decision**

$$S^\gamma = \left\{ A \in 2^\Omega \ \middle| \ |A| \in \lfloor 0, \gamma \rfloor \right\} \quad (10.13)$$

where $|.|$ is the cardinality function. The result $M_\gamma(.)$ of the **Partial Pignistic Transform** of order γ (noted γ^{th}-PPT) of $m(.)$ is defined on 2^Ω as:

$$M_\gamma(A) = \begin{cases} m(A) & \text{if } A = \emptyset \\ m(A) + \sum \left(\dfrac{m(B) \cdot |A|}{\sum_{k=1}^{\gamma} \binom{|B|}{k} \cdot k} \middle| \begin{array}{l} B \supseteq A \\ B \notin S^\gamma \end{array} \right) & \text{if } A \subseteq S^\gamma \\ 0 & \text{otherweise} \end{cases} \quad (10.14)$$

Then, the decision is made by simply choosing the element of the γth frame of decision which is the most believable, i. e. which gathers the highest score:

$$D^* = \underset{2^\Omega}{\operatorname{argmax}} \left(M_\gamma \right) \quad (10.15)$$

10.4.2.4 Application for Multimodal Fusion

Automatic clustering. A first classical method is to use the confusion matrix of the HMM based classifier to automatically identify sign clusters. The confusion matrix is converted to a sign cluster matrix by considering the confusions for each sign. Signs that are confused form a cluster. For example, assume that sign i is confused with sign j half of the time. Then the sign cluster of class i is $\{i,j\}$. The sign cluster of class j is separately calculated from its confusions in the estimation process. The disadvantage of this method is its sensitivity to odd mistakes which may result from the errors in the feature vector calculation as a result of bad segmentation or tracking.

We propose a more robust alternative which evaluates the decisions of the classifier and only consider the uncertainties of the classifier to form the sign clusters. For this purpose, we define a hesitation matrix. Its purpose is close to the classical confusion matrix, but it contains only the results of the uncertain decision, regardless with their correctness. Then, when a decision is certain (either true or false), it is not taken into account do define the hesitation matrix. On the contrary, when a decision is uncertain among sign i and sign j, it is counted in the hesitation matrix regardless with the ground truth of the sign being, i, j or even k. As a matter of fact, the confusion between a decision (partial or not) and the ground truth can be due to any other mistake (segmentation, threshold effect, etc. ...) whereas, on the contrary, the hesitation on the classification process only depends on the ambiguity at the level of the classification features with respect to the class borders. Then, it is more robust. In addition, it is not necessary to know the ground truth on the validation set on which the clusters are defined. This is a determining advantage in case of semi-supervised learning to adapt the system to the coder's specificity.

Partial Decision. Thanks to the PPT, it is possible to make partial decisions, which is particularly adapted to classification problems where the classes are defined in a hierarchical manner (dendrogram), such as explained in [433], where an illustration is given on the interest of the PPT to perform automatic lip-reading on French vowels. On classical problems where such a hierarchical does not exist (such as SL recognition), it is possible to simply let it appear by defining clusters based on the hesitation matrix described above. Then, during the decision making procedure, all the pieces of information are fused together and convert into an evidential formalism via the use of the Evidential Combination. Then, the format of the result of the Evidential Combination is naturally suitable to apply the PPT.

Optional sequential decision step. The only problem with such a method is that it does not guaranty that a decision is made: when the data are too uncertain, the PPT does not make any decision. Then, it can be fused with some other information, and finally, a last hesitation-free decision is taken. In [434], after a first decision step allowing some partial decisions, we propose to add some less conflictive non-manual information (that could not be taken into account earlier in the process without raising the amount of uncertainty) in order to perform a second decision step. The originality of the method is that this second step is optional: if no hesitation occurs at the first step, the good decision is not put back into question. This is possible thanks to the use of the PPT which automatically makes the most adapted decision (certain or not). We call this original method **sequential belief-based fusion**. Its comparison with classical methods demonstrates its interest for the highly conflictive and uncertain decision required in a gesture recognition system.

10.5 Applications

10.5.1 Sign Language Tutoring Tool

SignTutor is an interactive platform that aims to teach the basics of sign language. The interactivity comes from the automatic evaluation of the students' signing and visual feedback and information about the goodness of the performed sign. The system works on a low-cost vision based setup, which requires a single webcam, connected to a medium-level PC or a laptop that is able to meet the 25 fps in 640×480 camera resolution requirement.

To enable the system to work in different lighting conditions and environments, the system requires the user to wear two coloured gloves on each hand. With the gloves worn on the hands and no other similarly coloured objects in the camera view, there are no other restrictions.

The current system consists of 19 ASL signs that include both manual and non-manual components. The graphical user interface consists of four panels: Training, Information, Practice and Synthesis (Fig. 10.21). The training panel involves the

10.5 Applications

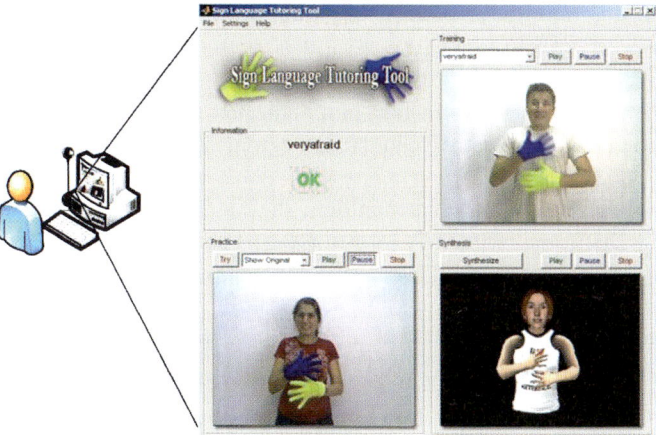

Fig. 10.21 SignTutor user interface

pre-recorded sign videos. These videos are prepared for the students' training. Once the student is ready to practice, and presses the try button, the program captures the students sign video.

The captured sign video is processed to analyze the manual and non-manual components. Here, we give a brief summary of the analysis, fusion and recognition steps. The techniques described here are also explained in the previous sections in detail so we only indicate the name of the technique and do not give the details. More details can be found in [370].

The analysis of the manual features starts with hand detection and segmentation based on the glove colours. Kalman filtering is used to smooth the hand trajectory and to estimate the velocity of each hand. The manual features consist of hand shape, position and motion features. Hand shape features are calculated from the ellipse fitted on each hand and a mask placed on the bounding box. Hand position at each frame is calculated by the distance of each hand center of mass to the face center of mass. As hand motion features, we used the continuous coordinates, and the velocity of each hand center of mass. The starting position of the hands are assumed as the (0,0) coordinate.

In this system, the head motions are analysed as the non-manual component. The system detects rigid head motions such as head rotations and head nods with the help of retina filtering as described in the previous sections. As a result, the head analyzer provides three features per frame: the quantity of motion and the vertical, horizontal velocity.

The recognition is applied via sequential belief-based fusion of manual and non-manual signs [434]. The sequential fusion method is based on two different classification steps: In the first step, we perform an inter-cluster belief-based classification, using a general HMM that receives all manual and non-manual features as input in a single feature vector. A BF is derived from this bank of

HMMs via the evidential combination. Then, the PPT is applied. This first step gives the final decision if there is no uncertainty at this level. Otherwise, a second optional step is applied. In this second step, we perform an intra-cluster classification and utilize the non-manual information in a dedicated model. The clusters are determined via the hesitation matrix automatically from the training set, prior to HMM training.

At the end of the sign analysis and recognition, the feedback about the students' performance is displayed in the information panel. There are three types of results: "ok" (the sign was confirmed), "false" (the sign was wrong) and "head is ok but hands are false". Possible errors are also shown in this field. The students can also watch a simple synthesised version of their performance on an avatar.

10.5.2 Cued Speech Manual Gesture Interpreter

In this chapter, we have presented several techniques in order to deal with FCS recognition:

- Hand segmentation
- Hand analysis: reduction of the variability of the shape and definition of the pointing finger
- Hand shape recognition (the shape descriptors are the FMD and the classification method is a 1vs1 Evidential Combination of SVMs followed by a PPT with an uncertainty parameter of 1 or 2)
- Face and feature detection
- Location of the pointing finger with respect to the face zones used in FCS.
- Lip segmentation
- Lip shape recognition
- Extraction of target image in case of static gestures
- Fusion of several static modalities (CS Hand shape and CS Location)

Then, the next step is to integrate all these functionalities into a global system in order to propose a French Cued Speech Translator. As the lip-reading functionality (based on the joint use of lip segmentation and lip shape recognition) as well as the fusion of manual and labial modalities (the manual gesture is static whereas the labial one is more complex [384]) are still open issues, we propose at the moment a system which is restricted to the manual part: the CS Manual Gesture Interpreter.

This system works as follows: a CS coder (it is important to be a skilled coder, in order to produce a code in which prosody is fluent, as the dedicated retina filter is tuned for such a rhythm) wearing a thin glove of uniform but unspecified colour is filmed at the frame rate of 50 images/s. The system is able to cope with unknown coder having different unknown morphology and glove. Once the video sequence is over, it is processed (this version of the interpreter works off-line), and the result is displayed. The screen is separated into two. On

10.6 Conclusion

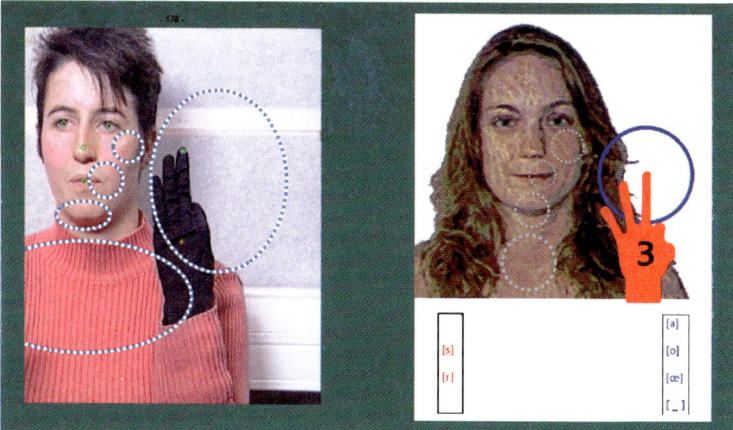

Fig. 10.22 User interface for the display of FCS manual gesture interpreter result

the left, the original video is played whereas on the right part, a virtual clone produces the gesture synchronously with the right part video (Fig. 10.22). Under the clone performing the recognised code, the corresponding potential phonemes are given. Note that, as no interpretation of higher level than the phonemic one is performed, the system is not restricted to any dictionary, and any French message can be processed.

10.6 Conclusion

Gestural interfaces can aid the hearing impaired to have more natural communication with either a computer or with other people. Sign language, the primary means of communication among the hearing impaired, and cued speech, which enriches lip reading with hand and facial cues, are inherently multimodal means of communication: They use gestures of the body, hands and face. Computer vision techniques to process and analyze these modalities have been presented in this chapter. These steps, as summarised below, are essential for an accurate and usable interface.

- A thorough analysis of each visual modality that is used to convey the message.
- The identification of static and temporal properties of each modality and their synchronization.
- Independent modelling and recognition of static/dynamic modalities.
- The integration of various modalities for accurate recognition.

We concentrated on sign languages and cued speech for two reasons: (1) Sign languages and cued speech are the two main media of hearing impaired communication; (2) they have different static and temporal characteristics, thus require different analysis and fusion techniques. After treating the problem in its most general form, we present two example applications: A sign language tutor that is aimed to teach signing to hearing people; and a cued speech manual gesture interpreter. The techniques discussed are general and can be used to develop other applications, either for the hearing impaired or for the general population, in a general modality replacement framework.

Chapter 11
Modality Replacement Framework for Applications for the Disabled

Savvas Argyropoulos[1,2] **Konstantinos Moustakas**[1,2]
[1] Electrical and Computer Engineering Dept., University of Thessaloniki, Hellas
[2] Informatics and Telematics Institute, Centre for Research and Technology, Hellas
savvas@ieee.org, moustak@iti.gr

11.1 Introduction

Recent technological advances have improved communication between disabled people. The emerging artificial intelligence techniques are starting to diminish the barriers for impaired people and change the way individuals with disabilities communicate. A quite challenging task involves the transformation of a signal to another perceivable form so that communication can be ensured. This chapter focuses on the effective processing of information conveyed in multiple modalities and their translation into signals that are more easily understood by impaired individuals. More specifically, the concept of modality replacement is introduced and two applications for the intercommunication of the disabled are thoroughly discussed.

In general, the major problem in intercommunication of impaired individuals is that they do not have access to the same modalities and the perceived information is limited by one's disabilities. For example, in a deaf-blind communication, a blind receiver can rely only on the audio modality to extract the transmitted message, whereas a non-impaired person can combine the audio signal with the lip shapes and achieve typically better recognition rates. This chapter discusses the development a of framework for the combination of all the incoming modalities from an individual, recognition of the transmitted message, and translation into another form that is perceivable by the receiver. In the blind receiver paradigm, for example, the recognised visual content can be transformed into a haptic or audio signal, which the blind receiver is able to understand, and thus communication is ensured. The key point in an automatic translation system is the efficient processing of the multimodal signals and their accurate fusion in order to enhance the performance of the recognition algorithms.

The first application for the demonstration of the modality replacement framework is the recognition of Cued Speech language [438], which is a specific gestural language (different from the sign language) used for communication between deaf people and other people. Cued Speech language consists of a combination of lip shapes and gestures. Since lip reading is not always adequate to understand speech, eight hand shapes in four positions near the face are used to disambiguate lip shapes that correspond to the same phonemes. Automatic real-time translation of gestural languages has been an active topic during the last years as it is expected to improve everyday life of impaired people [439]. An automatic cue generating system based on speech recognition is presented in [440]. In [441], 3D data for the characterization of the hand and face movements were recorded and a Cued Speech synthesizer was developed.

Ideally, an automatic Cued Speech recognition system should fuse the input of three modalities, (audio, lip shape, and hand shape) and take advantage of their complementary nature to deduce correctly the transmitted message. Multimodal approaches have been shown to be advantageous in continuous audio-visual speech processing. In [442], audio and visual features are integrated using a Coupled Hidden Markov Model (CHMM). Moreover, the complementary and supplementary information that is conveyed by speech and lip shape modalities is investigated in [443], where CHHMs with modified inference algorithms are also employed to exploit the interdependencies between these two modalities. CHMMs are also employed in [444] to model loosely-coupled modalities where only the onset of events is coupled in time. Moreover, the use of Dynamic Bayesian Networks (DBNs) is introduced to fuse the feature vectors extracted from lip shapes and the audio signal in [445].

An information theoretic approach for the estimation of the most informative features in audio-visual speech processing is presented in [446]. The most informative eigenlips for using mutual information criteria are selected and after the audio and visual features are obtained, the corresponding feature vectors are provided as input to a recognizer based on Hidden Markov Model (HMM). Additionally, in [447], an information theoretic framework is developed for multi-modal signal processing and joint analysis of multimedia sequences. An approach based on Markov chains is proposed and the existing correlation among modalities is exploited by maximizing the mutual information between audio and visual features.

Many of the aforementioned multimodal techniques have been employed for multimodal processing of sign language. However, the fact that hand shapes are made near the face and also that the exact number and orientation of fingers has to be determined in order to deduce the correct gesture differentiate Cued Speech from sign language. In sign language, the segmentation of the head and the hands is based on a skin colour technique [448]. Vision-based extraction of the hand features is prone to introducing errors when background colour is similar to the skin colour. Moreover, the skin colour depends on the user and it is affected by lighting, background and reflected light from user's clothes. In [449], the use of a sensor glove for automatic sign language recognition is presented to overcome these problems. The glove contains five sensors, one for each finger, plus two sensors to measure the tilt of the hand and the rotation of the hand, respectively.

Also, a Cued Speech hand gesture recognition tool is presented in [450]. A glove is employed in this method to improve hand shape segmentation and a set of parameters are used to construct a structural model. Specifically, a classification of fingers into Long, Small, Fat, and Wrong fingers is employed and a different value is attributed to each finger. Finally, classification into one of the eight possible hand shapes of the Cued Speech language is performed by computing the values according to the aforementioned structural model and intersecting with a set of admissible configurations. In [451], the limitation of using a glove is suppressed by a technique for automatic gesture recognition based on 2D and 3D methods.

In this chapter, the modality replacement concept and its applications are introduced. The basic idea is to exploit the correlation among modalities to enhance the information perceived by an impaired individual who can not access all incoming modalities. In that sense, a modality which would not be perceived due to a specific disability can be utilised to improve the information that is conveyed in the perceivable modalities and increase the accuracy rates of recognition. The results obtained by jointly fusing all the modalities outperform those obtained using only the perceived modalities since the inter-dependencies among them are modelled in an efficient manner and their correlation is exploited effectively.

Based on the aforementioned concept, a new multimodal fusion framework for continuous Cued Speech language recognition is developed. A rigorous feature extraction and representation method is presented for each of the modalities involved in Cued Speech; audio, lip shape, and gesture modalities. Furthermore, modality corruption by measurement noise or modelling errors is investigated and the advantages of modality replacements in such cases are demonstrated.

A critical feature of the recognition system is its ability to adaptively assess the reliability of each modality and assign a measure to weight its contribution. There exist different approaches to measure reliability, such as taking into account the noise level of the input signal. The common way of incorporating these reliability values into decision fusion is to use them as weighting coefficients and to compute a weighted average [452, 453]. The described scheme aims at maximizing the benefit of multimodal fusion so that the error rate of the system becomes than that of the cases where only the perceivable information is exploited. Modality reliability is also examined in [454], in the context of multimodal speaker identification. An adaptive cascade rule was proposed and the order of the classifiers was determined based on the reliability of each modality combination.

In this chapter, a modified CHMM is employed to model the complex interaction and inter-dependencies among the modalities and combine them efficiently to recognize correctly the transmitted message. Modality reliability is regarded as a means of giving priority to a single or combined modalities in the fusion process, rather than using it as a numerical weight. The main contribution of this chapter to the Cued Speech language recognition is the development of a fusion scheme for word recognition. To the best of our knowledge, this is the first approach to combine audio, lip shape, and hand shapes for the improvement of the recognition of the transmitted message in the context of Cued Speech language.

Additionally, another application for the disabled based on the concept of modality replacement is presented. A treasure hunting game for the collaboration of blind and deaf-mute people is developed. A novel human computer interaction

(HCI) interface integrating information from several different modalities, such as speech, gestures, haptics, and sign language, is thoroughly described. In each step, the transformation of signals into a perceivable form for the impaired users is required to ensure communication. Thus, the system combines various modules for the recognition of user input and the replacement of the missing modalities.

The chapter is structured as follows: in Section 2 the concept of modality replacement is discussed and a brief analysis of Cued Speech language is provided to familiarize the reader with the inherent problems of recognizing gestural languages accurately. Section 4 describes how the raw data from each modality can be represented and how robust feature vectors can be extracted. Section 5 presents the methodology for exploiting the correlation among the modalities and their fusion using CHMMs and experimental results discussed in detail in Section 6. The multimodal collaboration game for the intercommunication between blind and deaf-mute users is described in Section 7. Finally, Section 8 concludes the chapter with a discussion on the presented methods and future directions.

11.2 The Modality Replacement Concept

Many real-world processes in nature are products of complex interactions among a multitude of simple processes. Each of the simple processes may have its realization in a different observation domain or modality. The task is often to infer some information about the underlying global process. The most typical instance is the production of the human speech. People perceive the sound signal using their auditory instruments. However, it is well known that a part of the spoken information is transmitted visually, through images of lip movements. In the absence of reliable audio information, such as in a noisy vehicle or an airport, humans often disambiguate sound using the information conveyed in the lip movements. Thus, people combine information from different observation domains (audio signal and visual images) to acquire a better estimate of the transmitted word. This paradigm can be extended to the Cued Speech language, in which the cue listener combines spoken information, lip movements and hand shapes to model the interaction among the different modalities and reduce audio ambiguity using visual cues and perceive the underlying message.

The basic architecture of the modality replacement approach is depicted in Fig. 11.1. The performance of such a system is directly dependent on the efficient multi-modal processing of two or more modalities and the effective exploitation of their complementary nature and their mutual information to achieve accurate recognition of the transmitted content. After the recognition is performed effectively, either a modality translator can be employed to generate a new modality or the output can be utilised to detect and correct possibly erroneous feature vectors that may correspond to different modalities. The latter could be very useful in self-tutoring applications. For example, if an individual practices Cued Speech language, the automatic recognition algorithm could detect incorrect hand shapes

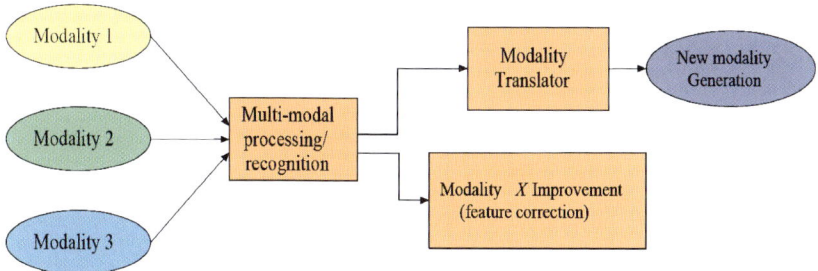

Fig. 11.1 The modality replacement system architecture

(based on the two other modalities) and indicate them so that the user can identify the wrong gestures and practice more on them.

11.3 Cued Speech

The Cued Speech paradigm has been selected to demonstrate the modality replacement concept since the modalities employed to convey information are highly inter-dependent. Cued Speech is a visual communication system, developed by Cornett. It uses eight hand shapes placed at four different positions near the face in combination with the natural lip movements of speech in order to make the sounds of spoken language look different from each other. In that sense, it can be considered as the visible counterpart of the spoken language, in which one uses visual information extracted from the speaker's lips to improve word recognition, especially in noisy environments.

Cued Speech is based on a syllabic decomposition: the message is formatted into a list of "Consonant-Vowel syllables", which is known as CV list. Each CV is coded using a different hand shape, which is combined with the lip shape, so that it is unique and understandable. The eight hand shapes incorporated in Cued Speech and the positions at the cue speaker's face are illustrated in Fig. 11.2. While talking, cued speakers execute a series of hand and finger gestures near the face closely related to what they are pronouncing. Thus, the transmitted message is contained into three modalities: audio, lip shapes, and hand shapes.

Speech recognition can be performed by extracting representative features from each modality and using a CHMM to model their asynchronous nature for the recognition of the transmitted message [455]. According to the modality replacement concept, the transmission and processing of the audio information conveyed in the speech signal can be utilised even if the cue listener can not perceive it, as illustrated in Fig. 11.3. Indeed, by modelling effectively the interdependency among the modalities used by the cue speaker the audio signal can be used to increase the accuracy of the recognition rates or to correct the other two modalities.

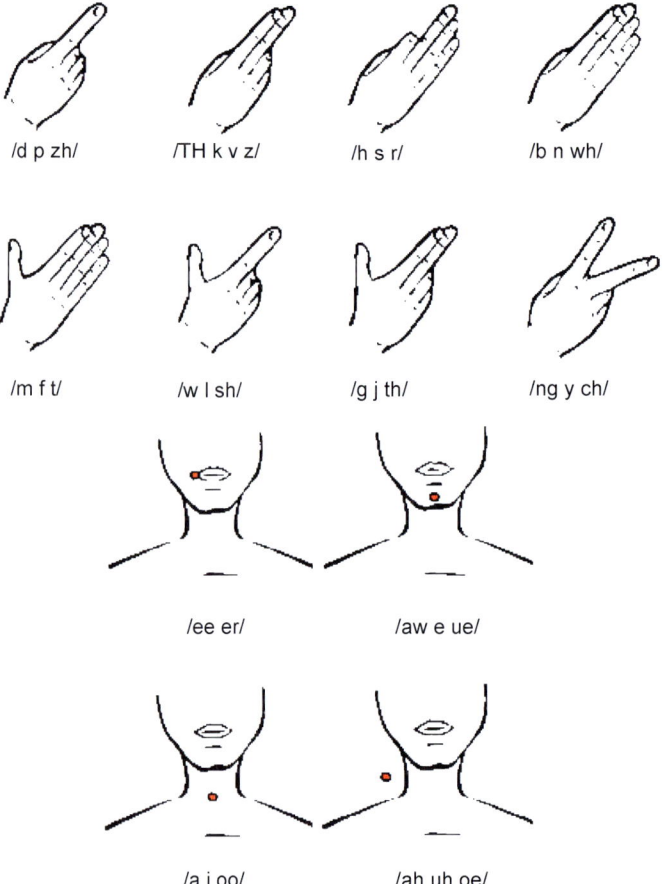

Fig. 11.2 The eight hand shapes and the four face positions in Cued Speech Language

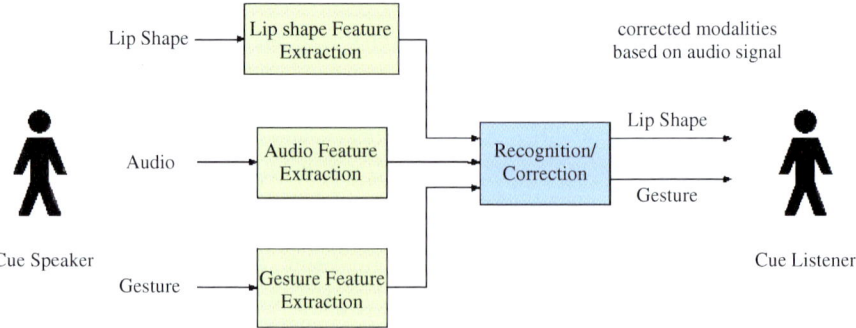

Fig. 11.3 The modality replacement system architecture in Cued Speech language

Since the scope of this chapter lies in the efficient combination of the modalities and the exploitation of their complementary and supplementary information and not in the feature extraction process, the feature extraction and representation procedure is constrained to a brief analysis which is given in the following section.

11.4 Feature Extraction and Representation for the Cued Speech Language

11.4.1 Audio Feature Extraction

The audio signal is considered as valuable information for the understanding of the conveyed message of an individual. However, robust features have to be extracted from the audio signal to ensure acceptable recognition rates under various environments. The most popular features used for speech processing are the Mel Frequency Cepstral Coefficients (MFCC) [456] since they yield good discrimination of the speech signal. The audio stream is processed over 15 *msec* frames centred on 25 *msec* Hamming window. Each analysis frame is first multiplied with a Hamming window and transformed to frequency domain using Fast Fourier Transform (FFT). Mel-scaled triangular filter-bank energies are calculated over the square magnitude of the spectrum and represented in logarithmic scale. The resulting MFCC features c_j, are derived using Discrete Cosine Transform (DCT) over log-scaled filter-bank energies e_i:

$$c_j = \frac{1}{N_M} \sum_{i=1}^{N_m} e_i \cos((i-0.5)\frac{j\pi}{N_M}), \quad j=1,2,\ldots,N. \tag{11.1}$$

where N_M is the number of mel-scaled filter banks and N is the number of MFCC features that are extracted. The MFCC feature vector for a frame comprises 13 MFCCs along with their first and second order derivatives.

11.4.2 Lip Shape Feature Extraction

The accuracy of a multimodal recognition system is closely related to the selected representation of each modality. For the lip shape modality, the robust location of facial features and especially the location of the mouth region is crucial. Then, a discriminant set of visual observation vectors have to be extracted. The process for the extraction of lip shape features is described in detail in [445] and is depicted in Fig. 11.4.

Initially, the speaker's face is located in the video sequence. Subsequently, the lower half of the detected face is selected as an initial candidate of the mouth

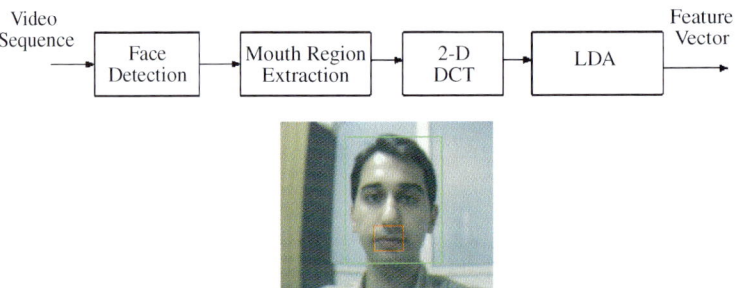

Fig. 11.4 Mouth location for lip shape representation and feature vector extraction

region and Linear Discriminant Analysis (LDA) is used to classify pixels into to classes: face and lip. After the lip region segmentation has been performed the contour of the lips is obtained using the binary chain encoding method and a normalised 64x64 region is obtained from the mouth region using an affine transform. In the following, this area is split into blocks and the 2D-DCT transform is applied to each of these blocks and the lower frequency coefficients are selected from each block, forming a vector of 32 coefficients. Finally, LDA is applied to the resulting vectors, where the classes correspond to the words considered in our application. A set of 15 coefficients, corresponding to the most significant generalised eigenvalues of the LDA decomposition is used as the lip shape observation vector. Apparently, other methods could also be employed to extract robust features for the representation of lip shapes, such as the methods in [457, 458].

11.4.3 Gesture Feature Extraction

Since Cued Speech is strongly dependent on the hand shapes made by the cue speaker while talking, it is essential for a multimodal speech recognition system to be able to detect and classify gestures in a correct and efficient manner. The main problem that has to be addressed is the segmentation of the hand and the fingers in front of the cue speaker's face and the possible occlusions, which pose limitations to the application of a skin colour mask. Moreover, since the exact orientation and the number of fingers is crucial the algorithm must be able to discriminate between fingers and estimate the position of the hand shape relative to the speaker's mouth. Furthermore, even the same person does not perform the same gesture in the same way every time (different speed, orientation of fingers, position, etc.) and sign boundaries are difficult to detect efficiently.

To overcome the aforementioned problems the presented system is based on gestures made with a glove consisting of six colours: one for each finger and one for the palm. In that way, a colour mask can be employed in order to determine which finger is present in each frame and estimate the position of that finger. The mass centre of each finger is computed and the absolute difference between the mass centre of the mouth (as computed in the lip shape extraction procedure) and the mass centre of each finger forms the feature vector for the hand shape modality.

11.5 Coupled Hidden Markov Models

The combination of multiple modalities for inference has proven to be a very powerful way to increase detection and recognition performance. By combining information provided by different models of the modalities, weakly incorrect evidence in one modality can be corrected by another modality. In general, the state spaces representing the modalities may be interdependent, which makes exact inference intractable.

HMMs are a popular probabilistic framework for modelling processes that have structure in time. They have a clear Bayesian semantics and efficient algorithms for state and parameter estimation. Especially, for the applications that integrate two or more streams of data, such as audio-visual speech processing, CHMMs have been developed. The structure of this model describes the state synchrony of the audio and visual components of speech while maintaining their natural correlation over time.

Since the modalities involved in Cued Speech language, namely audio, lip shape, and hand shape, contain supplementary and complementary information over time, CHMM is a logical selection to model the inter-dependencies among them and perform accurate word recognition. A CHMM can be considered as a collection of HMMs, one for each data stream, where the hidden backbone nodes at time t for each HMM are conditioned by the backbone nodes at time $t-1$ for all the related HMMs. It must be noted that CHMMs are very popular among the audio-visual speech recognition community, since they can model efficiently the endogenous asynchrony between the speech and lip shape modalities. The parameters of the CHMM are described below:

$$\pi_0^c(i) = P(q_t^c = i) \tag{11.2}$$

$$b_t^c(i) = P(\mathbf{O}_t^c \mid q_t^c = i) \tag{11.3}$$

$$a_{i|j,k,n}^c = P(q_t^c = i \mid q_{t-1}^A = j, q_{t-1}^L = k, q_{t-1}^G = n) \tag{11.4}$$

where q_t^c is the state of the coupled node in the c_{th} stream at time t, $\pi_0^c(i)$ is the initial state probability distribution for state i in c_{th} stream, $\mathbf{O}_t^c$ is the observation of the nodes at time t in the c_{th} stream, $b_t^c(i)$ is the probability of the observation given the i state of the hidden nodes in the c_{th} stream, and $a_{i|j,k,n}^c$ is the state transitional probability from node i to node j in the c_{th} stream, given the state of the nodes at time $t-1$ for all the streams. Figure 11.5 illustrates the CHMM employed in this work.

Square nodes represent the observable nodes whereas circle nodes denote the hidden (backbone) nodes.

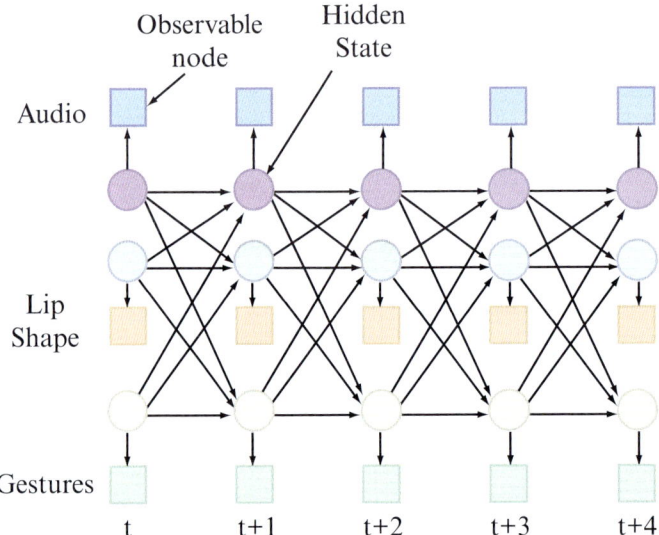

Fig. 11.5 Coupled Hidden Markov Model for fusion of three modalities

One of the most challenging tasks in automatic speech recognition systems is to increase robustness to environmental conditions. Although the stream weight needs to be properly estimated according to noise conditions, they cannot be determined based on the maximum likelihood criterion. Therefore, it is very important to build an efficient stream-weight optimization technique to achieve high recognition accuracy, as explained in the next subsection.

11.6 Modality Reliability

According to the above analysis, it is obvious that the fusion of information from different sources may reduce the overall uncertainty and increase the robustness of a recognition system. Ideally, the contribution of each modality to the overall output of the recognition system should be weighted according to a reliability measure. This measure denotes how each observation stream should be modified and acts as a weighting factor. In general, it is related to the environmental conditions (e. g., acoustic noise for the speech signal) or the modelling errors of the system.

The common way of incorporating these reliability values into decision fusion is to use them as weighting coefficients and to compute a weighted average. Thus, the probability $b_m(\mathbf{O}_t)$ of a feature $\mathbf{O}_t$ for a word m is given by:

$$b_m(\mathbf{O}_t) = w_A \cdot b_A(\mathbf{O}_t^A) + w_L \cdot b_L(\mathbf{O}_t^L) + w_G \cdot b_G(\mathbf{O}_t^G) \tag{11.5}$$

where $b_A(\mathbf{O}_t^A)$, $b_L(\mathbf{O}_t^L)$, $b_G(\mathbf{O}_t^G)$, are respectively the likelihoods for an audio feature $\mathbf{O}_t^A$, a lip shape feature $\mathbf{O}_t^L$, and a gesture feature $\mathbf{O}_t^G$. w_A, w_L, and w_G are audio, lip shape, and gesture weights, respectively, and $w_A + w_L + w_G = 1$.

In the presented method, a different approach is employed to determine the weights of each data stream. More specifically, for each modality, word recognition is performed using a mono-modal HMM for the training sequences. The results of the (mono-modal) word recognition indicate the noise levels in each modality and provide an approximation of their reliability. More specifically, when the mono-modal HHM classifier fails to identify the transmitted words it means that the observation features for the specific modality have been perturbed. On the other hand, a small word error rate using only one modality and the related HMM means that the corresponding feature vector is reliable and should be favoured in the CHMM. Additionally, the combinations of two modalities could be considered to perform recognition using a CHMM and provide another measure of the noise level in the corresponding signals. However, this approach was not selected because of the unacceptable computational cost. The weights estimated using this method are employed in the word inference stage as explained in the following section.

11.7 Modified Coupled Hidden Markov Model

11.7.1 Training

The Maximum Likelihood (ML) training of the dynamic Bayesian Networks in general and of the coupled HMMs in particular, is a well understood technique [175]. However, the iterative ML estimation of the parameters only converges to a local optimum, making the choice of the initial parameters of the model a critical issue. In this section, an efficient method for the initialization of the ML training that uses a Viterbi algorithm derived for the coupled HMM is presented. The Viterbi algorithm determines the optimal sequence of states for the coupled nodes of audio, lip shapes, and hand shapes that maximize the observation likelihood. The following steps describe the Viterbi algorithm for the three-stream CHMM used in the developed system.

- Initialization

$$\delta_0(i,j,\vartheta) = \pi_0^A(i)\pi_0^L(j)\pi_0^G(\vartheta)b_t^A(i)b_t^L(j)b_t^G(\vartheta) \tag{11.6}$$

$$\psi_0(i,j,\vartheta) = 0 \tag{11.7}$$

- Recursion

$$\delta_t(i,j,\vartheta) = \max_{k,l,m}\{\delta_{t-1}(k,l,m)a_{i|k,l,m}a_{j|k,l,m}a_{\vartheta|k,l,m}\}b_t^A(k)b_t^L(l)b_t^G(m) \tag{11.8}$$

$$\psi_t(i,j,\vartheta) = \arg\max_{k,l,m}\{\delta_{t-1}(k,l,m)a_{i|k,l,m}a_{j|k,l,m}a_{\vartheta|k,l,m}\} \tag{11.9}$$

- Termination

$$P = \max_{i,j,\vartheta}\{\delta_T(i,j,\vartheta)\} \quad (11.10)$$

$$q_T^A, q_T^L, q_T^G = \arg\max_{i,j,\vartheta}\{\delta_T(i,j,\vartheta)\} \quad (11.11)$$

- Backtracking

$$q_t^A, q_t^L, q_t^G = \psi_{t+1}(q_{t+1}^A, q_{t+1}^L, q_{t+1}^G) \quad (11.12)$$

The Viterbi algorithm determines the optimal sequence of states for the coupled nodes for each of the three modalities that maximize the observation sequence.

11.7.1.1 Recognition

The word recognition is performed using the Viterbi algorithm, described above, for the parameters of all the word models. It must be emphasised that the influence of each stream is weighted at the recognition process because, in general, the reliability and the information conveyed by each modality is different. Thus, the observation probabilities are modified as:

$$b_t^A(i) = b_t(\mathbf{O}_t^A \mid q_t^A = i)^{w_A} \quad (11.13)$$

$$b_t^L(i) = b_t(\mathbf{O}_t^L \mid q_t^L = i)^{w_L} \quad (11.14)$$

$$b_t^G(i) = b_t(\mathbf{O}_t^G \mid q_t^G = i)^{w_G} \quad (11.15)$$

where w_A, w_L, and w_G are respectively the weights for audio, lip shape, and gesture modalities and $w_A + w_L + w_G = 1$. The values of w_A, w_L, and w_G are obtained using the methodology of section 5.1.

11.8 Evaluation of the Cued Speech Recognition System

Since there is no available database for Cued Speech language, the multimodal speech recognition system was evaluated on multimodal database including gesture, lip and speech data corresponding to a small set of words. More specifically, the vocabulary consisted of digits from zero to nine. Each word of the considered vocabulary was repeated ten times; nine instances of each word were used for training and the remaining instance was used for testing. Five subjects were trained in order to perform the specified gestures.

Six states were considered for the coupled nodes in the CHMM, with no back transitions. The form of the feature vectors is described in Section 4. For audio only, lip shape only, and gesture only recognition, a HMM was used with six states and no back transitions. In the audio-only and all multimodal Cued Speech recognition experiments including audio, the audio sequences used in training were captured in clean acoustic conditions and the audio track of the testing sequences was altered by white noise at various SNR levels from 30 dB (clean) to 0 dB. The whole system was evaluated for different levels of noise, in terms of Signal to Noise Ratio (SNR), and the results are depicted in Fig. 11.6.

The experimental results indicate that the approach which exploits the audio, lip shape and gesture modalities (A-L-G) achieves smaller word error rate compared to the mono-modal approaches. More specifically, the system achieves consistently better results and the results are even more impressive compared to the audio-only method when the quality of the noise signal deteriorates (small SNR values). A reduction of approximately 20% in word error rate is achieved at 15 *dB*.

In order to demonstrate the merit of the presented scheme, the modality replacement scheme was also compared against a scheme which exploits only visual information; lip shapes and hand shapes. This scenario corresponds to an individual with hearing disabilities while trying to understand a cue speaker. Due to hearing loss, the impaired person can rely only on lip shapes and hand shapes to infer the transmitted message. However, much of the transmitted information is contained in the audio signal which the disabled person can not perceive. The results are illustrated in Fig. 11.6 where it can be seen that the Lip-Gesture (L-G) approach yields substantially lower recognition rates compared to the scheme which combines audio, lip shapes, and gestures (A-L-G).

The superior performance of the presented system can be attributed to the effective modelling of the interactions and the inter-dependencies between the three modalities. An instance of the word recognition results is given in Table 11.1 where the recognised word is presented for recognition in a noisy environment at

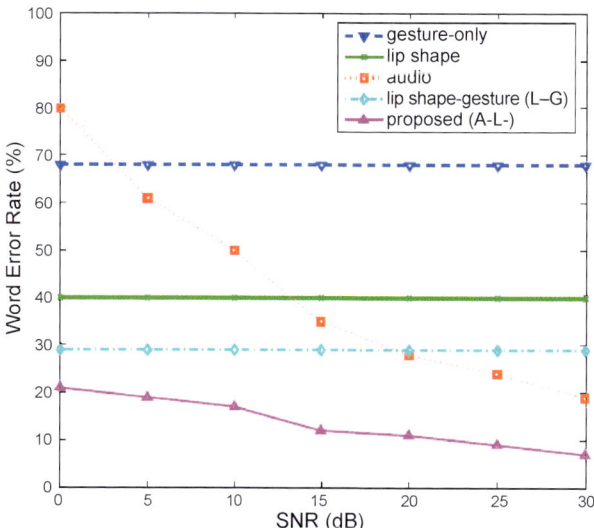

Fig. 11.6 Experimental Results

Table 11.1 Word recognition results at acoustic noise level 30 dB using different combinations of modalities for recognition

Method	zero	one	two	three	four	five	six	seven	eight	nine
A-L-G	zero	one	two	three	four	five	six	eight	eight	nine
L-G	zero	zero	two	three	five	five	six	eight	eight	nine
Audio	zero	one	five	three	five	five	six	eight	eight	nine
Lip	zero	zero	two	three	five	five	six	eight	eight	one
Gesture	one	zero	two	three	five	five	zero	one	eight	nine

noise levels 30 *dB*. It is worth noting that in the first case, where the voice signal is distorted by noise, the algorithm can correctly identify words even if all mono-modal recognizers fail to recognize the transmitted word. As depicted in Table 11.1 the three mono-modal recognizers misinterpret the word "four". However, the multimodal scheme can efficiently process the feature vectors from each modality and infer the correct word. This is a typical example that demonstrates the appropriateness of CHMMs for modelling processes that evolve in time.

11.9 Multimodal Human-computer Interfaces

Another area that has attracted much attention both from academia and industry during the last years is human-computer interaction (HCI) for multimodal interfaces. Since Sutherland's SketchPad in 1961 or Xerox' Alto in 1973, computer users have long been acquainted with more than the traditional keyboard to interact with a system. More recently, with the desire of increased productivity, seamless interaction and immersion, and e-inclusion of people with disabilities, multimodal interaction has emerged as a very active field of research.

Multimodal interfaces are those encompassing more input modalities than the traditional keyboard and mouse. Natural input modes are employed, such as voice, gestures, body movement, haptic interaction, facial expressions and physiological signals. As described in [459], multimodal interfaces should follow several guiding principles: multiple modalities that operate in different spaces need to share a common interaction space and be synchronised; multimodal interaction should be predictable and not unnecessarily complex and should degrade gracefully, for instance by providing for modality switching. Finally, multimodal interfaces should adapt to user's needs, abilities, and environment.

A key aspect in multimodal interfaces is the integration of information from several different modalities to assist in extracting high-level information non-verbally conveyed by users. Such high-level information can be related to the expressive and emotional content communicated by the users. In this framework, gesture has a dominant role as a primary non-verbal conveyor of expressive, emotional information. Research on gesture analysis, processing, and synthesis has received a growing interest from the scientific community in recent years and demonstrated its paramount importance for human machine interaction.

11.9 Multimodal Human-computer Interfaces

This section discusses the development of efficient tools and interfaces for the generation of an integrated platform for the intercommunication of blind and deaf-mute people. It is obvious that while multimodal signal processing is essential in such applications, specific issues like modality replacement and enhancement should be addressed in detail.

In the blind user's terminal the major modality to perceive a virtual environment is haptics while audio input is provided as supplementary side information. Force feedback interfaces allow blind and visually impaired users to access not only two-dimensional graphic information, but also information presented in 3D virtual reality environments (VEs) [460]. The greatest potential benefits from virtual environments can be found in applications concerning areas such as education, training, and communication of general ideas and concepts [461]. Several research projects have been conducted to assist visually impaired to understand 3D objects, scientific data and mathematical functions, by using force feedback devices [462].

PHANToM is the most commonly used force feedback device. Due to its hardware design, only one point of contact at a time is supported. This is very different from the way that people usually interact with surroundings and thus, the amount of information that can be transmitted through this haptic channel at a given time is very limited. However, research has shown that this form of exploration, although time consuming, allows users to recognize simple 3D objects. The PHANToM device has the advantage to provide the sense of touch along with the feeling of force feedback at the fingertip. Another device that is often used in such cases is the CyberGrasp that combines a data glove (CyberGlove) with an exoskeletal structure so as to provide force feedback to each of the fingers of the user (five degrees-of-freedom (DoF) force feedback, one DoF for each finger). The multimodal interface described in this chapter employs the PHANToM desktop device to enable haptic interaction of the blind user with the virtual environment.

Deaf-mute users have visual access to 3D virtual environments; however their immersion is significantly reduced by the lack of audio feedback. Furthermore, many efforts have been made to develop applications for the training of hearing impaired. Such applications include the visualization of the hand and body movements performed in order to produce words in sign language and applications based on computer vision techniques that aim to recognize such gestures in order to allow natural human machine interaction for the hearing impaired. In the context of the presented framework, the deaf-mute terminal incorporates sign-language analysis and synthesis tools to enable physical interaction of the deaf-mute user and the virtual environment.

11.9.1 Multimodal Collaborative Game

The effective communication between the impaired individuals is illustrated with the development of a collaborative treasure hunting game that exploits the aforementioned human-computer interfaces. The basic concept in multimodal interfaces for the disabled is the idea of modality replacement, defined as the use of information originating from various modalities to compensate for the missing input modality of the system or the users. The main objective of the described system is the development of tools, algorithms, and interfaces that employ modality replacement to enable the communication between blind or visually impaired and deaf-mute users. To achieve the desired result, the system combines the use of a set of different modules, such as *Gesture recognition*, *Sign language analysis and synthesis*, *Speech analysis and synthesis*, and *Haptics* into an innovative multimodal interface available to disabled users. Modality replacement is employed to enable information transformation between the various modalities used and, thus, enable the communication between the participating users.

Figure 11.7 presents the architecture of the developed system, the communication flow between the various modules used for the integration of the system, and the intermediate stages used for replacement between the various modalities. The left part of the figure refers to the blind user's terminal, while the right refers to the deaf-mute user's terminal. The interested reader is referred to [463, 464] for more information on the modules of the game. The different terminals of the treasure hunting game communicate through asynchronous TCP connection using TCP sockets. The following sockets are implemented in the context of the treasure hunting game:

- The SeeColor terminal which implements a server socket that receives queries for translating colour into sound. The code word consists of the following bytes, "$b;R;G;B$", where b is a boolean flag and R, G, B the colour values.
- The Blind user terminal which implements three sockets:
  - A client socket that connects to the SeeColor terminal.
  - A server socket to receive messages from the deaf-mute user terminal.
  - A client socket to send messages to the deaf-mute user terminal.

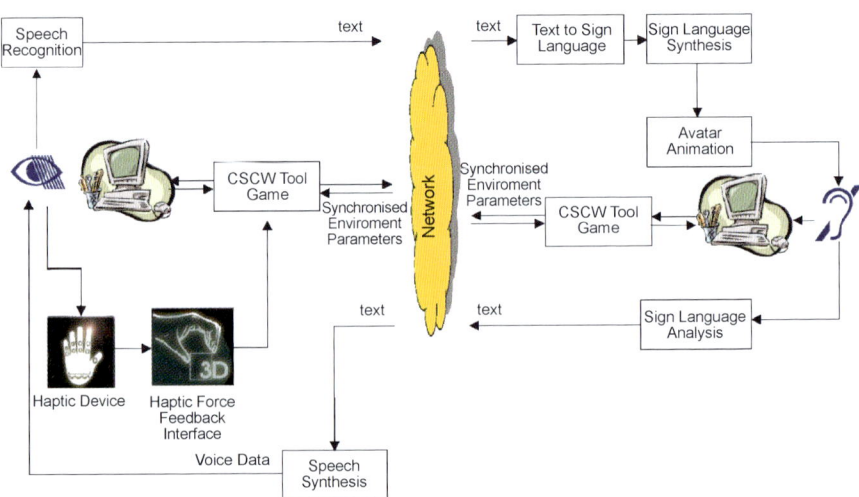

Fig. 11.7 Architecture of the multimodal collaborative game

11.9 Multimodal Human-computer Interfaces

- The Deaf-mute user terminal which implements two sockets:
  - A server socket to receive messages from the blind user terminal.
  - A client socket to send messages to the blind user terminal.

The aforementioned technologies are integrated in order to create an entertainment scenario. The scenario consists of seven steps, as illustrated in Fig. 11.8. In each step one of the users has to perform one or more actions in order to pass successfully to the next step. The storyboard is about an ancient city that is under

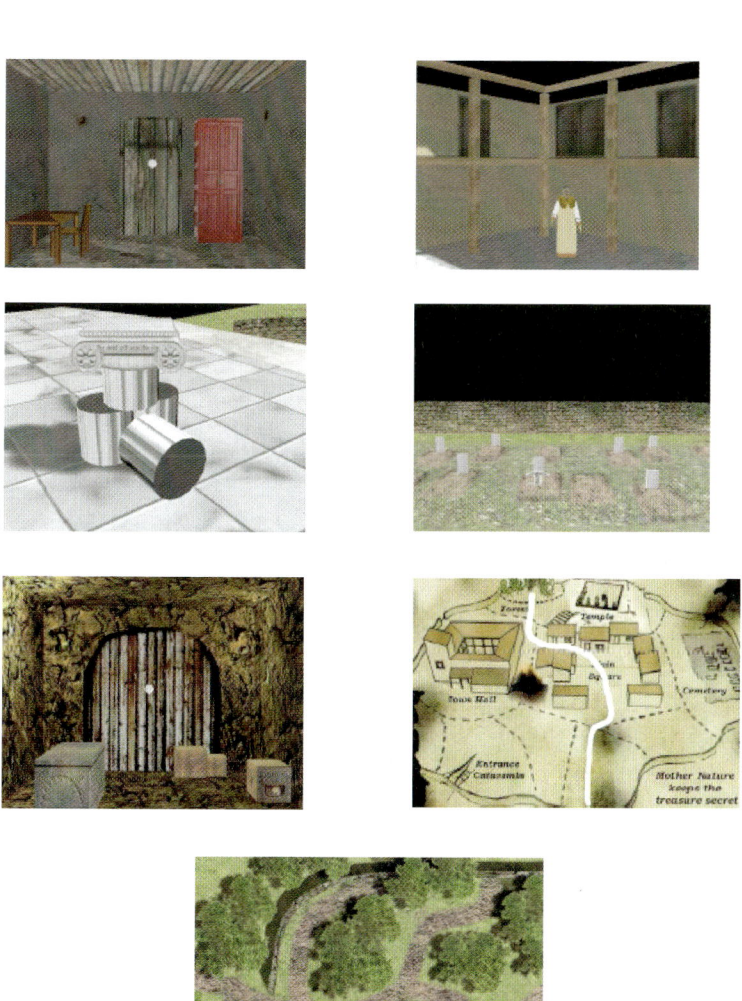

Fig. 11.8 The seven steps of the multimodal collaborative game

attack and citizens of the city try to find the designs in order to create high technology war machines.

In the first step, the blind user receives an audio message and is instructed to "find a red closet". Subsequently, the blind user explores the village using the haptic device. It is worth noting that audio modality replaces colour modality using the SeeColor module. Thus, the user can select the correct closet and receive further instructions which are transmitted to the other user. In the second step, the deaf-mute person receives the audio message which is converted to text using the speech recognition tool and then to sign language using the sign synthesis tool.

Subsequently, the user receives the message as a gesture through an avatar, as depicted in Fig. 11.9. This message guides the deaf-mute user to the town hall, where the mayor provides the audio message *"Go to the temple ruins"*.

The third step involves the blind user, who hears the message said by the mayor and goes to the temple ruins. In the temple ruins the blind user has to search for an object that has an inscription written on it. One of the columns has an inscription written on it that states *"The dead will save the city"*. The blind user is informed by an audio message whenever he finds this column and the message is sent to the deaf-mute user's terminal.

The fourth step involves again the deaf and mute user. The user receives the written text in sign language form. The text modality is translated to sign language symbols using the sign synthesis tool. Then the deaf and mute user has to under-

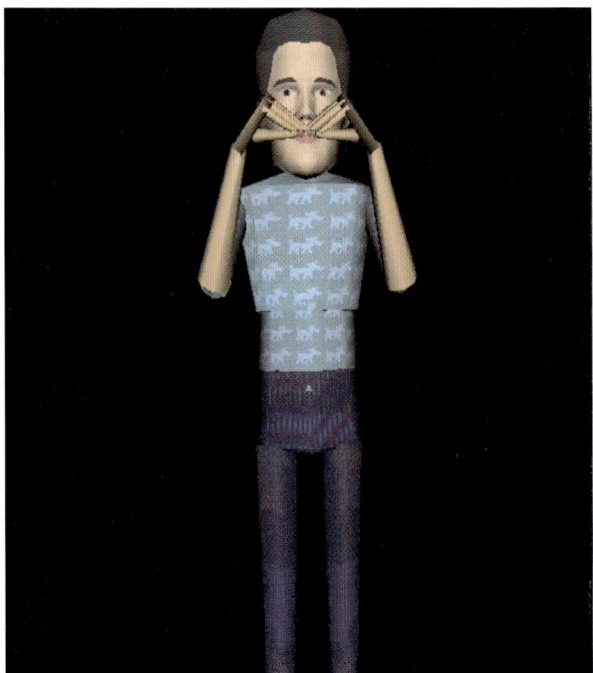

Fig. 11.9 Sign language synthesis using an avatar

stand the meaning of the inscription *"The dead will save the city"* and go to the cemetery using the mouse where he/she finds a key with the word *"Catacombs"* written on it.

In the fifth step, the text-to-speech (TTS) tool is employed to transform the instructions written on the key (*"CATACOMBS"*) to an audio signal that can be perceived by the blind user. The user has to search for the catacombs enter in them and find the box that contains a map. The map is then sent to the next level.

In the sixth step, the deaf user receives the map, and has to draw the route to the area where the treasure is hidden. The route is drawn on the map and the map is converted to a grooved line map, which is send to for the last level to the blind user.

In the seventh step, the blind user receives the grooved line map and has to find and follow the way to the forest where the treasure is hidden. Although the map is presented again as a 2D image the blind user can feel the 3D grooved map and follow the route to the forest. The 2D image and the 3D map are registered and this allows us to visualize the route that the blind user actually follows on the 2D image. The blind user is asked to press the key of the PHANToM device while he believes that the PHANTOM cursor lies in the path. Finally, after finding the forest he obtains a new grooved line map where the blind user has to search for the final location of the treasure. After searching in the forest streets the blind user should find the treasure.

11.10 Discussion

In this chapter, the concept of modality replacement for applications for the disabled was discussed. In the first application, a novel framework for continuous Cued Speech language recognition was presented based on multimodal signal processing techniques. Additionally, the representation and feature extraction of the multiple modalities was presented. The extracted features were fused using a modified CHMM to model the complementary information that is inherent among these modalities. The experimental results demonstrated the superiority of the proposed method over other schemes which exploit only visual information. Thus, it is evident that the effective processing of multiple modalities, even if they can not be perceived by the users, may increase the performance of recognition systems.

However, automatic Cued Speech language recognition remains a major challenge and an open research topic. Future work on such systems will concentrate on extending the vocabulary, which is at the moment limited to a small set. Furthermore, more sophisticated methods have to be employed for the extraction of feature vectors from hand shapes. Glove-based techniques are usually quite obtrusive and limit the user-friendliness of the system. Future research should focus on the use of depth data and gesture processing based on 3D images in order to extract a representative set of parameters.

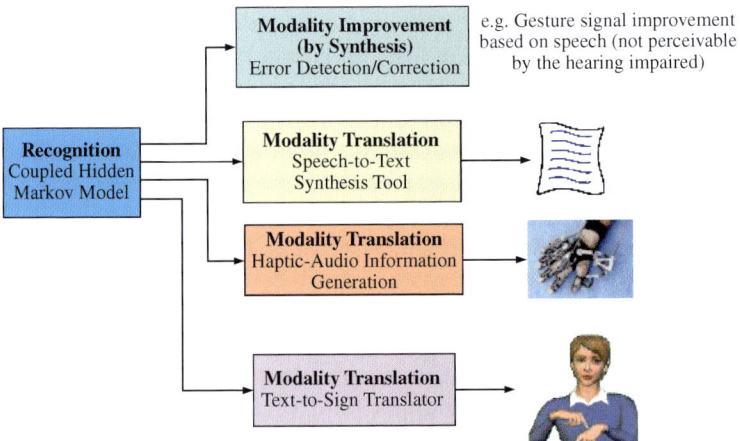

Fig. 11.10 Potential applications of the modality replacement framework

Moreover, future work will focus on implementing the modality replacement concept as depicted in Fig. 11.10. In particular, after the recognition of the transmitted message, a modality translation method should be applied in order to transform the conveyed information into a modality that can be perceived by the listener (e. g., text, haptic, audio, sign language, etc.).

Additionally, the overall performance of the automatic Cued Speech recognition system can be improved by more efficient fusion techniques. To this effect, the use of Dynamic Bayesian Networks, which are an extension of Hidden Markov Models, is expected to enable better modelling of the interactions between the modalities and provide a better understanding of their dependencies. Apart from this, an information theoretic framework could be developed in order to select optimally the most representative features in terms of mutual information.

The second application for the demonstration of the modality replacement idea in the intercommunication between impaired individuals was a treasure hunting game. The collaboration of a blind and deaf-mute user is required to solve a riddle and advance to the next levels. In every step, the information is communicated from one user to the other using various modules for the transformation of signals into a perceivable form. Thus, the system integrates a speech synthesis and analysis module, a sign language synthesis and analysis module, a haptics module, and a gesture recognition module. Multimodal fusion techniques assist in the recognition of the transmitted messages and improve the user-friendliness of the human-computer interface.

Chapter 12
A medical Component-based Framework for Image Guided Surgery

Daniela G. Trevisan[1,2], Vincent Nicolas[1], Benoit Macq[1], Luciana P. Nedel[2]
[1] Communications and Remote Sensing Laboratory, Université catholique de Louvain, (UCL) Place du Levant 2, Bat. Stevin, 1348 Louvain-le-Neuve, Belgium
[2] Instituto de Informática – Universidade Federal do Rio Grande do Sul (UFRGS) Caixa Postal 15.064 – 91.501-970 – Porto Alegre – RS – Brazil
vincent.nicolas@tele.ucl.ac.be, nedel@inf.ufrgs.br

12.1 Introduction

Advances in scanning technology and other data collection systems provide a wide spectrum of useful and complementary information about a patient's status to research and clinical domains. To combine these different types of information into a coherent presentation assuring usable and cognitively adequate interaction within the Operating Room becomes extremely difficult. More sophisticated methods of analysis, guidance and interaction need to be created to achieve these objectives.

Concerning some of the existents systems for surgical planning and intraoperative guidance, 3D Slicer [465] has been one of the first open-source application enabling data fusion and interventional imaging. Julius [466] is another extensible, cross-platform software framework providing a complete medical processing pipeline, but more targeted to visualization and analysis, whereas IGSTK [467] is the latest framework currently under development for open-source component-based rapid prototyping of image guided surgery applications.

Research efforts in Image Guided Surgery (IGS) systems and image processing are investigating how to leverage on such techniques to develop more effective, integrated solutions for the planning, simulation, and finally intra-operative guidance systems either based on a navigation concept or including human-computer interaction systems. As pointed out in [468], other problems regarding the use of new technologies are the lack of compatible and interchangeable equipments and limited communication among surgeons and others in the team especially during surgical procedures.

MedicalStudio proposes to address these issues by realizing a component-based framework targeted to medical interventions and equally attentive to research and therapeutic concerns. It consists of a unique framework for planning, simulating and performing interventional procedures assuring more compatibility within the surgical workflow. Its modular architecture can easily manage abstractions of hardware peripherals and directly make data available from them. Components developed in collaboration with several research centres and medical clinics have shown the promising dissemination and versatility of this medical framework in various disciplines.

The chapter is structured as follows. In section 2 we first describe the component-based architecture. In the General Purposes Components and Applications sections we illustrate examples of high level software components and applications implemented within the medical framework. The final section is devoted to the conclusions and the presentation of future plans.

12.2 MedicalStudio Framework

This section gives an overview on MedicalStudio architecture. It then details the core functionalities and implementation choices of the framework.

12.2.1 Architecture

A framework allowing the centralization of all tasks of assisted surgery must have a consistent and evolvable architecture. For that reason MedicalStudio is based on a component architecture. Such software architectures have been widely discussed either for multimodal interactions [469] or visualization systems [470]. The software architecture implemented in MedicalStudio is a modified version of the ARCH model [469] (Fig. 12.1a)) to add a specialization of the input and output modalities. The principal components of the architecture illustrated in Fig. 12.1(b) are detailed as follows:

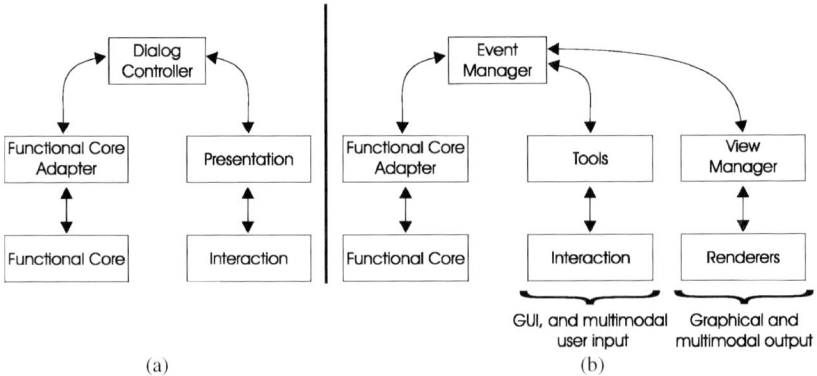

Fig. 12.1 (a) The ARCH software architecture (b) MedicaStudio architecture extending the ARCH model

1. **Functional Core** contains all data processing components such as registration algorithms, I/O filters, etc.
2. **Functional Core Adapter** is an abstraction layer that allows the event manager to communicate with the functional core.
3. **Event Manager** is the dialog controller between interactions and functional core. When interactions occur, the event manager will propagate them to the functional core, and if needed interpret them before.
4. **Interaction** represents user input components such as mouse, keyboard, magnetic pen, vocal recognition, etc.
5. **Tools** are components that map interactions into data processing or visualization modification. For example, a mouse click will be translated into a rotation (view manager) by the navigation tool, but it will be translated into a data modification (functional core) by a segmentation tool.
6. **View Manager** manages all view and knows how data can be visualized on which type of view.
7. **Renderers** are components which render specific data type onto specific views. An example is the rendering of an image on a 3D view with the raycasting algorithm. In order to clarify the development process and to fix where components should be in the visualization pipeline, [470] classified them into operators. MedicalStudio architecture groups the data operators into data components, the visualization operators into output components, and adds input components. There are plans to affine this subdivision to add more control on the visualization pipeline.

12.2.2 Framework Implementation

In order to assure cross-compatibility in terms of execution and development, MedicalStudio is written in C++. The language is commonly used by the signal and image processing community. Publicly available libraries are used to provide well known functionalities and avoid re-implementation of already validated methods. These libraries have been chosen in function of their specifications, their development language, their cross-compilation possibilities and the community working with them. VTK[1] is used for all visualization tasks. Gtk[2] is used for all graphical user interfaces. ITK[3] is used for signal and image processing. Dcmtk[4] is used for Dicom 3.0 standard compliance. Figure 12.2 is showing how these libraries are inserted into the architecture pipeline. This list of libraries is not fixed.

---

[1] VTK: http://www.vtk.org
[2] GTK: http://www.gtk.org
[3] ITK: http://www.itk.org
[4] Dcmtk: http://www.dicom.offis.de/dcmtk.php.en/

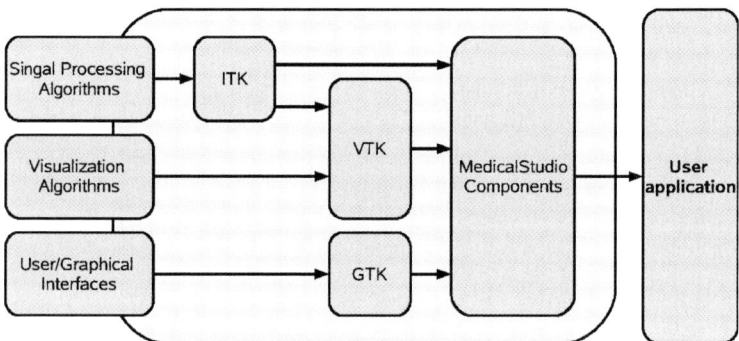

Fig. 12.2 MedicalStudio component-based architecture pipeline

Thanks to the modular architecture any other library can be linked into a new component providing high extensibility to MedicalStudio. We define a plug-in as a group of components compiled together into one shared library. These plug-ins can group components by any criteria but by convention it is preferable to group them either by type or procedure context. An xml file is created for each plug-in describing its content. With this, the kernel is able to create a repository of all available components without loading them into memory, plug-ins will be loaded lately, only when required letting resources available for data or processing. This facilitates the distribution process as only two files will be needed to add this procedure to the basic MedicalStudio platform. This allows easy customization of final applications. For example all the components for a specific neurosurgery procedure will be grouped into one plug-in. For applications that do not need the neurosurgery plug-in (e.g. maxillo-facial surgery), removing the two files is enough to eliminate the unwanted functionalities.

12.3 General Purpose Components

In the following part we will present examples of generic software components that can be found in MedicalStudio to develop more complex scenarios for the comprehensive support of interventions (planning, simulation, intraoperative guidance) and clinical research. This includes components for registration, segmentation and 3D reconstruction and augmented visualization. Thanks to the component-based architecture, while developing a new plug-in in MedicalStudio the programmer does not need to care about implementing basic tools such as 2D and 3D rendering, view organization and coordination, colours settings, transparency settings or performing operations such as scaling, rotations and zooming. All these functionalities as well as other plug-ins integrated into MedicalStudio can easily share all the data required for their execution.

12.3 General Purpose Components

12.3.1 Multimodal Registration

Currently MedicalStudio supports two kinds of multimodal registration: rigid and non-rigid.

1) Rigid Registration: the rigid registration algorithm relies on the Insight Registration and Segmentation Toolkit. It makes use of a Simultaneous Perturbation Stochastic Approximation of the cost function gradient to seek for the optimum. This method has first been introduced by Spall [471]. The estimate of the gradient is fast and more robust to the presence of local minima than classical gradient descent schemes. Such registration algorithm was validated in collaboration with the Radiation Oncology Dept., St-Luc University Hospital, Belgium. A group of fifteen patients with pharyngo-laryngeal tumours were imaged by CT scan, MRI (T1-and T2 weighed) and PET (transmission and FDG emission) with constrain masks. All these images were automatically registered using Mutual Information as criterion. Four of the fifteen patients were registered manually. The results from both methods were compared in terms of accuracy, reproducibility, robustness and speed. Maximum Euclidean deviations to the reference transformations were smaller than 1.7 mm for PET-CT registration and smaller than 4.7 mm for MR-CT registration. Furthermore, the automated method converged to validated results for clinical cases where experts failed using a manual registration. Automated registration needs 2 to 8 min on standard platforms. Figure 12.3 illustrates this method into the MedicalStudio interface.

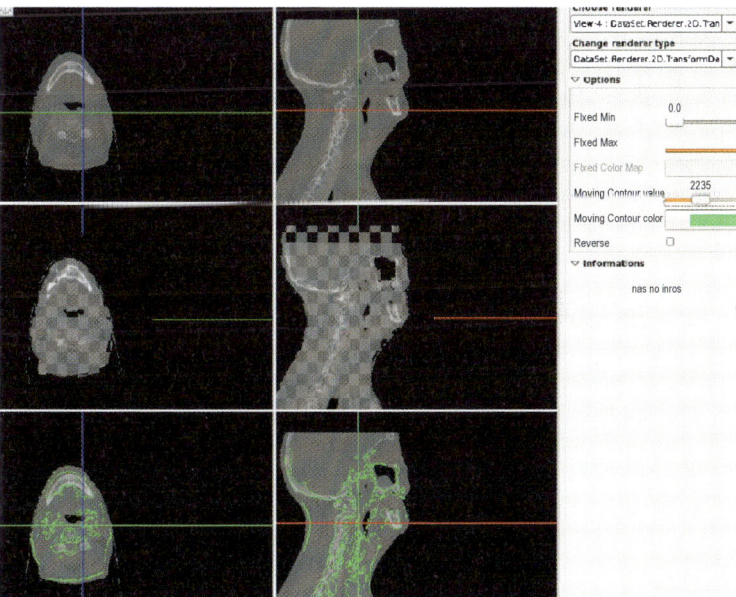

Fig. 12.3 Multimodal rigid-registration: CT data (upper viewer) PET data (middle viewer) Result (bottom viewer)

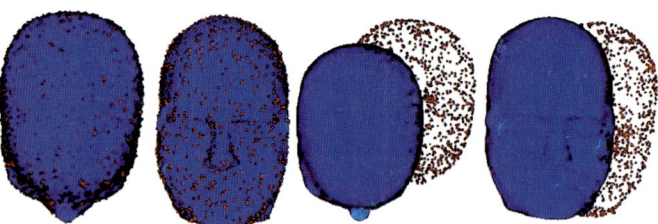

Fig. 12.4 Rigid registration using the surface-based algorithm described in [472]

Another kind of rigid registration uses the surface-based algorithm. It minimizes the mean square distance between the points representing the real object and the surface from segmented MRI images. For that we applied the implementation of Saito's EDT (Euclidean Distance transform) found in [472]. Results of this method are illustrated in Fig. 12.4. This method is reused by other specialized medicalstudio plug-ins such as the Transcranial Magnetic Stimulation (TMS) application described in [473] and the ACROGuide application described in section 4.

2) Non-rigid Registration: The use of the SPSA (Simultaneous Perturbation Stochastic Approximation) method has been investigated for optimizing a large set of parameters characterizing a non-rigid deformation. We use volumetric tetrahedral meshes as non-rigid deformation models. Their main advantage is the capability to deal with non-uniform sampling of the image domain and to allow multigrid representation of the deformation. The deformation is constrained by the linear elastic energy acting as a regularization term. This regularization term can allow more flexibility in some regions than in others by giving different mechanical properties to elements in different regions. The implementation allows running this algorithm in parallel on Symmetric Multi Processor architectures (SMP) by distributing independent computations of the cost functions to different threads. This work has been developed in collaboration with Brigham and Womens Hospital, Harvard Medical School, Computational Radiology Laboratory [474].

12.3.2 Segmentation and 3D Reconstruction

This plug-in contains various components allowing the segmentation and appropriate labelling of anatomical structures for 3D reconstruction. As well as methods for manual segmentation where borders are drawn directly onto the raw image dataset, one of the classic methods for performing the task is intensity based filtering from MRI or CT dataset using the Marching Cubes algorithm [475] and acting in the same way as thresholding segmentation. Figure 12.5 illustrates the result of this method. The correct reconstruction also requires connectivity filtering to extract cells that share common points and satisfy a scalar threshold criterion. Another algorithm implements automatic, atlas-guided segmentation [476] which is suitable in presence of deformed anatomy caused by tumours and operates through

12.4 Applications

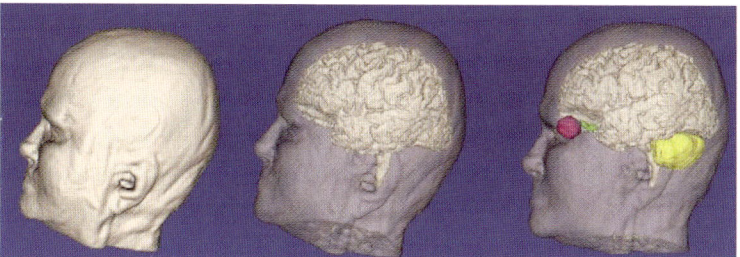

Fig. 12.5 Reconstructed 3D model from MRI data with segmented structures

a combination of rigid and non rigid registration components. The computed transformations map the atlas segmented structures onto the subject volume. The plug-in includes level set segmentation [477] with active contour modelling for boundary object detection letting an initial interface evolve towards the object boundary. The level set implementation automatically manages the topology changes of the evolving interface, allowing detection of several objects.

12.4 Applications

This section illustrates examples of medical applications developed within the MedicalStudio framework by making use of all the available plug-ins for IGS generic components as well as the basics tools for view management, visualization and manipulation of images and volumes.

The flowchart that represents the system designed by us to support maxillofacial surgery including pre-operative CT scanning and planning to surgical guidance is shown in Fig. 12.6. The process begins with the images acquisition using a CT scanner. Then, a threshold is used to filter the images, separating the bones and the other soft tissues. After this processing, the images are segmented and a 3D model of the skull is reconstructed from the segmented images. Using the reconstructed 3D model and the 3D Medical Assistant for Orthognatic Computer Surgery, a path-line representing the osteotomy path is then designed. The next step involves the hardware calibration and the registration of real and virtual objects in the ACROGuide component. This step consists of picking points over patient mandible using an optical tracker system and then registers them with the 3D surface reconstructed before from patient CT images.

12.4.1 3D Medical Assistant for Orthognatic Computer Surgery

This application contributes to the realization of a comprehensive orthognatic image guided surgery system aim to improve accuracy and performance of surgi-

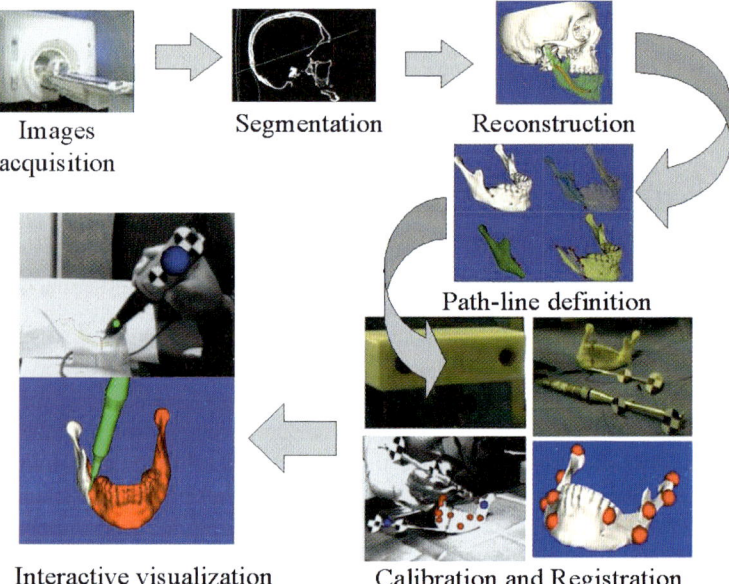

Fig. 12.6 System flowchart supported by MedicalStudio to provide surgical guidance during maxillofacial surgery

cal acts throughout the different phases involved (diagnosis, planning, simulation and surgery). A wizard is being developed to guide the surgeon during the 3D cephalometric analysis [478] (see Fig. 12.7). The analysis is divided into 16 modules which determine and quantify facial asymmetry, dentomaxillofacial dysmorphosis and cranio maxillo facial deformations. As the whole analysis uses 3D concepts, the interface allows the visualization and navigation in 2D (using the three orthogonal axes) and in 3D (using a polygonal reconstruction of the bones). Figure 12.7 shows one of the steps that the surgeon should do. It emphasizes the region of interest where the action is to take place. Effects of completed actions are highlighted and should be easy for the user to distinguish.

12.4.2 ACROGuide

In collaboration with the Service de Stomatologie et Chirurgie Maxillo-faciale at Saint Luc Hospital, in Brussels, we proposed the development of an application using a mixed reality interface to guide surgeons in this type of surgeries. The goal of this application is to increase the first surgery success, avoiding a second intervention. To achieve this objective, we are mixing real and virtual images by projecting a virtual guidance path-line on the patient mandible live video images. Then, the surgeon should cut the patient mandible paying attention to follow this path-line and avoiding touching the dental nerve [479].

12.5 Conclusions and Future Works

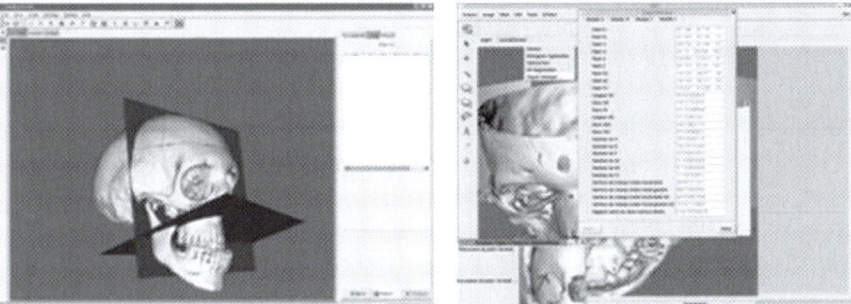

Fig. 12.7 Maxillofacial computer assisted planning. (left) Selected planes (right) Wizard for assistance

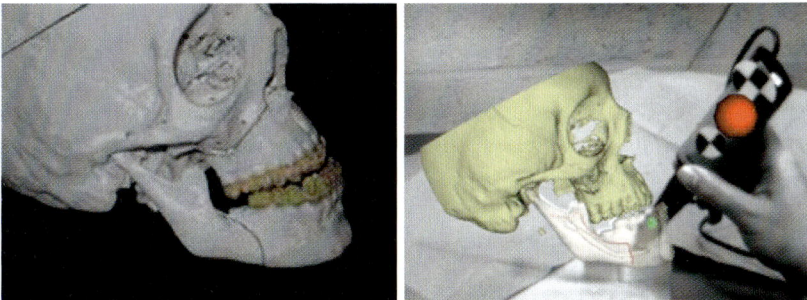

Fig. 12.8 (a) Head patient mock-up reproduced from 3D CT images. (b) Augmented reality view with guidance information projected into video image in MedicalStudio framework

When the markers are recognized by the system, two spheres are displayed on the screen: one attached to the tool; and the other, bigger, displayed in the middle of the tool (see Fig. 12.8). The smaller one indicates if the tooltip is placed (i.e. assuming the green colour) or not (i.e. assuming the blue colour) on the correct position indicated by the path-line (red line) projected into the image video. The big one indicates the distance between the tooltip and dental nerve using three colours: gray means go ahead; blue means pay attention (i.e., you are less than 3mm from the nerve); red means stop (i.e., you will probably touch the nerve).

12.5 Conclusions and Future Works

In this work we have presented MedicalStudio[5] which is a medical framework supporting several surgical guidance aspects. The system is suited for the processing of multimodal image data registration as well as augmented visualization. The

---

[5] MedicalStudio is actually available for download at http://www.medicalstudio.org.

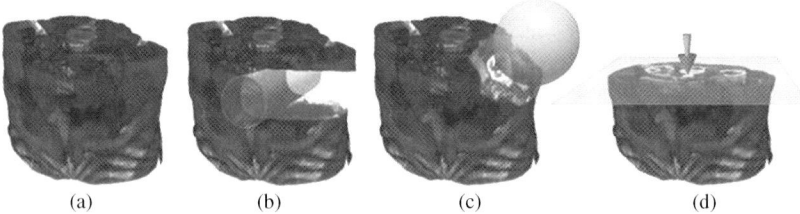

Fig. 12.9 Example of an interactive sculpting session, where the volume of interest (VOI) is a torso dataset. Visualization is shown for each tool: (a) original VOI, (b) VOI carved by the eraser tool, (c) VOI carved by the digger tool, and (d) VOI carved by the clipper tools

main advantages of this medical framework can be summarized as following: 1.Flexibility: the component-based architecture allows the integration of new components sharing resources offered by other plug-ins furnished by a large community of developers. In this case only the plug-ins used are loaded. 2. Interoperability: basic resources such as reading and visualizing support to several medical images formats guarantee the exchange of data between pre and intraoperative phases. 3. Benchmarking: having different approaches integrated as plug-ins allows to researchers compare results. 4. Cross-platform: the same applications scenarios can be compiled currently to Windows, Linux and MacOSX platforms. 5. Independent Validation: the component-based architecture favours independent validation of single modules. It is an advantage to reuse already validated components when building a new medical application.

On going work is focused on the integration of a new component for rendering using hardware graphic in collaboration with the Computer Graphics Laboratory at University Federal of Rio Grande do Sul. The support of graphics hardware for texture-based visualization allows efficient implementation of rendering techniques that can be combined with interactive sculpting tools to enable interactive inspection of 3D datasets (see Fig. 12.9) [480]. This integration will test the flexibility of MedicalStudio to support such kind of application.

Subsequent versions of MedicalStudio will include integration with PACS providing DICOM network compatibility and extend image support format such as electrocardiogram and ultrasound images which require a differentiated image processing treatment. Finally future works will address how advanced technologies such as further intra-operative devices, robotics and telecollaboration can be supported by this framework and integrated seamlessly into the operation room.

12.6 Acknowledgment

This work was supported by the SIMILAR Network of Excellence, by the Painter project (Waleo II: 0516051). The authors would also like to thank CNPq (Brazilian Council for Research and Development) for the financial support.

Chapter 13
Multimodal Interfaces for Laparoscopic Training

Pablo Lamata, Carlos Alberola, Francisco Sánchez-Margallo,
Miguel Ángel Florido and Enrique J. Gómez.
On behalf of the SINERGIA Project[1]

13.1 Introduction

Multimodal technologies are offering promising simulation paradigms for training different practitioners as surgeons. Traditional methods of acquiring surgical skill, using the apprenticeship model, haven't been able to accommodate the new skills required for laparoscopic surgery [480]. There is a need of training outside the operating room that can be satisfied with multimodal interactive virtual interfaces.

Surgical simulators offer a virtual environment similar to the operating theatre that enables a surgeon to practice different surgical procedures. Virtual environments offer numerous advantages over the traditional learning process: they can display a wide variety of different pathologic conditions; they allow a trainee to practice at any time and as many times as required; and finally they can monitor the skill in real-time, providing constructive feedback to the student as the teacher does. In short, multimodal virtual reality simulators are a valuable tool for training and skills assessment [480]. In addition, their permit direct comparison of performance between surgeons since different users can complete exactly the same task, without the effects of patient variability or disease severity.

As an example of the importance of multimodal interfaces in laparoscopic training this chapter describes a specific development: the SINERGIA simulator [481, 482]. This simulator is being developed by the SINERGIA Spanish cooperative network, which is composed by clinical and technical Spanish partners around Spanish geography. We describe in this Chapter the first prototype (PT1) of the system, which is being developed. This research departs from the work of MedICLab (Polytechnic University of Valencia), where an early prototype was developed [483].

13.2 Functionality

Surgical trainees interact with the simulator using two sensorial channels: visual and haptic. A monitor displays the virtual surgical scene, as it may be captured with a laparoscope, providing the visual interface. On the other hand a haptic feedback device, the Laparoscopic Surgical Workstation (Immersion Corp., San Jose, U.S.A.), simulates the haptic interaction of the surgeon. This device provides force feedback to the user, who can feel the consistency of virtual organs.

The simulator is task-oriented, i. e. there are different particular tasks which the user is meant to perform with the system. These tasks are grouped in seven didactic units: hand-eye co-ordination, navigation, grasping, pulling, cutting, dissection and suturing.

Each didactic unit has been designed in based on a specific learning goal, i. e. the co-ordination didactic unit has its main purpose the training of the tools handling while the surgeon is watching the monitor (in contrast with open surgery, in which the surgeon views directly the surgical scene). In the case of the suture didactic unit, the implemented task helps the surgeon to learn the knotting process, task considered essential by the clinicians. Other tasks will be explained in detail section 13.3.

13.3 Technical Issues

To simulate the interaction of the different elements in the virtual scene, new collision detection and handling algorithms have been development. The imple-

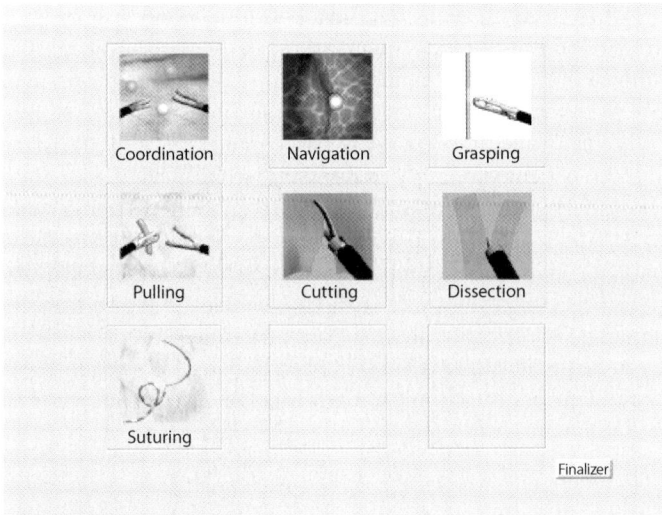

Fig. 13.1 Simulator User Interface

13.3 Technical Issues

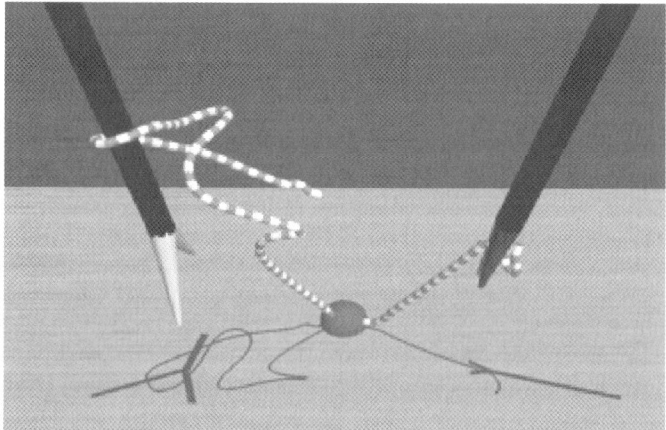

Fig. 13.2 Example of multimodal interface: Suture Simulation

mented method uses volumetric information of the tool for the collision response of organ-tool collision, in contrast to most surgical simulators, in which the interaction is assumed to occur at a single point [484]. In case of simulating the dynamics elements, a superficial model models the soft tissue behavior. For the thread, the model proposed in [485] for microsurgery has been implemented.

13.3.1 Simulator Architecture

13.3.1.1 Hardware Architecture

The hardware components of the systems and the relations between them can be observed in Fig. 13.3:

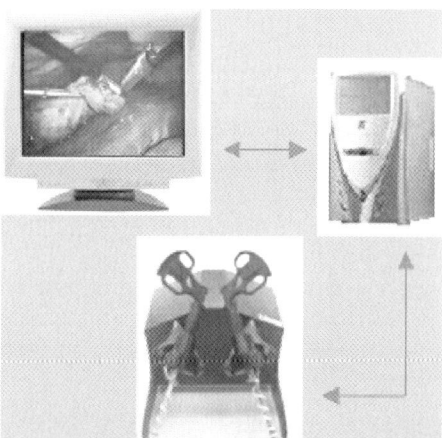

Fig. 13.3 Components of the simulator: haptic feedback device, the monitor and workstation

- Haptic Feedback devices (LSW of Immersion): This device provides force feedback to the user.
- Monitor: displays the virtual surgical scene, as it may be capture with a laparoscope in the operating room.
- Workstation: with controls both user interfaces and their interaction.

13.3.1.2 Software Architecture

The simulator is composed by four main modules controlled by the main program, the architecture can be observed in Fig. 13.4:

- Biomechanical model: it calculates the deformation of organs in the virtual scene. A T2-Mesh model [486] is used, which is a kind of spring-mass model that approaches objects by a triangular surface mesh.
- Collision module: calculates the interaction between the virtual models.
- Visual motor: represents the geometry in the visual device (screen).
- Haptic motor: reads the positions of the haptic device and returns the haptic forces to the user.

All these components are controlled by the simulator main program that executes the following steps is its main simulation loop:

1. Read the positions of the tools represented by the haptic device.
2. Calculate the collisions between the elements in the scene and the response of these collisions.
3. Calculate de deformation of the deformable models that represents the organs.
4. Display the new geometry and the reaction forces resulted from the deformation process.

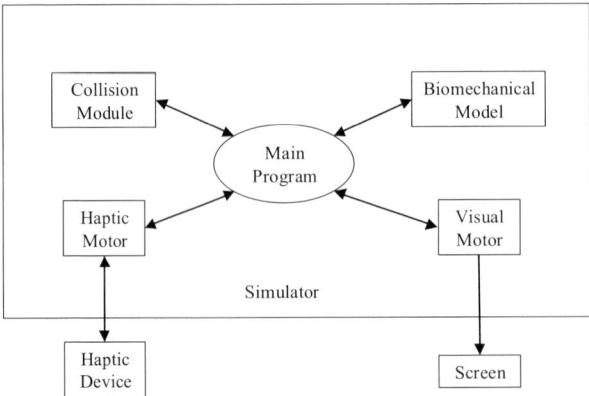

Fig. 13.4 Sinergia simulator architecture

13.3.2 Collision Detection and Handling

Surgical simulators are complex computer applications in which many agents (organs, tools etc.) must represent scenes that look like those that a surgeon may encounter in daily practice. These scenes, however, are not static snapshots of reality, but dynamical scenarios in which the above mentioned agents move according to its own moving patterns and this motion has to be realistic. This statement may seem trivial at first glance; we take for granted that if a virtual environment depicts a scissor with its two cutting edges approaching each other, and a string is found in the way of this closing scissor, the string will be cut apart. Nevertheless, unless a collision is found and it is properly dealt with, i. e., unless the system detects that an object is located between the two edges, the objects is declared softer than the edges, and the model of the object allows for topological changes (the single string becomes now two independent strings) such a cutting operation will not be carried out, and both the scissor and the string would gently share a virtual space, but with no interaction at all between them. Therefore, in order for realistic interactions to be attained, a module called collision detection and management performs an outstanding operation.

In Sinergia we have relied on available software [487] to deal with the collision detection problem. However, this software does not work efficiently if topological changes occur in the deformable model, a situation that arises when cutting or dissection operations are carried out. In order for these operations to be manageable with our simulator without loss of efficiency, a collision detection library has been developed by one of the partners (MedicLab) that tests geometrically the interaction between tools (rigid objects) and deformable objects, and does manage topological changes. Our major effort has been focused on collision management. Basically, collision management can be divided into methods that rely on the biomechanical model of the objects to be handled and those that do not. As for the former the object new position and shape, as a result of a collision, is controlled by the model equations, with restrictions (or penalties) given by the kinematics of

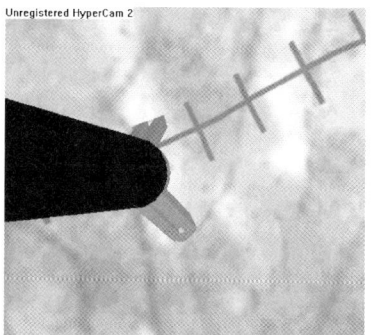
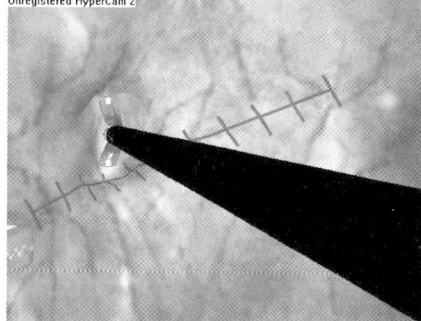

Fig. 13.5 Sinergia collision detection example

the collision. As for the latter, and assuming that the objects are modeled by means of a triangular mesh, the problem is posed as finding the new positions of the vertices of the triangles involved in the collision. Then, the biomechanical model will calculate how the deformations that stem from the collision propagate throughout the object.

If real time constraints are an issue, collision management based on biomechanical models may be somewhat slow, due to the amount of calculations to be carried out. Since this is our case, we have resorted to a management that finds the new vertices positions. Within this approach many proposals found in the literature [488, 489] deal with only a single point in the surgical tool. This may lead to unrealistic effects; for instance, the leftmost image below, has been obtained using this simplistic management. Clearly, the upper part of the tool has apparently gone through the simulated tissue.

This is not the case when more points are handled (see rightmost image). In this case the whole tool is pushing the tissue down, which is the effect sought in the simulation.

Details of the algorithm will not be described here; however it is worth mentioning that the algorithm is based on the tool kinematics (the tool velocity vector) and the normals to the triangles involved in the collision. As an illustration we show how we handle a common situation, which coincides with the one indicated above: the surgical tool pushes a tissue down (without tearing it apart). In this case **Vm** indicates the velocity vector of the tool. Clearly from the picture, vertices of the triangles involved in the collision are moved to the face of the tool whose normal vector is parallel to the surface vector of the tissue. Then, the biomechanical model of the organ recalculates how this deformation affects nearby triangles. It could also deal with possible break-ups of the tissue.

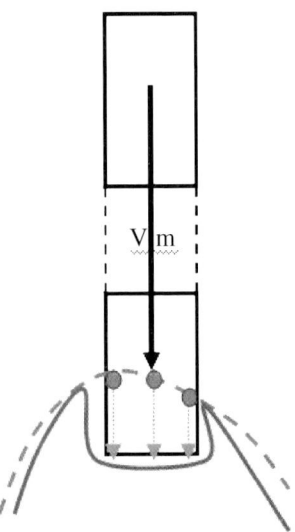

Fig. 13.6 Collision detection algorithm

13.3 Technical Issues

Many other situations like the one encounter here may be found during the simulation. Those identified by us are properly dealt with in a similar manner as the one just described.

13.3.3 Design of Multimodal Interface Scenarios and Surgical Simulation Tasks

Hands-on practice in a virtual environment is an effective way of building and maintaining laparoscopic surgery skills. Surgical scenarios suppose an important contribution in that training, and contribute towards a highly realistic environment that familiarizes novice surgeons with the real surgical scene [490].

Moreover, the design of simple scenarios to enhance or teach some surgical skills (dissection, suturing, etc.) allows training surgeons to develop some abilities before a laparoscopic intervention. In our project we have developed some simple scenarios to enhance some abilities:

a) Camera Navigation: Handling the camera is the first skill the aspiring surgeon has to learn. In this exercise we develop some basic navigation skills and an initial feeling for how laparoscopic instruments have to be used. The student has to find the green ball in the scene, zoom up and match its size to the on-screen circle. Then the ball will disappear, and another ball will be presented to repeat the exercise. In Fig. 13.7a, a screenshot of this exercise is represented.

b) Coordinated Navigation: Using both hands in a coordinated manner is a very important ability for laparoscopic. In this exercise, the student has to focus, with one hand, the green ball that appears in the virtual environment, and touch it with the right-hand instrument. The figure that represents this scenario is similar to the previous exercise (Fig. 13.7b).

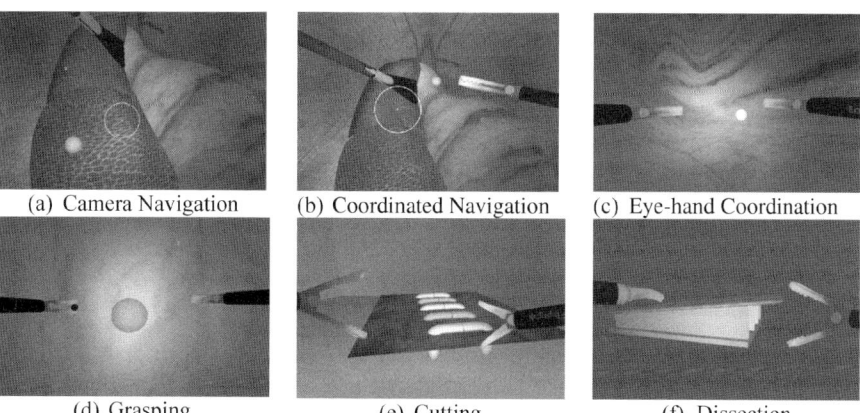

(a) Camera Navigation   (b) Coordinated Navigation   (c) Eye-hand Coordination

(d) Grasping   (e) Cutting   (f) Dissection

Fig. 13.7 Multimodal Interface Surgical Scenarios

c) Eye-hand coordination: Coordinate the scene that the surgeon sees, and use the laparoscopic instruments, is another important skill. In this exercise, the student has to touch color balls with the left-hand, right-hand, or both instruments based on theirs colors (Fig. 13.7c).
d) Grasping: In this exercise the student practices the grasp coordinated ability. The objective of the exercise is to grasp, stretch and move the objects that appear into a sphere (Fig. 13.7d).
e) Cutting: This exercise develops the ability of the student to cut with the laparoscopic instruments. The objective is to select a cylinder, stretch, and cut it (Fig. 13.7e).
f) Dissection: The objective is to teach the dissection procedure. The student has to take a piece of the object with one instrument, stretch and dissect it with the other tool (Fig. 13.7f).

13.4 Research Issues

Virtual Reality (VR) technologies are opening up new ways of training laparoscopic surgeons. Skills learnt with simple VR laparoscopic simulators can be transferred to the Operating Room (OR) environment [491]. But little is known about the actual requirements of simulators in order to be an effective training tool. It is widely believed that a higher fidelity in simulation will improve training [492], but increased fidelity may not always lead to an enhanced training capability [493]. Studies are needed to clarify the relationship between fidelity and training effectiveness [494]. This research issue is being addressed in two complementary ways, studying the sensory interaction of surgeons and proposing a conceptual framework for the analysis, design and evaluation of simulators.

13.4.1 Study of Laparoscopic Sensory Interaction

A methodology for surgeon sensory interaction characterization has been defined and applied to a surgical skill: tissue consistency perception. The aim is to determine the relative importance of three components of a perceptual surgical skill: visual cues, haptic information, and previous surgical knowledge and experience [495, 496].

Defined methodology for surgeon sensory interaction characterization has provided consistent and useful results about tissue consistency perception. Statistical differences were found when perceiving consistency depending on the tissue, but not depending on the expertise level of surgeons. Nevertheless results suggest that surgeons improve their sensory perception capabilities in an early stage, but also build a mental representation of tissue consistencies after years of experience. This is interpreted to be the "visual haptics" concept which could affect a lost in attention on visual and tactile stimuli. It has been assessed the higher importance of tactile stimuli over visual cues in this skill, which suggest that only little sensory substitution might be present.

All these results contribute to the definition of simulation requirements. They suggest that VR simulators need haptic devices with force feedback capability if consistency information is to be delivered. Moreover, a simple model with only some discrete levels of forces seems to be enough to calculate these forces. This model could be parameterized by two main variables: the kind of tissue and the degree of fixation to other organs or to the abdominal cavity.

13.4.2 *Laparoscopic Simulation Conceptual Framework*

A conceptual framework for the design, analysis and evaluation of simulation technologies is being developed in collaboration with the Dept. of Surgical Oncology and Technology (Imperial College of London). The aim is to propose a methodology to identify which VR didactic resources are important to achieve predefined training objectives [497].

VR didactic resources have been defined and classified in three main categories: Fidelity, Virtual and Evaluation resources. Fidelity resources refer to the different levels of realism offered by a simulator in its interaction and behavior. They can be further divided into sensorial, mechanical and physiological. Virtual resources are features unique to a computer simulated environment that can enhance training, like cues and instructions given to the user to guide a task, or to manage a training program. Evaluation resources are metrics to evaluate performance, follow up progress and ways to deliver constructive feedback to the user.

Each category is defined, studied and compared in different commercial laparoscopic simulators using a pre-defined criterion. Results contribute to the definition of simulation requirements and offers guidelines to formulate hypotheses about the importance of different didactic resources.

13.5 Conclusion

VR surgical simulation is useful for training surgical skills without requiring high level of realism and it could be considered a good method to evaluate surgeon's performance, by means of objective metrics which allow them to follow up progress giving a constructive feedback to the user.

Multimodal interfaces are essential in surgical simulation due to the multiple source of information used by surgeon (visual, haptic) in the interventions. An appropriate training in surgical techniques requires a complete learning and understanding of all these sources from early training stages.

13.6 Acknowledgement

Authors would like to express their gratitude to all members of the SINERGIA Thematic Collaborative Network (G03/135) funded by the Spanish Ministry of Health. Special thanks to all the people that contributed in this chapter: Samuel Rodríguez, Pablo J. Figueras from GBT-UPM, Verónica García, Emma Muñoz from LPI-UVA, Diana Suárez, Manuel Bernal, Juan Ruiz from ULPGC, Óscar López, Carlos Montserrat, Mariano Alcañiz from MediClab, J. Blas Pagador, Jeús Usón from CCMI, Salvador Pascual from HSMP.

References

1. T. Butz and J.-Ph. Thiran, "From error probability to information theoretic (multi-modal) signal processing", Signal Processing, vol. 85, no. 5, pp. 875–902, 2005.
2. Bolt, R. A: Put-that-there: Voice and Gesture at the Graphics Interface. Proceedings of the 7th Annual Conference on Computer Graphics and Interactive Techniques, Seattle, 1980, 262–270.
3. Hovy, E. and Arens, Y.: When is a Picture Worth a Thousand Words ? Allocation of Modalities in Multimedia Communication. Paper presented at the AAAI Symposium on Human-Computer Interfaces, Stanford, 1990.
4. Oviatt, S. and Cohen, P.: Multimodal Interfaces That Process What Comes Naturally. Communications of the ACM, 43/3, 2000, 45–53.
5. Bernsen, N. O., and Dybkjær, L. (in press (a)): Multimodal Usability.
6. Bernsen, N. O.: Foundations of Multimodal Representations. A Taxonomy of Representational Modalities. Interacting with Computers 6.4, 1994, 347–71.
7. Bernsen, N. O.: Multimodality in Language and Speech Systems – From Theory to Design Support Tool. In Granström, B., House, D., and Karlsson, I. (Eds.): Multimodality in Language and Speech Systems. Dordrecht: Kluwer Academic Publishers 2002, 93–148.
8. Jurafsky, D. and Martin, J. H.: Speech and Language Processing. An Introduction to Natural Language Processing, Computational Linguistics, and Speech Recognition. Prentice Hall, 2000.
9. Bernsen, N. O. Why are Analogue Graphics and Natural Language Both Needed in HCI? In Paterno, F. (Ed.), Design, Specification and Verification of Interactive Systems. Proceedings of the Eurographics Workshop, Carrara, Italy, 1994, 165–179. Focus on Computer Graphics. Springer Verlag, 1995: 235–51.
10. Lakoff, G.: Women, Fire, and Dangerous Things: What Categories Reveal about the Mind. Chicago: University of Chicago Press 1987.
11. Rosch, E. Principles of Categorization. In Rosch, E. and Lloyd, B. B. (Eds.). Cognition and Categorization. Hillsdale, NJ: Erlbaum, 1978.
12. Bernsen, N. O., and Dybkjær, L. (in press (b)): Annotation Schemes for Verbal and Nonverbal Communication: Some General Issues.
13. McNeill, D: Hand and Mind. University of Chicago Press, 1992.
14. Bernsen, N. O. Towards a Tool for Predicting Speech Functionality. Speech Communication 23, 1997, 181–210.
15. Baber, C., and Noyes, J. (Eds.). Interactive Speech Technology. London: Taylor & Francis, 1993.

16. Bernsen, N. O., and Dybkjær, L.: Working Paper on Speech Functionality. Esprit Long-Term Research Project DISC Year 2 Deliverable D2.10. University of Southern Denmark, 1999a. See www.disc2.dk.
17. Bernsen, N. O., and Dybkjær, L.: A Theory of Speech in Multimodal Systems. In Dalsgaard, P., Lee, C.-H., Heisterkamp, P., and Cole, R. (Eds.). Proceedings of the ESCA Workshop on Interactive Dialogue in Multi-Modal Systems, Irsee, Germany. Bonn: European Speech Communication Association, 1999b 105–108.
18. Bernsen, N. O., and Dybkjær, L.: Report on Iterative Testing of Multimodal Usability and Evaluation Guide. SIMILAR Deliverable D98, October 2007.
19. Moustakas, K., Nikolakis, G., Tzovaras, D., Deville, B., Marras, I. and Pavlek, J.: Multimodal Tools and Interfaces for the Intercommunication between Visually Impaired and "Deaf and Mute" People. eNTERFACE'06, July 17th – August 11th, Dubrovnik, Croatia, Final Project Report, 2006.
20. Martin, J.-C.: Cooperations between Modalities and Binding Through Synchrony in Multimodal Interfaces. PhD Thesis (in French). ENST, Orsay, France 1995.
21. Nigay, L. and Coutaz, J.: A Design Space for Multimodal Systems: Concurrent Processing and Data Fusion. International Conference on Human-Computer Interaction. ACM Press, 1993, 172–178.
22. J.W. Fisher III and J.C. Principe, "A methodology for information theoretic feature extraction," in World Congress on Computational Intelligence, Anchorage, USA, March 1998.
23. J.C. Principe, D. Xu, and J.W. Fisher III, "Learning from examples with information theoretic criteria," Journal of VLSI Signal Processing Systems, vol. 26, pp. 61–77, 2000.
24. T. Darrell, J.W. Fisher III, P. Viola, and W. Freeman, "Audio-visual segmentation and the 'cocktail party effect'," in International Conference on Multimodal Interfaces, Beijing, China, pp. 32–40, 2000.
25. A.O. Hero, B. Ma, and O. Michel, "Imaging applications of stochastic minimal graphs," in IEEE International Conference on Image Processing, Thessaloniki, Greece, pp. 573–576, 2001.
26. J.C. Principe, J.W. Fisher III, and D. Xu, Unsupervised Adaptive Filtering, chapter 7: "Information Theoretic Learning," John Wiley & Sons, 2000.
27. W.M. Wells III, P. Viola, H. Atsumi S. Nakajima, and R. Kikinis, "Multimodal volume registration by maximization of mutual information," Medical Image Analysis, vol. 1, no. 1, pp. 35–51, March 1996.
28. F. Maes, A. Collignon, D. Vandermeulen, G. Marchal, and P. Suetens, "Multimodality image registration by maximization of mutual information," IEEE Transactions on Medical Imaging, vol. 16, no. 2, pp. 187–198, April 1997.
29. C. Studholme, D.J. Hawkes, and D.L.G. Hill, "A normalised entropy measure for multimodality image alignment," in SPIE Medical Imaging: Image Processing, San Diego, USA, 1998, pp. 132–142.
30. Roche, G. Malandin, X. Pennec, and N. Ayache, "The correlation ratio as a new similarity measure for multimodal image registration," in Medical Image Computing and Computer-Assisted Intervention, Cambridge, USA, pp.1115–1124, 1998.
31. E. Gokcay and J.C. Principe, "Information theoretic clustering," IEEE Transactions on Pattern Analysis and Machine Intelligence, vol.24, no.2, pp.158–171, February 2002.
32. T. Butz and J.-Ph. Thiran, "From error probability to information theoretic (multi-modal) signal processing," Signal Processing, vol.85, no.5, pp.875–902, 2005.
33. T.M. Cover and J.A. Thomas, Elements of Information Theory, John Wiley & Sons, 1991.
34. R.M. Fano, Transmission of Information: A Statistical Theory of Communication, MIT Press and John Wiley & Sons, 1961.
35. Renyi, Probability Theory, Elsevier Publishing Company, 1970.
36. D. Erdogmus and J.C. Principe, "Information transfer through classifiers and its relation to probability of error," in International Joint Conference on Neural Networks, Washington D.C., 2001.
37. L. Devroye and L. GyÄorfi, Non-parametric Density Estimation, John Wiley & Sons, 1985.

38. L. Devroye, A Course in Density Estimation, BirkhÄauser, 1987.
39. T. Butz and J.-Ph. Thiran, "Multi-modal signal processing: An information theoretical framework," Tech. Rep. 02.01, Signal Processing Institute (ITS), Swiss Federal Institute of Technology (EPFL), 2002.
40. T. Butz, O. Cuisenaire, and J.-Ph. Thiran, "Multi-modal medical image registration: From information theory to optimization objective," in IEEE International Conference on Digital Signal Processing, Santorini, Greece, vol.I, pp.407–414, July 2002.
41. Studholme, D.J. Hawkes, and D.L.G. Hill, "An overlap invariant entropy measure of 3D medical image alignment," Pattern Recognition, vol. 32, pp. 71–86, 1999.
42. M. Holden, D.L.G. Hill, E.R.E. Denton, J.M. Jarosz, T.C.S. Cox, and D.J. Hawkes, "Voxel similarity measures for 3D serial MR brain image registration," in Information Processing in Medical Imaging, Visegrd, Hungary, vol. 1613, pp. 466–471, 1999.
43. Roche, G. Malandain, and N. Ayache, "Unifying maximum likelihood approaches in medical image registration," Tech. Rep. 3741, Inst. National de Recherche en Informatique et en Automatique, Sophia Antipolis, July 1999.
44. D.E. Goldberg, Genetic Algorithms in Search, Optimization, and Machine Learning, Addison-Wesley Publishing Company, 1989.
45. M. Wall, GAlib 2.4.5: A C++ Library of Genetic Algorithm Components, Massachusetts Institute of Technology, 1999.
46. W. Gropp, E. Lusk, and A. Skjellum, Using MPI: Portable Parallel Programming With the Message-Passing Interface, MIT Press, 2nd edition, 1999.
47. MPI Forum, A Message-Passing Interface Standard, University of Tennessee, 1999.
48. J. West, J.M. Fitzpatrick, M.Y. Wang, B.M. Dawant, C.R. Maurer Jr., R.M. Kessler, R.J. Maciunas, Ch. Barillot, D. Lemoine, A. Collignon, F. Maes, P. Suetens, D. Vandermeulen, P.A. van den Elsen, S. Napel, T.S. Sumanaweera, B. Harkness, P.F. Hemler, D.L.G. Hill, D.J. Hawkes, C. Studholme, J.B.A. Maintz, M.A. Viergever, G. Malandain, X. Pennec, M.E. Noz, G.Q. Maguire Jr., M. Pollack, Ch.A. Pelizzari, R.A. Robb, D. Hanson, and R.P. Woods, "Comparison and evaluation of retrospective intermodality brain image registration techniques," Journal of Computer Assisted Tomography, vol. 21, pp. 554–566, 1997.
49. D.L. Collins, A.P. Zijdenbos, V. Kollokian, J.G. Sled, N.J. Kabani, C.J. Holmes, and A.C. Evans, "Design and construction of a realistic digital brain phantom," IEEE Transactions on Medical Imaging, vol. 17, no. 3, pp. 463–468, June 1998.
50. R.K.-S. Kwan, A.C. Evans, and G.B. Pike, "MRI simulation-based evaluation of image-processing and classification methods," IEEE Transactions on Medical Imaging, vol. 18, no. 11, pp. 1085–1097, November 1999.
51. P. Viola and W.M. Wells III, "Alignment by maximization of mutual information," in 5th International Conference on Computer Vision, 1995, pp. 16–23.
52. T. Butz and J.-Ph. Thiran, "Affine registration with feature space mutual information," in Medical Image Computing and Computer-Assisted Intervention, Utrecht, The Netherlands, pp. 549–556, 2001.
53. G. Gerig, O. KÄubler, R. Kikinis, and F.A. Jolesz, "Nonlinear anisotropic Filtering of MRI data," IEEE Transactions on Medical Imaging, vol. 11, no. 2, pp. 221–232, June 1992.
54. E. Solanas and J.-Ph. Thiran, "Exploiting voxel correlation for automated MRI bias field," in Medical Image Computing and Computer-Assisted Intervention, Utrecht, The Netherlands, pp. 1220–1221, 2001.
55. Likar, M.A. Viergever, and F. Pernus, "Retrospective correction of MR intensity inhomogeneity by information minimization," in Medical Image Computing and Computer-Assisted Intervention, Pittsburgh, PA, pp. 375–384, 2000.
56. J.W. Fisher III, T. Darrell, W.T. Freeman, and P. Viola, "Learning joint statistical models for audio-visual fusion and segregation," in Advances in Neural Information Processing Systems, Denver, USA, November 2000.
57. T. Butz and J.-Ph. Thiran, "Feature space mutual information in video-speech sequences," in IEEE International Conference on Multimedia and Expo, Lausanne, Switzerland, vol. II, pp. 361–364, August 2002.

58. M. Unser, A. Aldroubi, and M. Eden, "B-spline signal processing: Part I – Theory," IEEE Transactions on Signal Processing, vol. 41, no. 2, pp. 821–833, February 1993.
59. M. Unser, A. Aldroubi, and M. Eden, "B-spline signal processing: Part II – E±cient design and applications," IEEE Transactions on Signal Processing, vol. 41, no. 2, pp. 834–848, February 1993.
60. T. Butz and J.-Ph. Thiran, "Feature-space mutual information for multi-modal signal processing, with application to medical image registration," in XI. European Signal Processing Conference, Toulouse, France, vol. I, pp.3–10, September 2002.
61. J.N. Kapur and H.K. Kesavan, Entropy Optimization Principles with Applications, Academic Press, 1992.
62. Oviatt, S. and Wahlster, W (1997). Special issue on Multimodal Interfaces. Human Computer Interaction, 12, 1&2.
63. Bézivin, J., Dupé, G., Jouault, F., Pitette, G., and Rougui, J. (2003) First Ex-periments with the ATL Transformation Language: Transforming XSLT into Xquery. In: Proc. of OOPSLA Workshop on Transformations (Anaheim, 2003), http://www.softmetaware.com/oopsla2003/bezivin.pdf.
64. Oviatt, S., Darrell, T., Flikner, M. (2004) Multimodal Interfaces that Flex, Adapt, and Persist. Communication of the ACM, 47, 1: 30–33.
65. Browne, D., Totterdell, P., and Norman, M. (1990) Adaptive User Interfaces, Academic Press, Computer and People Series.
66. Furtado, E., Furtado, J.J.V., Silva, W.B., Rodrigues, D.W.T., Taddeo, L.S., Limbourg, Q., and Vanderdonckt, J. (2001) An Ontology-Based Method for Universal Design of User Interfaces. In: Proc. of Workshop on Multiple User Interfaces over the Internet: Engineering and Applications Trends MUI'2001 (Lille, 10 September 2001), Seffah A., Radhakrishnan T., and Canals G. (eds.), http://www.cs.concordia.ca/~faculty/seffah/ihm2001/papers/furtado.pdf.
67. Berti, S. and Paternò, F. (2000) Migratory multimodal interfaces in multidevice environments. In: Proc. of 7th ACM Int. Conf. on Multimodal Interfaces ICMI'2005 (Trento, October 4–6, 2005), ACM Press, New York, pp. 92–99.
68. Thevenin, D. and Coutaz, J. (1999) Plasticity of User Interfaces: Framework and Research Agenda. In: Proceedings of IFIP TC 13 Conf. on Human-Computer Interaction Interact'99 (Edinburgh, September 1999), Sasse A & Johnson Ch. (eds.), IOS Press, Amsterdam, pp. 110–117.
69. Calvary, G., Coutaz, J., Thevenin, D., Limbourg, Q., Bouillon, L., and Vanderdonckt, J. (2003) A Unifying Reference Framework for Multi-Target User Interfaces. Interacting with Computers, 15,3: 289–308.
70. Bouchet, J., Nigay, L. (2004) ICARE: A Component-Based Approach for the Design and Development of Multimodal Interfaces. In: Proceedings of the ACM Conference on Human factors in computing systems CHI'2004 (Vienna, April 24–29, 2004), Extended abstracts, ACM Press, New York, pp. 1325–1328.
71. Coutaz, J., Nigay, L., Salber, D., Blandford, A., May, J., and Young, R. (1995) Four Easy Pieces for Assessing the Usability of Multimodal Interaction: The CARE properties. In: Proc. of the IFIP TC 13 Int. Conf. on Human-Computer Interaction INTERACT'95 Lillehammer, 27–29 June 1995), Nordbyn, K., Helmersen, P.H., Gilmore, D.J., Arnesen S.A. (eds.), Chapman & Hall, London, pp. 115–120.
72. Coutaz, J. (2006) Meta-User Interfaces for Ambient Spaces. In: Proc. of 4th Int. Workshop on Task Models and Diagrams for User Interface Design Tamodia'2006 (Hasselt, October 23–24, 2006). Lecture Notes in Computer Science, Vol. 4385, Springer, Heidelberg, pp. 1–15.
73. Dowell, J. and Long, J. (1989) Toward a conception for an engineering discipline of human factors. Ergonomics 32, 11: 1513–1535.
74. Stanciulescu, A., Limbourg, Q., Vanderdonckt, J., Michotte, B., and Montero, F. (2006) A Transformational Approach for Multimodal Web User Interfaces based on UsiXML. In:

Proc. of 7th ACM Int. Conf. on Multimodal Interfaces ICMI'2005 (Trento, 4–6 October 2005), ACM Press, New York, 2005, pp. 259–266.
75. Nigay, L. and Coutaz, J. (1993) A Design Space for Multimodal Systems: Concurrent Processing and Data Fusion. In: Proceedings of ACM Conf. on Human Factors in Computing Systems InterCHI'93 (Amsterdam, April 24–29, 1993), ACM Press, New York, pp. 172–178.
76. Rousseau, C., Bellik, Y., Vernier, F., and Bazalgette, D. (2006) A Framework for the Intelligent Multimodal Presentation of Information, Signal Processing, 88, 12: 3696–3713.
77. Rousseau, C., Bellik, Y., and Vernier, F. (2005) Multimodal output specification/simulation platform: In: Proc. of 7th ACM Int. Conf. on Multimodal Interfaces ICMI'2005 (Trento, October 4–6, 2005), ACM Press, New York, pp. 84–91.
78. Myers, B., Hudson, S.E., and Pausch, R. (2000) Past, Present, and Future of User Interface Software Tools. ACM Transactions on Computer-Human Interaction, 7,1: 3–28.
79. Bolt, R.A. (1980) Put-that-there: Voice and gesture at the graphics interface. Computer Graphics, 14(3): 262–270.
80. Balme, L., Demeure, A., Barralon, N., Coutaz, J., and Calvary, G. (2004) CAMELEON-RT: A Software Architecture Reference Model for Distributed, Migratable, and Plastic User Interfaces. In: Proceedings of 2nd European Symposium on Ambient Intelligence EUSAI'2004 (Eindhoven, November 8–11, 2004), Lecture Notes in Computer Science, Vol. 3295, Springer, Heidelberg, pp. 291–302.
81. Dragicevic, P. and Fekete, J.-D. (2004) ICON: The Input Configurator Toolkit: towards High Input Adaptability in Interactive Applications. In: Proceedings of the ACM Working Conference on Advanced Visual Interfaces AVI 2004 (Gallipoli, May 25–28, 2004), ACM Press, New York, pp. 244–247.
82. Oviatt, S. (2002) Multimodal interfaces. In: Handbook of human-computer interaction, Jacko J. and Sears A. (eds.), Lawrence Erlbaum, Mahwah.
83. Oviatt, S. (1999) Ten myths of multimodal interaction, Communications of the ACM, 42, 11: 74–81.
84. Palanque, Ph. and Schyn, A. (2003) A Model-Based Approach for Engineering Multimodal Interactive. In: Proc. of 9th IFIP TC13 Int. Conf. on Human-Computer Interaction Interact'2003 (Zurich, September 1–5, 2003), IOS Press, Amsterdam, pp. 543–550.
85. Sinha, A.K. and Landay, J.A. (2003) Capturing User Tests in a Multimodal, Multidevice Informal Prototyping Tool. In: Proceedings of the 5th ACM International Conference on Multimodal Interfaces ICMI'2003 (Vancouver, November 5–7, 2003), ACM Press, New York, pp. 117–124.
86. Stanciulescu, A. and Vanderdonckt, J. (2006) Design Options for Multimodal Web Applications. In: Proc. of 6th Int. Conf. on Computer-Aided Design of User Interfaces CADUI'2006 (Bucharest, June 6–8, 2006), Springer, Berlin, pp. 41–56.
87. Vanderdonckt, J. (2005) A MDA-Compliant Environment for Developing User Interfaces of Information Systems. In: Proc. of 17th Conf. on Advanced Information Systems Engineering CAiSE'05 (Porto, June 13–17, 2005), Pastor O. & Falcão e Cunha J. (eds.), Lecture Notes in Computer Science, Vol. 3520, Springer, Heidelberg, pp. 16–31.
88. Fitzmaurice, G., Ishii, H., and Buxton, W. (1995) Bricks: Laying the foundations for graspable user interfaces. In: Proceedings of the ACM Conference on Human Factors in Computing Systems CHI'95 (Denver, May 07–11, 1995). ACM Press, New York, pp. 442–449.
89. Kurtev, I., Bézivin, J., and Aksit, M. (2002) Technological Spaces: an Initial Appraisal. CoopIS, DOA'2002 Federated Conferences, Industrial track, Irvine, 2002, http://www.sciences.univ-nantes.fr/lina/atl/www/papers/Position PaperKurtev.pdf.
90. Cockton, G. (2004) From Quality in Use to Value in the World. In: Proceedings of the ACM Conference on Human factors in computing systems CHI'2004 (Vienna, April 24–29, 2004), Extended abstracts, ACM Press, New York, pp. 1287–1290.
91. Bass, L. (1992) A Metamodel for the Runtime Architecture of an Interactive System, The UIMS Developers Workshop. ACM SIGCHI Bulletin, 24, 1: 32–37.

92. Rekimoto, J. (1997) Pick and Drop: A Direct Manipulation Technique for Multiple Computer Environments. In: Proc. of 10th ACM Symposium on User Interface Software Technologies UIST'97 (Banff, October 14–17, 1997), ACM Press, New York, pp. 31–39.
93. Katsurada, K., Nakamura, Y., Yamada, H., Nitta, T. (2003) XISL: A Language for Describing Multimodal Interaction Scenarios. In: Proceedings of the 5th ACM International Conference on Multimodal Interfaces ICMI'2003 (Vancouver, November 5–7, 2003), ACM Press, New York, pp. 281–284.
94. Stephanidis, C. and Savidis, A. (2001) Universal Access in the Information Society: Methods, Tools, and Interaction Technologies, Journal of the Universal Access in Information Society, 1, 1: 40–55.
95. Coutaz, J. and Calvary, G. (2007) HCI and Software Engineering: Designing for User Interface Plasticity. In: Sears A. and Jacko J. (eds.), The Human-Computer Interaction Handbook: Fundamentals, Evolving Technologies and Emerging Applications, 2nd ed., Taylor & Francis Publishers.
96. Martin, J.-C. (1998) TYCOON: Theoretical Framework and Software Tools for Multimodal Interfaces. In: Lee J (ed.), Intelligence and Multimodality in Multimedia Interfaces, AAAI Press.
97. Nigay, L. and Coutaz, J. (1995) A Generic Platform for Addressing the Multimodal Challenge. In: Proceedings of the ACM Conference on Human Factors in Computing Systems CHI'95 (Denver, May 07–11, 1995), ACM Press, New York, pp. 98–105.
98. André, E. (2000) The generation of multimedia presentations. In: Dale R., Mois L.H. & Somers H. (eds.), A handbook of natural language processing: techniques and applications for the processing of language as text, Marcel Dekker Inc., pp. 305–327.
99. Gram, C. and Cockton, G. (1996) Design Principles for Interactive Software, Chapman & Hall, London.
100. Oreizy, P., Gorlick, M., Taylor, R., Heimbigner, D., Johnson, G., Medvidovic, N., Quilici, A., Rosenblum, D., and Wolf, A. (1999) An Architecture-Based Approach to Self-Adaptive Software, IEEE Intelligent Systems, 14,3: 54–62.
101. G. Monaci, O. Divorra-Escoda, and P. Vandergheynst, "Analysis of multimodal sequences using geometric video representations," Signal Processing, vol. 86, 2006.
102. R. Chellappa, C. Wilson, and S. Sirohey, "Human and machine recognition of faces: A survey," Proceedings of the IEEE, vol.83, no.5, pp.705–740, May 1995.
103. M. Yang, D. Kriegman, and N. Ahuja, "Detecting faces in images: A survey," IEEE Trans. PAMI, vol.24, no.1, pp.34–58, January 2002.
104. E. Hjelmas and B. Low, "Face detection: A survey," Computer Vision and Image Understanding, vol.83, pp.236–274, 2001.
105. D. Chai and K.Ngan, "Locating facial region of a head-and-shoulders colour image," in Third International Conference on Automatic Face and Gesture Recognition. Nara, Japan:IEEE, pp.124–129, April 1998.
106. D. Chai and K. Ngan, "Face segmentation using skin-colour map in videophone applications," IEEE Transactions on Circuits and Systems for Video Technology, vol.9, no.4, pp.551–564, June 1999.
107. C. Garcia and G. Tziritas, "Face detection using quantised skin colour regions merging and wavelet packet analysis," IEEE Transactions on Multimedia, vol.1, no.3, pp.264–277, September 1999.
108. J. C. Terrillon, M. David, and S. Akamatsu, "Automatic detection of human faces in natural scene images by use of a skin colour model and of invariant moments," in Third International Conference on Automatic Face and Gesture Recognition. Nara, Japan: IEEE, pp.112–117, April 1998.
109. J. Terrillon, M. Shirazi, and M. Sadek, "Invariant face detection with support vector machines," in Proc. Of the 15th. International Conference on Pattern Recognition, Barcelona, Spain, pp. 210–217, 2000.

110. E. Saber and A. M. Tekalp, "Frontal-view face detection and facial feature extraction using colour, shape and symmetry based cost functions," Pattern Recognition Letters, vol.19, pp.669–680, 1998.
111. K. Sobottka and I. Pitas, "A novel method for automatic face segmentation, facial feature extraction and tracking," Signal Processing: Image Communication, vol.12, pp.263–281, 1999.
112. N. Herodotou, K. Plataniotis, and A. Venetsanopoulos, "Automatic location and tracking of the facial region in colour video sequences," Signal Processing: Image Communication, vol.14, pp.359–388, 1999.
113. H. Sahbi and N. Boujemaa, "Coarse to fine face detection based on skin colour adaptation," in Workshop on Biometric Authentication, ECCV's 2002. Copenhagen, Denmark.: Springer Verlag, 2002.
114. R. Hsu, M. A. Mottaleb, and A. Jain, "Face detection in colour images," IEEE Transactions on Pattern Analysis and Machine Intelligence, vol.24, no.5, pp.696–706, May 2002.
115. Y. Li, S. Gong, J. Sherrah, and H. Liddell, "Support vector machine based multiview face detection and recognition," Image and Vision Computing, vol.22, no.5, pp.413–427, 2004.
116. H. Rowley, S. Baluja, and T. Kanade, "Neural network-based face detection," IEEE Transactions on Pattern Analysis and Machine Intelligence, vol.20, no.1,pp.23–38, Jan 1998.
117. H. Rowley, S. Baluja and T. Kanade, "Rotation invariant neural network-based face detection," in Proc. IEEE Conf. Computer Vision and Pattern Recognition, pp.38–44, June 1998.
118. E. Osuna, R. Freund, and F. Girosi, "Training support vector machines: An application to face detection," in Proc. Of Computer Vision and Pattern Recognition, Puerto Rico, June 1997.
119. R. Féraud, O. Bernier, J. Viallet, and M. Collobert, "A fast and accurate face detector based on neural networks," IEEE Trans. on Pattern Analysis and Machine Intelligence, vol.23, no.1, pp.42–53, January 2001.
120. S. Lin, S. Y. Kung, and L. Lin, "Face recognition/detection by probabilistic decision-based neural network," IEEE Transactions on Neural Networks, vol.8, pp.114–132, 1997.
121. D. Roth, M. Yang, and N. Ahuja, "A SNoW-based face detector," Advances in Neural Information Processing Systems 12 (NIPS12), 2000.
122. B. Moghaddam and A. Pentland, "Probabilistic visual learning for object representation," IEEE Transactions on Pattern Analysis and Machine Intelligence, vol.19, no.7, pp.696–710, Jul 1997.
123. K. K. Sung and T. Poggio, "Example-based learning for view-based human face detection," IEEE Transactions on Pattern Analysis and Machine Intelligence, vol. 20, no. 1, pp. 39–51, Jan 1998.
124. M. H. Yang, N. Ahuja, and D. Kriegman, "Mixtures of linear subspaces for face detection," in Proc. Fourth Int. Conf. Automatic Face and Gesture Recognition, pp.70–76, 2000.
125. V. Popovici and J. Thiran, "Face detection using SVM trained in eigenfaces space," in 4th International Conference on Audio and Video-Based Biometric Person Authentication, ser. Lecture Notes in Computer Science, vol. 2688. Berlin: Springer-Verlag, 2003, pp. 190–198.
126. C. Kervrann, F. Davoine, P. Pérez, H. Li, R. Forchheimer, and C. Labit, "Generalised likelihood ratio-based face detection and extraction of mouth features," in Proc. Third International Conference on Automatic Face and Gesture Recognition, pp.103–108, 1998.
127. L. Meng and T. Nguyen, "Frontal face detection using multi-segment and wavelet," in Conference on Information Science and Systems, 1999.
128. H. Schneiderman, "Learning statistical structure for object detection," in Computer Analysis of Images and Patterns (CAIP). Springer Verlag, August 2003.
129. C. Papageorgiou and T. Poggio, "A trainable system for object detection," International Journal of Computer Vision, vol. 38, no. 1, pp. 15–33, 2000.
130. H. Schneiderman and T. Kanade, "Probabilistic modelling of local appearance and spatial relationships for object recognition," in Proc. IEEE Int. Conf. On Computer Vision and Pattern Recognition (CVPR'98), pp.45–51, July 1998.

131. A. Colmenarez and T. S. Huang, "Pattern detection with information-based maximum discrimination and error bootstrapping," in Int. Conference on Pattern Recognition, 1998.
132. F. Fleuret and D. Geman, "Coarse-to-fine face detection," International Journal of Computer Vision, vol.41, pp.85–107, 2001.
133. P. Viola and M. Jones, "Rapid object detection using a boosted cascade of simple features," in Proc. Of IEEE Conference on Computer Vision and Pattern Recognition, Kauai, HI, December 2001.
134. Y. Freund and R. Schapire, "A decision-theoretic generalization of on-line learning and an application to boosting," Journal of Computer and System Sciences, vol.55, no.1, pp.119–139, August 1997.
135. M. Jones and P. Viola, "Fast multi-view face detection," in Proc. IEEE Conference on Computer Vision and Pattern Recognition (CVPR), June 2003.
136. H. Sahbi, D. Geman, and N. Boujemaa, "Face detection using coarse-to-fine support vector classifiers," in Proc. Of the IEEE International Conferene on Image Processing (ICIP'2002), Rochester, New York, USA, 2002.
137. D. Zhang, S. Li, and D. Gatica-Perez, "Real-time face detection using boosting learning in hierarchical feature spaces," IDIAP, Tech. Rep. IDIAP-RR 03-70, December 2003.
138. S. Romdhani, P. Torr, B. SchÄolkopf, and A. Blake, "Efficient face detection by a cascaded support vector machine expansion," in Proceedings of the Royal Society, Series A (Accepted), 2004.
139. D. Keren, M. Osadchy, and C. Gotsman, "Antifaces: A novel, fast method for image detection," IEEE Transactions on Pattern Analysis and Machine Intelligence, vol.23, no.7, pp.747–761, 2001.
140. H. Schneiderman, "Learning a restricted bayesian network for object detection," in Proc. IEEE Int. Conf. On Computer Vision and Pattern Recognition (CVPR'04), June 2004.
141. H. Schneiderman, "Feature-centric evaluation for efficient cascaded object detection," in IEEE Conference on Computer Vision and Pattern Recognition (CVPR). IEEE, June 2004.
142. H. Wang and S. Chang, "A highly efficient system for automatic face region detection in MPEG video," IEEE Transactions on Circuits and Systems for Video Technology, vol.7, no.4, pp.615–628, Aug 1997.
143. H. Luo and A. Eleftheriadis, "On face detection in the compressed domain," in ACM Multimedia Conference, pp. 285–294, 2000.
144. P. Fonseca and J. Nesvadba, "Face detection in the compressed domain," in IEEE Proc. Int. Conf. On Image Processing (ICIP 2004). Singapore: IEEE, pp. 2015–2018, October 2004.
145. I. Ozer and W. Wolf, "Human detection in compressed domain," in Proc. International Conference on Image Processing, pp. 274–277, October 2001.
146. P. Salembier and F. Marques, "Region-based representation of image and video: Segmentation tools for multimedia services," IEEE Transactions on Circuits and Systems for Video Technology, vol.9, no.8, pp.1147–1167, December 1999.
147. M. H. Yang and N. Ahuja, "Detecting human faces in colour images," in IEEE International Conference on Image Processing, Chicago, USA, pp.127–130, October 1998.
148. N. Ahuja, "A transform for multiscale image segmentation by integrated edge and region detection," IEEE Transactions on Pattern Analysis and Machine Intelligence, vol.18, no.12, pp.1211–1235, 1996.
149. F. Marques and V. Vilaplana, "Face segmentation and tracking based on connected operators and partition projection," Pattern Recognition, vol.35, no.3, pp.601–614, 2002.
150. V. Vilaplana and F. Marques, "Face detection and segmentation on a hierarchical image representation," in EUSIPCO'07, September 2007.
151. P. Salembier and L. Garrido, "Binary partition tree as an efficient representation for image processing, segmentation and information retrieval," IEEE Transactions on Image Processing, vol.9, no.4, pp.561–575, April 2000.
152. V. Govindaraju, "Locating human faces in photographs," International Journal on Computer Vision, vol.19, no.2, pp.129–146, 1996.

153. K. Yow and R. Cipolla, "Feature-based human face detection," Image and Vision Computing, vol.15, no.9, pp.713–735, 1997.
154. D. Maio and D. Maltoni, "Real-time face location on gray-scale static images," Pattern Recognition, vol.33, no.9, pp.1525–1539, 2000.
155. G. Feng and P. Yuen, "Variance projection function and its application to eye detection for human face recognition," Pattern Recognition Letters, vol.19, no.9, pp.899–906, 1998.
156. Z. Zhou and X. Geng, "Projection functions for eye detection," Pattern Recognition, vol.37, no.5, pp.1049–1056, 2004.
157. T. F. Cootes, D. H. Cooper, C. J. Taylor, and J. Graham, "A trainable method of parametric shape description," Image and Vision Computing, vol.10, no.5, p.289–294, 1992.
158. T. F. Cootes, G. J. Edwards, and C. J. Taylor, "Active appearance models," in Proc. 5th European Conference on Computer Vision, vol.2, pp.484–498, 1998.
159. T. F. Cootes and G. J. Edwards and C. J. Taylor, "Interpreting face images using active appearance models," in Proc. 3rd International Conference on Automatic Face and Gesture Recognition, pp.300–305, 1998.
160. M. B. Stegmann, B. K. Ersbll, and R. Larsen, "Fame – a Flexible appearance modelling environment," IEEE Transactions on Medical Imaging, vol.22, no.10, pp.1319–1331, 2003.
161. J. Hershey and J. R. Movellan, "Audio vision: Using audio-visual synchrony to locate sounds," Neural Information Processing Systems, pp.813–819, 1999.
162. M. Slaney and M. Covell, "Facesync: A linear operator for measuring synchronization of video facial images and audio tracks," Neural Information Processing Systems, pp.814–820, 2000.
163. H. J. Nock, G. Iyengar, and C. Neti, "Assessing face and speech consistency for monologue detection in video," Proc. ACM Multimedia, 2002.
164. H. J. Nock, G. Iyengar and C. Neti, "Speaker localization using audio-visual synchrony: An empirical study," Proceedings of the International Conference on Image and Video Retrieval, 2003.
165. J. W. Fisher III, T. Darrell, W. T. Freeman, and P. Viola, "Learning joint statistical models for audio-visual fusion and segregation," Advances in Neural Information Processing Systems, 2000.
166. J. W. Fisher III and T. Darrell, "Speaker association with signal-level audio-visual fusion," IEEE Transactions on Multimedia, vol.6, no.3, pp.406–413, 2004.
167. T. Butz and J. P. Thiran, "Feature space mutual information in speech-video sequences," Proceedings of the IEEE International Conference on Multimedia and Expo, vol.2, pp.361–364, 2002.
168. T. Butz and J. P. Thiran, "From error probability to information theoretic (multi-modal) signal processing," Signal Processing, no.85, pp.875–902, 2005.
169. P. Besson, M. Kunt, T. Butz, and J. P. Thiran, "A multimodal approach to extract optimised audio features for speaker detection," Proceedings of European Signal Processing Conference (EUSIPCO), 2005.
170. C. Bishop, Neural networks for Pattern Recognition. Oxford University Press, 1995.
171. E. K. Patterson, S. Gurbuz, Z. Tufekci, and J. N. Gowdy, "Moving-talker, speaker-independent feature study and baseline results using the CUAVE multimodal speech corpus," EURASIP JASP, vol. 2002(11), pp.1189–1201, 2002.
172. P. Besson, G. Monaci, P. Vandergheynst, and M. Kunt, "Experimental framework for speaker detection on the CUAVE database, EPFL-ITS Tech. Rep. 2006-003," EPFL, Lausanne, Switzerland, Tech. Rep., 2006.
173. H. McGurk and J. MacDonald, "Hearing lips and seeing voices," Nature, vol.264, pp.746–748, 1976.
174. G. Potamianos, C. Neti, J. Luettin, and I. Matthews, "Audio-visual automatic speech recognition: an overview," in Issues in audio-visual speech processing, G. Bailly, E. Vatikiotis-Bateson, and P. Perrier, Eds. MIT Press, 2004.
175. L. R. Rabiner, "A tutorial on Hidden Markov Models and selected applications in speech recognition," Proceedings of the IEEE, vol. 77(2), 1989.

176. A. K. Jain, "Fundamentals of Digital Image Processing," Prentice-Hall, 1989.
177. G. Potamianos, H. P. Graf, and E. Cosatto, "An image transform approach for HMM based automatic lipreading," in Proceedings of the International Conference on Image Processing, vol.3, pp.173–177, 1998.
178. R. Reilly and P. Scanlon, "Feature analysis for automatic speech reading," Proc. Workshop on Multimedia Signal Processing, pp.625–630, 2001.
179. A. Adjoudani and C. Benoît, "On the integration of auditory and visual parameters in an HMM-based ASR," in Speechreading by humans and machines, D. G. Stork and M. E. Hennecke, Eds. Springer, pp.461–471, 1996.
180. H. Bourlard and S. Dupont, "A new asr approach based on independent processing and recombination of partial frequency bands," Proc. International Conference on Spoken Language Processing, pp.426–429, 1996.
181. H. Liu and H. Motoda, "Feature selection for knowledge discovery and data mining," Kluwer Academic Publishers, 1998.
182. R. Battiti, "Using mutual information for selecting features in supervised neural net working," IEEE Transactions on Neural Networks, vol.5(4), 1994.
183. H. Peng, F. Long, and C. Ding, "Feature selection based on mutual information: Criteria of max-dependency, max-relevance, and min-redundancy," IEEE Transactions on Pattern Analysis and Machine Intelligence, vol.27(8), 2005.
184. F. Fleuret, "Fast binary feature selection with conditional mutual information," Journal of Machine Learning Research, vol. 5), pp.1531–1555, 2004.
185. T.Cover and J.Thomas, "Elements of Information Theory," Wiley Series in Telecommunications, New York, 1991.
186. D. Ellis and J. Bilmes, "Using mutual information to design feature combinations," in Proceedings of ICSLP, vol.3, pp.79–82, 2000.
187. P. Scanlon, G. Potamianos, V. Libal, and S. M. Chu, "Mutual information based visual feature selection for lipreading," ICSLP, pp. 2037–2040, 2004.
188. S. Young, D. Kershaw, J. Odell, D. Ollason, V. Valtchev, and P. Woodland, "The HTK Book. Cambridge," Entropic Ltd., 1999.
189. H. Misra, H. Bourlard, and V. Tyagi, "New entropy based combination rules in hmm/ann multi-stream asr," Proceedings of the 2003 IEEE International Conference on Acoustics, Speech, and Signal Processing (ICASSP '03), 2003.
190. R. W. Picard, Affective Computing, MIT Press, 1997.
191. A. Jaimes, Human-Centred Multimedia: Culture, Deployment, and Access, IEEE Multimedia Magazine, Vol. 13, No.1, 2006.
192. A. Pentland, Socially Aware Computation and Communication, Computer, vol. 38, no. 3, pp. 33–40, 2005.
193. R. W. Picard, Towards computers that recognize and respond to user emotion, IBM Syst. Journal, 39 (3–4), 705–719, 2000.
194. P. Cohen, Multimodal Interaction: A new focal area for AI. IJCAI 2001, pp. 1467–1473.
195. P. Cohen, M. Johnston, D. McGee, S. Oviatt, J. Clow, I. Smith, The efficiency of multimodal interaction: A case study, in Proceedings of International Conference on Spoken Language Processing, ICSLP'98, Australia, 1998.
196. S. Oviatt, Ten myths of multimodal interaction, Communications of the ACM, Volume 42, Number 11 (1999), Pages 74–81.
197. S. Oviatt, A. DeAngeli, K. Kuhn, Integration and synchronization of input modes during multimodal human-computer interaction, In Proceedings of Conf. Human Factors in Computing Systems CHI'97, ACM Press, NY, 1997, pp. 415 – 422.
198. A. Jaimes and N. Sebe, Multimodal Human Computer Interaction: A Survey, IEEE International Workshop on Human Computer Interaction, ICCV 2005, Beijing, China.
199. A. Mehrabian, Communication without words, Psychology Today, vol. 2, no. 4, pp. 53–56, 1968.
200. Z. Zeng, J. Tu, M. Liu, T. S. Huang, B. Pianfetti, D. Roth, S. Levinson, Audio-Visual Affect Recognition, IEEE Trans. Multimedia, vol. 9, no. 2, Feb. 2007.

201. M. Pantic and L.J.M. Rothkrantz, Towards an affect-sensitive multimodal human-computer interaction, Proc. of the IEEE, vol. 91, no. 9, pp. 1370–1390, 2003.
202. S. Ioannou, A. Raouzaiou, V. Tzouvaras, T. Mailis, K. Karpouzis, S. Kollias, Emotion recognition through facial expression analysis based on a neurofuzzy network, Special Issue on Emotion: Understanding & Recognition, Neural Networks, Elsevier, Volume 18, Issue 4, May 2005, Pages 423–435.
203. L.C. De Silva and P.C. Ng, Bimodal emotion recognition, in Proc. Face and Gesture Recognition Conf., 332–335, 2000.
204. A. Kapoor, R. W. Picard and Y. Ivanov, Probabilistic combination of multiple modalities to detect interest, Proc. of IEEE ICPR, 2004.
205. N. Fragopanagos and J. G. Taylor, Emotion recognition in human computer interaction, Neural Networks, vol. 18, pp. 389–405, 2005.
206. H. Gunes and M. Piccardi, Fusing Face and Body Gesture for Machine Recognition of Emotions, 2005 IEEE International Workshop on Robots and Human Interactive Communication, pp. 306 – 311.
207. G. Caridakis, L. Malatesta, L. Kessous, N. Amir, A. Raouzaiou, K. Karpouzis, Modelling naturalistic affective states via facial and vocal expressions recognition, International Conference on Multimodal Interfaces (ICMI'06), Banff, Alberta, Canada, November 2–4, 2006, pp. 146–154.
208. K. Karpouzis, G. Caridakis, L. Kessous, N. Amir, A. Raouzaiou, L. Malatesta, S. Kollias, Modelling naturalistic affective states via facial, vocal, and bodily expressions recognition, in T. Huang, A. Nijholt, M. Pantic, A. Pentland (eds.), AI for Human Computing, LNAI Volume 4451/2007, Springer.
209. M. Wallace, S. Ioannou, A. Raouzaiou, K. Karpouzis, S. Kollias, Dealing with Feature Uncertainty in Facial Expression Recognition Using Possibilistic Fuzzy Rule Evaluation, International Journal of Intelligent Systems Technologies and Applications, Volume 1, Number 3–4, 2006.
210. M. Pantic, Face for interface, in The Encyclopedia of Multimedia Technology and Networking, M. Pagani, Ed. Hershey, PA: Idea Group Reference, 2005, vol. 1, pp. 308–314.
211. L. Wu, Sharon L. Oviatt, Philip R. Cohen, Multimodal Integration – A Statistical View, IEEE Transactions on Multimedia, vol. 1, no. 4, December 1999.
212. P. Teissier, J. Robert-Ribes, and J. L. Schwartz, Comparing models for audiovisual fusion in a noisy-vowel recognition task, IEEE Trans. Speech Audio Processing, vol. 7, pp. 629–642, Nov. 1999.
213. A. Rogozan, Discriminative learning of visual data for audiovisual speech recognition, Int. J. Artif. Intell. Tools, vol. 8, pp. 43–52, 1999.
214. S. Haykin, Neural Networks: A Comprehensive Foundation, Prentice Hall International, 1999.
215. G. Littlewort, M. S. Bartlett, I. Fasel, J. Susskind, J. Movellan, Dynamics of facial expression extracted automatically from video, Image and Vision Computing 24 (2006) 615–625.
216. P. Ekman and W. V. Friesen. Felt, false, and miserable smiles. Journal of Nonverbal Behaviour, 6:238–252, 1982.
217. M. G. Frank and P. Ekman. Not all smiles are created equal: The differences between enjoyment and other smiles. Humour: The International Journal for Research in Humour, 6:9–26, 1993.
218. G. Littlewort, M. S. Bartlett, I. Fasel, J. Susskind, J. Movellan, Dynamics of facial expression extracted automatically from video, Image and Vision Computing 24 (2006) 615–625.
219. P. Ekman and W. V. Friesen, The facial Action Coding System: A Technique for the Measurement of Facial Movement. San Francisco: Consulting Psychologists Press, 1978.
220. P. Ekman, Facial expression and Emotion. Am. Psychologist, Vol. 48 (1993) 384–392.
221. R. Cowie and R. Cornelius, Describing the Emotional States that are Expressed in Speech, Speech Communications, 40(5–32), 2003.
222. R. Cowie, E. Douglas-Cowie, S. Savvidou, E. McMahon, M. Sawey, M. Schroeder, 'FeelTrace': An Instrument for recording Perceived Emotion in Real Time, Proceedings of ISCA Workshop on Speech and Emotion, pp 19–24, 2000.

223. ERMIS, Emotionally Rich Man-machine Intelligent System IST-2000-29319 http://www.image.ntua.gr/ermis.
224. C.M Whissel, (1989) 'The dictionary of affect in language', in Plutchnik, R. and Kellerman, H. (Eds.): Emotion: Theory, Research and Experience: The Measurement of Emotions, Academic Press, New York, Vol. 4, pp.113–131.
225. R. Cowie, E. Douglas-Cowie, S. Savvidou, E. McMahon, M. Sawey, and M. Schröder, 'Feeltrace': An instrument for recording perceived emotion in real time, in Proc. ISCA-Workshop on Speech and Emotion, 2000, pp. 19–24.
226. R. Bertolami,H. Bunke, Early feature stream integration versus decision level combination in a multiple classifier system for text line recognition, 18th International Conference on Pattern Recognition (ICPR'06).
227. C. Tomasi, T. Kanade, Detection and Tracking of Point Features, Carnegie Mellon University Technical Report CMU-CS-91-132, April 1991.
228. Ying-li Tian, Takeo Kanade, J. F. Cohn, Recognizing Action Units for Facial Expression Analysis, IEEE Transactions on PAMI, Vol.23, No.2, February 2001.
229. A. Murat Tekalp, Joern Ostermann, Face and 2-D mesh animation in MPEG-4, Signal Processing: Image Communication 15 (2000) 387–421.
230. I.A. Essa and A.P. Pentland, "Coding, Analysis, Interpretation, and Recognition of Facial Expressions," IEEE Trans. Pattern Analysis and Machine Intelligence, vol. 19, no. 7, pp. 757–763, July 1997.
231. S. H. Leung, S. L. Wang SL, W. H. Lau, Lip image segmentation using fuzzy clustering incorporating an elliptic shape function, IEEE Trans. on Image Processing, vol.13, No.1, January 2004.
232. N. Sebe, M.S. Lew, I. Cohen, Y. Sun, T. Gevers, T.S. Huang, Authentic Facial Expression Analysis, International Conference on Automatic Face and Gesture Recognition (FG'04), Seoul, Korea, May 2004, pp. 517–522.
233. T. Cootes, G. Edwards, C. Taylor, Active appearance models, IEEE PAMI 23 (6), 2001, pp. 681–685.
234. M. Pantic, L.J.M Rothkrantz, Expert system for automatic analysis of facial expressions, Image and Vision Computing vol 18, 2000, pp. 881–905.
235. M. H. Yang, D. Kriegman, N. Ahuja, "Detecting Faces in Images: A Survey", PAMI, Vol.24(1), 2002, pp. 34–58.
236. C. Papageorgiou, M. Oren and T. Poggio, A general framework for object detection. In international Conference on Computer Vision, 1998.
237. P. Viola, M. Jones, Rapid Object Detection using a Boosted Cascade of Simple Features. Computer Vision and Pattern Recognition, 2001. CVPR 2001. Proceedings of the 2001 IEEE Computer Society Conference on , Volume: 1 , 8–14 Dec. 2001 Pages:I-511–I-518 vol.1.
238. I. Fasel, B. Fortenberry, and J. R. Movellan, A generative framework for real-time object detection and classification, Computer Vision and Image Understanding, Volume 98 , Issue 1 (April 2005), pp. 182 – 210.
239. R. Fransens, Jan De Prins, SVM-based Nonparametric Discriminant Analysis, An Application to Face Detection, Ninth IEEE International Conference on Computer Vision Volume 2, October 13–16, 2003.
240. M. T. Hagan, and M. Menhaj, Training feedforward networks with the Marquardt algorithm, IEEE Trans. Neural Networks, vol. 5, no. 6, 1994, pp. 989–993.
241. ERMIS, Emotionally Rich Man-machine Intelligent System IST-2000-29319 (http://www.image.ntua.gr/ermis).
242. C. Lee, S. Narayanan, Emotion recognition using a data-driven fuzzy inference system. Proc. Eurospeech 2003, Geneva.
243. O. Kwon, K. Chan, J. Hao, Emotion recognition by speech signals. Proc. Eurospeech 2003, Geneva, pp. 125–128, 2003.
244. G. Zhou, J. Hansen, J. Kaiser, Methods for stress classification: Nonlinear TEO and linear speech based features. Proc. IEEE International Conference on Acoustics, Speech, and Signal Processing, vol. IV, pp. 2087–2090, 1999.

245. J. Ang, R. Dhillon, A. Krupski, E. Shriberg, A. Stolcke, Prosody-based automatic detection of annoyance and frustration in human-computer dialog. Proc. ICSLP 2002, Denver, Colorado, Sept. 2002.
246. S. Yacoub, S. Simske, X. Lin, J. Burns, Recognition of emotions in interactive voice response systems, in Proeedings of Eurospeech 2003, Geneva, 2003.
247. A. Batliner, K. Fischer, R. Huber, J. Spilker, E. Noeth, How to find trouble in communication. Speech Communication 40, pp. 117–143, 2003.
248. S. McGilloway, R. Cowie, E. Douglas-Cowie, S. Gielen, M. Westerdijk, S. Stroeve, Automatic recognition of emotion from voice: a rough benchmark. Proc. ISCA ITRW on Speech and Emotion, 5–7 September 2000, Textflow, Belfast, pp. 207–212, 2000.
249. R. Nakatsu, N. Tosa, J. Nicholson, Emotion recognition and its application to computer agents with spontaneous interactive capabilities. Proc. IEEE International Workshop on Multimedia Signal Processing, pp. 439–444, 1999.
250. A. Batliner, S. Steidl, B. Schuller, D. Seppi, K. Laskowski, T. Vogt, L. Devillers, L. Vidrascu, N. Amir, L. Kessous, V. Aharonson, Combining efforts for improving automatic classification of emotional user states, In Erjavec, T. and Gros, J. (Ed.), Language Technologies, IS-LTC 2006, pp. 240–245, Ljubljana, Slovenia, 2006.
251. P. Mertens, The Prosogram: Semi-Automatic Transcription of Prosody based on a Tonal Perception Model. in B. Bel & I. Marlien (eds.), Proc. of Speech Prosody, Japan, 2004.
252. J. L. Elman, Distributed representations, simple recurrent networks, and grammatical structure. Machine Learning, 7, 195–224, 1991.
253. A. M. Schaefer, H. G. Zimmermann, Recurrent Neural Networks Are Universal Approximators, ICANN 2006, pp. 632–640.
254. B. Hammer, P. Tino, Recurrent neural networks with small weights implement definite memory machines, Neural Computation 15(8), 1897–1929, 2003.
255. E. Douglas-Cowie, N. Campbell, R. Cowie, P. Roach, Emotional speech: Towards a new generation of databases, Speech Communication 40, pp. 33–60, 2003.
256. R. Banse, K. Scherer, K., Acoustic profiles in vocal emotion expression. J. Pers. Social Psychol. 70 (3), 614–636, 1996.
257. P. Ekman and W. V. Friesen. Felt, false, and miserable smiles. Journal of Nonverbal Behaviour, 6:238–252, 1982.
258. M. G. Frank and P. Ekman. Not all smiles are created equal: The differences between enjoyment and other smiles. Humour: The International Journal for Research in Humour, 6:9–26, 1993.
259. M.F. Valstar, M. Pantic, Z. Ambadar and J.F. Cohn, 'Spontaneous vs. posed facial behaviour: Automatic analysis of brow actions (pdf file)', in Proceedings of ACM Int'l Conf. Multimodal Interfaces (ICMI'06), pp. 162–170, Banff, Canada, November 2006.
260. http://emotion-research.net/toolbox/toolboxdatabase.2006-09-26.5667892524.
261. J.F. Kelley, Natural Language and computers: Six empirical steps for writing an easy-to-use computer application. Unpublished doctoral dissertation, The Johns Hopkins University, 1983.
262. J. W. Young, Head and Face Anthropometry of Adult U.S. Civilians, FAA Civil Aeromedical Institute, 1963–1993 (final report 1993).
263. I. Cohen, N. Sebe, A. Garg, L.S. Chen, T.S. Huang, Facial expression recognition from video sequences: temporal and static modelling, Computer Vision and Image Understanding 91 (2003) 160–187.
264. M. Pantic and I. Patras, 'Dynamics of Facial Expression: Recognition of Facial Actions and Their Temporal Segments from Face Profile Image Sequences', IEEE Transactions on Systems, Man and Cybernetics – Part B, vol. 36, no. 2, pp. 433–449, April 2006.
265. Rabiner L, Juang ▯ (1993) Fundamentals of Speech Recognition New Jersey: Prentice-Hall, Englewood Cliffs.
266. Turk M (2002) Gesture recognition. In K. M. Stanney (Ed.), Handbook of virtual environments: Design, implementation, and applications. Mahwah, NJ: Lawrence Erlbaum Associates, Inc. pp 223–238.

267. Feyereisen P, De Lannoy J (1991) Gesture and Speech : Psychological Investigations. Cambridge University Press.
268. Poggi I (2001) From a typology of gestures to a procedure for gesture production. In: Gesture Workshop, pp 158–168.
269. Oviatt S (2003) Multimodal Interfaces. In Human-Computer Interaction Handbook: Fundamentals, Evolving Technologies and Emerging Applications. Lawrence Erlbaum Assoc. Mahwah, NJ, pp 286–304.
270. Bolt RA (1980) "Put-that-there": voice and gesture at the graphics interface, In: 7-th Annual Conference on Computer Graphics and Interactive Techniques, Washington, Seattle, WA, USA, pp 262–270.
271. Malerczyk C, Dähne P, Schnaider M (2005) Pointing Gesture-Based Interaction for Museum Exhibits. In: HCI International 2005. Mahwah, New Jersey: Lawrence Erlbaum Associates, Inc.
272. Wilson A, Cutrell E (2005) FlowMouse: A computer vision-based pointing and gesture input device In: Human-Computer Interaction INTERACT'05, Rome, Italy, pp 565–578.
273. de Silva GC, Lyons MJ, Kawato S, Tetsutani N (2003) Human Factors Evaluation of a Vision-Based Facial Gesture Interface. Workshop on Computer Vision and Pattern Recognition for Computer Human Interaction, vol. 5, Madison, USA, pp 52.
274. Kaiser E, Olwal A, McGee D, Benko H, Corradini A, Li X, Cohen P, Feiner S (2003) Mutual disambiguation of 3D multimodal interaction in augmented and virtual reality. In: Proceedings of the Fifth International Conference on Multimodal, Vancouver, Canada, pp 12–19.
275. Stiefelhagen R, Fuegen C, Gieselmann P, Holzapfel H, Nickel K, Waibel A (2004) Natural human–robot interaction using speech, gaze and gestures. In: IEEE/RSJ International Conference on Intelligent Robots and Systems, Sendai, Japan, pp 2422–2427.
276. Sharma R, Yeasin M, Krahnstoever N, Rauschert I, Cai G, Brewer I, MacEachren AM, Sengupta K (2003) Speech-gesture driven multimodal interfaces for crisis management. Proceedings of the IEEE, vol. 91(9) : 1327–1354.
277. Schapira E, Sharma R (2001) Experimental evaluation of vision and speech based multimodal interfaces. In: PUI '01 workshop on Perceptive user interfaces, New York, USA, pp 1–9.
278. Chen F, Choi E, Epps J, Lichman S, Ruiz N, Shi Y, Taib R, Wu M (2005) A study of manual gesture-based selection for the PEMMI multimodal transport management interface. In: International Conference on Multimodal Interfaces (ICMI), Trento, Italy, pp 274–281.
279. Wang SB, Demirdjian D (2005) Inferring body pose using speech content. In: International Conference on Multimodal Interfaces ICMI, Trento, Italy, pp 53–60.
280. Dossogne S, Macq B (2005) SIMILAR Dreams – Multimodal Interfaces in Our Future Life, Presses Universitaires de Louvain, Louvain la Neuve.
281. Argyropoulos S, Moustakas K, Karpov A, Aran O, Tzovaras D, Tsakiris T, Varni G, Kwon B (2007) A multimodal framework for the communication of the disabled. In: Summer Workshop on Multimodal Interfaces eNTERFACE'07, Istanbul, Turkey.
282. Corno F, Farinetti L, Signorile I (2002) An eye-gaze input device for people with severe motor disabilities. In: SSGRR-2002 International Conference on Advances in Infrastructure for e-Business, e-Education, e-Science, and e-Medicine on the Internet, L'Aquila, Italy.
283. LC Technologies, Inc. Eyegaze Systems. http://www.eyegaze.com.
284. Garcia-Moreno F (2001) Eye Gaze Tracking System Visual Mouse Application Development. Technical Report, Ecole Nationale Supériere de Physique de Strasbourg (ENSPS) and School of Computer Science Queen's University Belfast.
285. Grauman K, Betke M, Lombardi J, Gips J, Bradski G (2003) Communication via Eye Blinks and Eyebrow Raises: Video-Based Human-Computer Interfaces. Universal Access in the Information Society, vol. 4 : 359–373.
286. Bates R, Istance HO (2003) Why are eye mice unpopular? A detailed comparison of head and eye controlled assistive technology pointing devices. Universal Access in the Information Society. Springer-Verlag Heidelberg. vol. 2 : 280–290.

287. Karpov A, Ronzhin A, Cadiou A (2006) A multi-modal system ICANDO: Intellectual Computer AssistaNt for Disabled Operators. In: 9-th International Conference on Spoken Language Processing INTERSPEECH-ICSLP'2006, Pittsburgh, PA, USA, pp 1998–2001.
288. Richter J (1999) Programming Applications for Microsoft Windows, Microsoft Press.
289. Ronzhin A, Karpov A (2007) Russian Voice Interface. Pattern Recognition and Image Analysis, 17 (2) : 321–336.
290. Ronzhin A, Karpov A, Timofeev A, Litvinov M (2005) Multimodal human-computer interface for assisting neurosurgical system. In: 11-th International Conference on Human-Computer Interaction HCII'2005, Las Vegas, USA.
291. Young S, Evermann G, Gales M, Hain T, Kershaw D, Liu X, Moore G, Odell J, Ollason D, Povey D, Valtchev V, Woodland P (2006) The HTK Book (for HTK Version 3.4), Cambridge University Engineering Department.
292. Ward D, Blackwell A, MacKay D (2000) Dasher: A data entry interface using continuous gestures and language models, In Proceedings of the ACM Symposium on User Interface Software and Technology UIST'2000, New York: ACM, pp 129–137.
293. Lienhart R, Maydt D (2002) An Extended Set of Haar-like Features for Rapid Object Detection. In: IEEE International Conference on Image Processing ICIP'2002, Rochester, New York, USA, pp 900–903.
294. Gorodnichy D, Roth G (2004) Nouse 'Use your nose as a mouse' perceptual vision technology for hands-free games and interfaces. Image and Vision Computing, 22 : 931–942.
295. Bouguet JY (2000) Pyramidal implementation of the Lucas-Kanade feature tracker. Technical Report, Intel Corporation, Microprocessor Research Labs.
296. Carbini S, Viallet JE, Bernier O, Bascle B (2005a) Tracking body parts of multiple persons for multi-person multimodal interface. In: IEEE International ICCV Workshop on Human-Computer Interaction, Beijing, China, pp 16–25.
297. Carbini S, Delphin-Poulat L, Perron L, Viallet JE (2006a) From a Wizard of Oz Experiment to a Real Time Speech and Gesture Multimodal Interface, Special issue of Signal Processing (ELSEVIER) on Multimodal Human-Computer Interfaces, vol. 86(12) : 3559–3577.
298. Atienza R, Zelinsky A (2003) Interactive skills using active gaze tracking. In: International Conference on Multimodal Interfaces (ICMI), Vancouver, Canada, pp 188–195.
299. Nickel K, Seemann E, Stiefelhagen R (2004) 3D-tracking of head and hands for pointing gesture recognition in a human robot interaction scenario. In: IEEE International Conference on Automatic Face and Gesture Recognition, Seoul, Korea, pp 565–570.
300. Kato T, Kurata T, Sakaue K (2002) Vizwear-active: distributed Monte Carlo face tracking for wearable active camera, In: International Conference on Pattern Recognition, Quebec, Canada, vol. 1, pp 395–400.
301. Jojic N, Brumitt B, Meyers B, Harris S (2000) Detecting and estimating pointing gestures in dense disparity maps. In: IEEE International Conference on Face and Gesture Recognition, Grenoble, France pp 468–475.
302. Yamamoto Y, Yoda I, Sakaue K (2004) Arm-pointing gesture interface using surrounded stereo cameras system. In: International Conference on Pattern Recognition (ICPR), Cambridge, UK, pp 965–970.
303. Demirdjian D, Darrell T (2002) 3-D articulated pose tracking for untethered diectic reference. In: International Conference on Multimodal Interfaces (ICMI), Pittsburgh, PA, USA, pp 267.
304. Viallet JE, Carbini S (2007) A simple hand posture recognition for efficient single handed pointing and selection, In: 7-th International Gesture Workshop, Lisbon, Portugal, pp 26–27.
305. Moustakas K, Tzovaras D, Carbini S, Bernier O, Viallet JE, Raidt S, Mancas M, Dimiccoli M, Yagci E, Balci S, Ibanez Leon E, Strintzis MG (2006) MASTERPIECE: Experiencing Physical Interaction in VR Applications, IEEE Multimedia, vol. 13, Issue 3, pp 92–100.
306. Carbini S, Viallet JE, Delphin-Poulat L (2005b) Context dependent interpretation of multimodal speech pointing gesture interface. In: ICMI International Conference on Multimodal Interfaces, Doctoral Spotlight and Demo Papers, Trento, Italy, pp 1–4.

307. Morris MR, Huang A, Paepcke A, Winograd T (2006) Cooperative Gestures: Multi-User Gestural Interactions for Co-located Groupware, In: Proceedings of CHI 2006, Montreal, Canada, pp 1201–1210.
308. ISO 9241-9:2000(E) (2000) Ergonomic Requirements for Office Work with Visual Display Terminals (VDTs), Part 9: Requirements for Non-Keyboard Input Devices, International Standards Organization.
309. Fitts M (1954) The information capacity of the human motor system in controlling amplitude of movement, Journal of Experimental Psychology, 47 : 381–391.
310. Soukoreff RW, MacKenzie IS (2004) Towards a standard for pointing device evaluation, perspectives on 27 years of Fitts' law research in HCI, Int. J. Hum. Comput. Stud., 61(6): 751–789.
311. Carbini S, Viallet JE (2006b) Evaluation of contactless multimodal pointing devices. In: Second IASTED International Conference on Human-Computer Interaction, Chamonix, France, pp 226–231.
312. Sukaviriya, P., Foley, J.D., and Griffith, T. A Second Generation User Interface Design Environment: The Model and The Runtime Architecture, in INTERCHI'93: 375–382.
313. Neches, R., Foley, J., Szekely, P., Sukaviriya, P., Luo, P., Kovacevic, S., Hudson, S. Knowledgeable development environments using shared design models, Proceedings IUI'93, pp. 63 – 70, ACM Press.
314. Szekely, P., Luo, P., Neches, R., "Beyond Interface Builders: Model-Based Interface Tools". Proceedings INTERCHI'93, pp. 383–390, ACM Press, 1993.
315. Puerta A., Eisenstein J., XIML: A Common Representation for Interaction Data. In: Proceedings of IUI2002: Sixth International Conference on Intelligent User Interfaces, ACM, January 2002.
316. Abrams, M., Phanouriou, C., Batongbacal, A., Williams, S., Shuster, J., UIML: An Appliance-Independent XML User Interface Language, Proceedings of the 8th WWW conference, 1994.
317. Mori G., Paternò F., Santoro C. Design and Development of Multidevice User Interfaces through Multiple Logical Descriptions. IEEE Transactions on Software Engineering August 2004, Vol 30, No 8, IEEE Press, pp.507–520.
318. Limbourg, Q., Vanderdonckt, J., UsiXML: A User Interface Description Language Supporting Multiple Levels of Independence, in Matera, M., Comai, S. (Eds.), Engineering Advanced Web Applications, Rinton Press, Paramus, 2004.
319. Chen, Y., Ma, W.-Y., Zhang, H.-J. Detecting Web page structure for adaptive viewing on small form factor devices. In Proceedings of the twelfth international conference on World Wide Web (WWW'03) (May 20–24, 2003, Budapest, Hungary), ACM 1-58113-680-3/03/0005, pp 225–233.
320. MacKay, B., Watters, C. R. Duffy, J.: Web Page Transformation When Switching Devices. In Proceedings of Sixth International Conference on Human Computer Interaction with Mobile Devices and Services (Mobile HCI'04) (Glasgow, September 2004), LNCS 3160. Springer-Verlag, 228–239.
321. Nichols, J., Myers B. A., Higgins M., Hughes J., Harris T. K., Rosenfeld R., Pignol M. "Generating remote control interfaces for complex appliances". Proceedings ACM UIST'02, pp.161–170, 2002.
322. Gajos, K., Christianson, D. , Hoffmann, R., Shaked, T. , Henning, K. , Long, J. J. and Weld D. S. Fast and robust interface generation for ubiquitous applications. In Seventh International Conference on Ubiquitous Computing (UBICOMP'05), September 2005, pp. 37–55.
323. Ponnekanti, S. R. Lee, B. Fox, A. Hanrahan, P. and Winograd T. ICrafter: A service framework for ubiquitous computing environments. In Proceedings of UBICOMP 2001. (Atlanta, Georgia, USA., 2001). LNCS 2201, ISBN:3-540-42614-0, Springer Verlag London UK Pp 56–75.
324. Garlan, D., Siewiorek, D., Smailagic, A., Steenkiste, P. Project Aura: Toward Distraction Free Pervasive Computing. IEEE Pervasive Computing, Vol 21, No 2 (April–June 2002), 22–31.

325. Bharat K. A. and Cardelli. L., Migratory Applications. In proceedings of User Interface Software and Technology (UIST '95). Pittsburgh PA USA. November 15–17, 1995. pp. 133–142.
326. Kozuch M., Satyanarayanan M., Internet Suspend/Resume, Proceedings of the Fourth IEEE Workshop on Mobile Computing Systems and Applications (WMCSA'02) IEEE Press, 2002.
327. Chung G., Dewan P., A mechanism for Supporting Client Migration in a Shared Window System, Proceedings UIST'96, pp.11–20, ACM Press.
328. Balme, L. Demeure, A., Barralon, N., Coutaz, J., Calvary, G. CAMELEON-RT: a Software Architecture Reference Model for Distributed, Migratable, and Plastic User Interfaces. In Proceedings the Second European Symposium on Ambient Intelligence (EUSAI '04), LNCS 3295, Markopoulos et al. Springer-Verlag, Berlin Heidelberg, 2004, 291–302.
329. Bandelloni, R., Berti, S., Paternò, F. Mixed-Initiative, Trans-Modal Interface Migration. In Proceedings of Sixth International Conference on Human Computer Interaction with Mobile Devices and Services, Mobile HCI'04, Glasgow, 2004, LNCS 3160. Springer-Verlag, 216–227.
330. Lieberman, H.; Paternò, F.; Wulf, V. (Eds.), End User Development, Springer Verlag, Human-Computer Interaction Series , Vol. 9 2006, XVI, 492 p., Hardcover, ISBN: 978-1-4020-4220-1.
331. André, E., Finkler, W., Graf, W., Rist, T., Schauder, A., Wahlster, W. (1993) WIP: The Automatic Synthesis of Multimodal Presentations. In: Maybury M. (eds), Intelligent Multimedia Interfaces, AAAI Press, pp. 75–93.
332. Maybury, M.T. (ed) (1993) Intelligent Multimedia Interfaces. AAAI Press/MIT Press.
333. Oviatt, S., Cohen, P. (2000) Multimodal interfaces that process what comes naturally. Communications of the ACM, 43, 3: 45–53.
334. Sinha, A. K., Landay, J. A. (2003) Capturing User Tests in a Multimodal, Multidevice Informal Prototyping Tool. In: Proceedings of ICMI'2003, the Fifth International Conference on Multimodal Interfaces, ACM Press, pp. 117–124.
335. Harrison, B. et al. (1998) Squeeze me, Hold me, Tilt Me! An exploration of Manipulative User Interface. In: Karat, C-M et al. (eds) Proceedings of the SIGCHI conference on Human factors in computing systems, ACM Press, pp. 17–24.
336. Coutaz, J., Nigay, L., Salber, D., Blandford, A., May, J. & Young, R. (1995) Four Easy Pieces for Assessing the Usability of Multimodal Interaction: The CARE properties. In: Arnesen S. A., Gilmore D. (eds) Proceedings of the INTERACT'95 conference, Chapman&Hall Publ., pp. 115–120.
337. Martin, J. C. (1997) TYCOON: Theoretical Framework and Software Tools for Multimodal Interfaces. In: Lee J. (ed) Intelligence and Multimodality in Multimedia Interfaces, AAAI Press.
338. Mansoux, B., Nigay, L., Troccaz, J. (2006) Output Multimodal Interaction: The Case of Augmented Surgery. In: Blandford A, Bryan-kinns, N., Curzon, P., Nigay, L., (eds) Proceedings of HCI'06, Human Computer Interaction, People and Computers XX, the 20th BCS HCI Group conference in co-operation with ACM, Springer Publ., pp. 177–192.
339. Bouchet, J., Nigay, L., & Ganille, T. (2005). The ICARE Component-Based Approach for Multimodal Input Interaction: Application to real-time military aircraft cockpits. In Conference Proceedings of HCI International 2005, the 11th International Conference on Human-Computer Interaction, Lawrence Erlbaum Associates.
340. Zouinar, M. et al. (2003) Multimodal Interaction on Mobile Artifacts. In Kintzig, C, Poulain, G., Privat, G., Favennec, P-N. (eds) Communicating with smart objects-developing technology for usable pervasive computing systems, Kogan Page Science.
341. Myers, B. (1994) Challenges of HCI Design and Implementation. Interactions, 1,1: 73–83.
342. Little, T.D.C, Ghafoor, A., Chen, C.Y.R., Chang, C.S., Berra, P.B. (1991) Multimedia Synchronization. IEEE Data Engineering Bulletin, 14, 3: 26–35.

343. Henry, T.R., Hudson, S.E., Newell, G.L. (1990) Integrating Gesture and Snapping into a User Interface Toolkit. In: Proceedings of the 3rd annual ACM SIGGRAPH symposium on User Interface Software and Technology, ACM Press, pp. 112–121.
344. Nigay, L., Coutaz, J. (1993) A Design Space for Multimodal Systems: Concurrent Processing and Data Fusion. In: Ashlung, S., Mullet, K., Henderson, A., E. Hollnagel, E., White, T. (eds), Proceedings of the SIGCHI conference on Human factors in computing systems, ACM Press, pp. 172–178.
345. Nigay, L., Coutaz, J. (1995) A Generic Platform for Addressing the Multimodal Challenge. In: Katz, I., Mack, R., Marks, L. (eds), Proceedings of the SIGCHI conference on Human factors in computing systems, ACM Press, pp. 98–105.
346. Coutaz J. (1987) PAC: an Implementation Model for Dialog Design. In Bullinger, H-J., Shackel, B. (eds) Proceedings of the INTERACT'87 conference, North Holland Publ., pp. 431–436.
347. Myers, B., Hudson, S.E., Pausch, R. (2000) Past, Present, and Future of User Interface Software Tools, ACM Transactions on Computer-Human Interaction TOCHI, 7(1): 3–28.
348. Dragicevic, P., Fekete, J.-D. (2004) ICON: The Input Configurator Toolkit: towards High Input Adaptability in Interactive Applications. In: Proceedings of the working conference on Advanced Visual Interfaces AVI 2004, ACM Press, pp. 244–247.
349. Oviatt, S. (2000) Taming recognition errors with a multimodal interface. Communications of the ACM, 43,9: 45–51.
350. Westeyn, T., Brashear, H., Atrash, A., Starner, T. (2003) Georgia Tech Gesture Toolkit: Supporting Experiments in Gesture Recognition. In: Proceedings of ICMI'2003, the Fifth International Conference on Multimodal Interfaces, ACM Press, pp. 85–92.
351. Glass, J., Weinstein, E., Cyphers, S., Polifroni, J., Chung, G., Nakano M. (2004) A Framework for Developing Conversational User Interfaces. In: Jacob, R., Limbourg, Q., Vanderdonckt, J. (eds.) Proceedings of the Fifth International Conference on Computer-Aided Design of User Interfaces CADUI 2004, pp.354–65.
352. Krahnstoever, N., Kettebekov, S., Yeasin, M., Sharma, R. (2002) A Real-time Framework for Natural Multimodal Interaction with Large Screen Displays,. In: Proceedings of ICMI'2002, Proceedings of the Fourth International Conference on Multimodal Interfaces, ACM Press, pp. 349.
353. Johnston, M., Cohen, P. R., McGee, D., Oviatt, S. L., Pittman, J. A., Smith, I. (1997) Unification-based Multimodal Integration. In: Proceedings of the 35th Annual Meeting of the Association for Computational Linguistics and of the 8th Conference of the European Chapter of the Association for Computational Linguistics, pp. 281–288.
354. Moran, D. B., Cheyer, A. J., Julia L. E., Martin, D. L., Park, S. (1997) Multimodal User Interfaces in the Open Agent Architecture. In: Proceedings of IUI'97, the 2nd international conference on Intelligent User Interfaces, ACM Press, pp. 61–68.
355. Beaudoin-Lafon, M. (2004) Designing Interaction, not Interfaces. In: Proceedings of the working conference on Advanced Visual Interfaces AVI 2004, pp. 15–22.
356. Elting, C., Rapp, S., Mölher, G., Strube M. (2003) Architecture and Implementation of Multimodal Plug and Play. In: Proceedings of ICMI'2003, the Fifth International Conference on Multimodal Interfaces, ACM Press, pp. 93–100.
357. Rousseau, C., Bellik, Y., Vernier, F., Bazalgette, D. (2004) Architecture Framework For Output Multimodal Systems Design. In: Proceedings of OZCHI 2004.
358. Bouchet, J., Nigay, L. (2004) ICARE: A Component-Based Approach for the Design and Development of Multimodal Interfaces. In: Dykstra-Erickson, E., Tscheligi, M. (eds) Proceedings of the SIGCHI conference on Human factors in computing systems, Extended abstracts, ACM Press, pp. 1325–1328.
359. Bouchet, J., Nigay, L., Ganille, T. (2004). ICARE Software Components for Rapidly Developing Multimodal Interfaces. In: Proceedings of ICMI'2004, the Sixth International Conference on Multimodal Interfaces, ACM Press, pp. 251–258.
360. Gaines, B.R. (1991) Modelling and Forecasting the Information Sciences. Information Sciences, 57-58: 3–22.

References

361. Bolt, R. (1980) Put that there: Voice and gesture at the graphics interface. Computer Graphics: 262–270.
362. Gray, P., Ramsay, A., Serrano, M. (2007) A Demonstration of the OpenInterface Interaction Development Environment. Demonstration at the 20th annual ACM SIGGRAPH symposium on User Interface Software and Technology.
363. Stokoe WC (1960) Sign Language Structure: An Outline of the Visual Communication Systems of the American Deaf, Studies in Linguistics: Occasional papers 8.
364. Liddell SK (2003) Grammar, Gesture, and Meaning in American Sign Language, Cambridge University Press.
365. Ong SCW, Ranganath S (2005) Automatic Sign Language Analysis: A Survey and the Future beyond Lexical Meaning., IEEE Transactions on Pat-tern Analysis and Machine Intelligence 27(6):873–891.
366. Parton BS (2006) Sign language recognition and translation: A multidisciplined approach from the field of artificial intelligence, Journal of deaf studies and deaf education 11(1):94–101.
367. Keskin C, Balci K, Aran O, Sankur B, Akarun L (2007) A Multimodal 3D Healthcare Communication System. In: 3DTV Conference, Greece.
368. Fang G, Gao W, Zhao D (2007) Large-Vocabulary Continuous Sign Language Recognition Based on Transition-Movement Models, IEEE Transactions on Systems, Man and Cybernetics, Part A 37(1):1–9.
369. Holden E, Lee G, Owens R (2005) Australian sign language recognition, Machine Vision and Applications 16(5):312–320.
370. Aran O, Ari I, Benoit A, Campr P, Carrillo AH, Fanard F, Akarun L, Caplier A, Rombaut M, Sankur B (2006) Sign Language Tutoring Tool. In: eNTERFACE 2006, The Summer Workshop on Multimodal Interfaces, Dubrovnik, Croatia, pp 23–33.
371. Feris R, Turk M, Raskar R, Tan K, Ohashi, G (2004) Exploiting Depth Dis-continuities for Vision-Based Fingerspelling Recognition. In: CVPRW 04: Proceedings of the 2004 Conference on Computer Vision and Pattern Recog-nition Workshop (CVPRW04), IEEE Computer Society, Washington, DC, USA, pp 155.
372. Wu J, Gao W (2001) The Recognition of Finger-Spelling for Chinese Sign Language. In: Gesture Workshop, pp 96–100.
373. Cornett RO (1967) Cued Speech. American Annals of the Deaf 112:3–13.
374. Habili N, Lim C, Moini A (2004) Segmentation of the face and hands in sign language video sequences using colour and motion cues. IEEE Trans. Circuits Syst. Video Techn. 14(8):1086–1097.
375. Awad G, Han J, Sutherland A (2006) A Unified System for Segmentation and Tracking of Face and Hands in Sign Language Recognition. In: ICPR 06: Proceedings of the 18th International Conference on Pattern Recognition, IEEE Computer Society, Washington, DC, USA, pp 239–242.
376. Aran O, Akarun L (2006) Recognizing Two Handed Gestures with Generative, Discriminative and Ensemble Methods via Fisher Kernels, Lecture Notes in Computer Science: Multimedia Content Representation, Classification and Security International Workshop, MRCS 2006, Istanbul, Turkey, pp 159–166.
377. Jayaram, S., Schmugge, S., Shin, M.C., Tsap, L.V., 2004. Effect of colorspace transformation, the illuminance component, and color modeling on skin detection, Paper presented at the Comput. Vision and Pattern Recognition, 2004 (CVPR 2004), Proc. 2004 IEEE Comput. Soc. Conf. on.
378. Hjelmäs H, Low B (2001) Face detection: a survey. Computer Vision and Image Understanding, 83:236–274.
379. Yang MH, Kriegman D, Ahuja N (2002) Detecting face in images: a survey. IEEE Trans on PAMI, 24(1):34–58.
380. Viola P, Jones J (2004) Robust Real Time Face Detection. International Journal of Computer Vision, 57(2):137–154.
381. Bullier J (2001) Integrated model of visual processing. Brain Research, 36(2-3):96–107.

382. Benoit A, Caplier A (2005) Biological approach for head motion detection and analysis. EUSIPCO 2005, Antalya, Turkey.
383. Benoit A, Caplier A (2005) Head Nods Analysis : Interpretation Of Non Verbal Communication Gestures. IEEE ICIP 2005, Genova.
384. Burger T (2007) Reconnaissance automatique des gestes de la Langue française Parlée Complétée. Thèse de Doctorat, France.
385. M. Isard and A. Blake (1998). CONDENSATION "Conditional density propagation for visual tracking." Int. J. Computer Vision, 29, 1, 5–28.
386. Burger T, Urankar A, Aran O, Akarun L, Caplier A. (2007b) Cued Speech Hand Shape Recognition. In:2nd International Conference on Computer Vision Theory and Applications (VISAPP07), Spain.
387. Norkin CC, Levangie PK (1992) Joint structure and function. (2nd ed.). Philadelphia: F.A. Davis.
388. Y. Wu, Lin J. Y., and T. S. Huang. Capturing natural hand articulation. In Proc. 8th Int. Conf. on Computer Vision, volume II, pages 426–432, Vancouver, Canada, July 2001. http://citeseer.ist.psu.edu/wu01capturing.html.
389. Zhang D, Lu G (2003) Evaluation of MPEG-7 shape descriptors against other shape descriptors. Multimedia Systems 9(1).
390. Hu MK (1962) Visual pattern recognition by moment invariants. IRE Trans. on Information Theory, 8:179–187.
391. Caplier A, Bonnaud L, Malassiotis S, Strintzis MG (2004) Comparison of 2D and 3D analysis for automated Cued Speech gesture recognition. In:SPECOM.
392. Adam S, Ogier JM, Cariou C, Mullot R, Gardes J, Lecourtier Y (2001) Utilisation de la transformée de Fourier-Mellin pour la reconnaissance de formes multi-orientées et multi-échelles : application à l'analyse de documents techniques". Traitement du Signal, 18(1).
393. Burger T, Caplier A , Perret P (2007a) Cued Speech Gesture Recognition: a First Prototype Based on Early Reduction. International Journal of Image and Video Processing, Special Issue on Image & Video Processing for Disability.
394. Burger T, Aran O, Caplier A. (2006a) Modelling Hesitation and Conflict: A Belief-Based Approach for Multi-class Problems. In: ICMLA 06: Proceedings of the 5th International Conference on Machine Learning and Applications, IEEE Computer Society, Washington, DC, USA, pp 95–100.
395. Benoit A, Caplier A (2005) Hypo-vigilance Analysis: Open or Closed Eye or Mouth? Blinking or Yawning Frequency?. IEEE AVSS 2005, Como, Italy.
396. Kass M, Witkin A, Terzopoulos D (1988) Snakes: active contour models. Int. Journal of Computer Vision, 1(4):321–331.
397. Terzopoulos D, Waters K (1993) Analysis and Synthesis of Facial Image Sequences Using Physical and Anatomical Models. IEEE Trans. On Pattern Analysis and Machine Intelligence, 15(6):569–579.
398. Aleksic P, Williams J, Wu Z, Katsaggelos A (2002) Audio-Visual Speech Recognition using MPEG-4 Compliant Features. Eurasip Journal on Applied Signal Processing, Special Issue on Joint Audio-visual speech processing, pp.1213–1227.
399. Cootes TF, Hill A, Taylor CJ, Haslam J (1994) Use of Active Shape Models for Locating structures in Medical Images, Image and Vision Computing, 12(6):355–365.
400. Eveno N, A. Caplier A, Coulon PY (2004) Automatic and Accurate Lip Tracking. IEEE Transactions on Circuits and Systems for Video technology, 14(5)706–715.
401. Zhang L (1997) Estimation of the mouth features using deformable templates. Int. Conf. on Image Processing (ICIP'97), Santa Barbara, CA, October, pp. 328–331.
402. Beaumesnil B, Chaumont M, Luthon F (2006) Liptracking and MPEG4 Animation with Feedback Control. IEEE International Conference On Acoustics, Speech, and Signal Processing.
403. Luettin J, Thacker N, Beet S (1996) Statistical Lip Modelling for Visual Speech Recognition. In Proceedings of the 8th European Signal Processing Conference (Eusipco'96).

404. Gacon P, Coulon PY, Bailly G (2005) Non-Linear Active Model for Mouth Inner and Outer Contours Detection. European Signal Processing Conference, Antalya, Turkey.
405. Pantic M, Rothkrantz M. (2000) Automatic Analysis of Facial Expressions: The State of the Art. IEEE Trans. on Pattern Analysis and Machine Intelligence, 22(12).
406. Fasel B, Luettin J (2003) Automatic Facial Expression Analysis: A Survey. Pattern Recognition, 1(30):259–275.
407. Yacoob Y, Davis LS. (1996) Recognizing Human Facial Expressions from Long Image Sequences Using Optical Flow, IEEE Trans. on Pattern Analysis and Machine Intelligence, 18(6):636–642.
408. Black MJ, Yacoob Y (1997) Recognizing Facial Expression in Image Sequences Using Local Parameterised Models of Image motion. International Journal of Computer Vision, 25(1):23–48.
409. Essa IA, Pentland AP (1997) Coding, Analysis, Interpretation, and Recognition of Facial Expressions. IEEE Trans. on Pattern Analysis and Machine Intelligence, 19(7)757–763.
410. Cohn JF, Zlochower AJ, Lien JJ, Kanade T (1998) Feature-Point Tracking by Optical Flow Discriminates Subtles Differences in Facial Expression, Proc. IEEE International Conference on Automatic Face and Gesture Recognition, April, Nara, Japan, pp. 396–401.
411. Zhang Z, Lyons L, Schuster M, Akamatsu S. (1998) Comparison between Geometry-Based and Gabor Wavelets-Based Facial Expression Recognition Using Multi-Layer Perceptron. Proc. IEEE International Conference on Automatic Face and Gesture Recognition, pp. 454–459.
412. Gao Y, Leung MKH, Hui SC, Tananda MW (2003) Facial Expression Recognition From LineBased Caricatures, IEEE Trans. on System Man and Cybernetics – PART A: System and Humans, 33(3).
413. Oliver N, Pentland A, Bérard F. (2000) LAFTER: A real-time face and tracker with facial expression recognition. Pattern Recognition, 33:1369–1382.
414. Abboud B, Davoine F, Dang M (2004) Facial expression recognition and synthesis based on appearance model. Signal Processing: Image Communication, 19(8)723–740.
415. Lien JJ, Kanade T, Cohn JF, Li C. (1998) Subtly different facial expression recognition and expression intensity estimation, Proc. IEEE Computer Vision and Pattern Recognition, Santa Barbara, CA, pp. 853–859.
416. Tian Y, Kanade T, Cohn JF. (2001) Recognizing Action Units for Facial Expression Analysis, IEEE Trans. Pattern Analysis and Machine Intelligence, 23(2)97–115.
417. Cohen I, Cozman FG, Sebe N, Cirelo MC, Huang TS (2003) Learning Bayesian network classifiers for facial expression recognition using both labelled and unlabeled data, Proc. IEEE Computer Vision and Pattern Recognition.
418. Tsapatsoulis N, Karpouzis K, Stamou G, Piat F, Kollias SA (2000) A fuzzy system for emotion classification based on the MPEG-4 facial definition parameter set. Proc. 10th European Signal Processing Conference, Tampere, Finland.
419. Hammal Z, Couvreur L, Caplier A, Rombaut M (2007) Facial Expression Classification: An Approach based on the Fusion of Facial Deformation unsung the Transferable Belief Model. Int. Jour. of Approximate Reasoning.
420. Smets P and Kennes R (1994) The transferable belief model, Artificial Intelligence, 66(2): 91–234.
421. Attina V (2005) La Langue française Parlée Complétée : production et perception. Thèse de Doctorat en Sciences Cognitives, Institut National Polytechnique de Grenoble, France.
422. Rabiner LR (1989) A Tutorial on Hidden Markov Models and Selected Applications in Speech Recognition. In:Proceedings of the IEEE, pp 257–285.
423. Brand M, Oliver N, Pentland A (1997) Coupled hidden Markov models for complex action recognition. In: IEEE Computer Society Conference on Computer Vision and Pattern Recognition (CVPR97), pp 994.
424. Vogler C, Metaxas D (1999) Parallel Hidden Markov Models for American Sign Language Recognition. In: International Conference on Computer Vision, Kerkyra, Greece, pp 116–122.

425. Bengio Y, Frasconi P (1996) Input-output HMM's for sequence processing. IEEE Transactions on Neural Networks, 7(5):1231–1249.
426. Dempster AP (1968) A generalization of Bayesian inference. Journal of the Royal Statistical Society, Series B, 30(2):205–247.
427. Shafer G. (1976) A Mathematical Theory of Evidence, Princeton University Press.
428. Vannoorenberghe P and Smets P (2005) Partially Supervised Learning by a Credal EM Approach. Symbolic and Quantitative Approaches to Reasoning with Uncertainty.
429. Denoeux T (1995) A k-nearest neighbour classification rule based on Dempster-Shafer theory. IEEE Transactions on Systems, Man and Cybernetics, 25(5):804–813.
430. Denoeux T (1997) Analysis of evidence-theoretic decision rules for pattern classification. Pattern Recognition, 30(7): 1095–1107.
431. Denoeux T (2000) A neural network classifier based on Dempster-Shafer theory. IEEE Transactions on Systems, Man and Cybernetics A, 30(2):131–150.
432. Cox RT (1946) Probability, Frequency, and Reasonable Expectation. American Journal hysique, 14:1–13.
433. Burger T, Caplier A (2007) Partial Pignistic Transform". International Journal of Approximate Reasoning, Submitted.
434. Aran O, Burger T, Caplier A, Akarun L (2007) Sequential Belief-Based Fusion of Manual and Non-Manual Signs". Gesture Workshop, Lisbon, Portugal.
435. Burger T, Benoit A, Caplier A (2006b) Extracting static hand gestures in dynamic context. Proceeding of ICIP'06, Atlanta, USA.
436. Hérault J, Durette B (2007) Modelling Visual Perception for Image Processing. F. Sandoval et al. (Eds.): IWANN 2007, LNCS 4507, Springer-Verlag Berlin Heidelberg, pp.662–675.
437. MPT:Machine Perception Toolbox, face detection algorithm: http://mplab.ucsd.edu/grants/project1/free-software/MPTWebSite/introductionframe.html.
438. R. O. Cornett, "Cued speech," American Annals for the Deaf, vol. 112, pp. 3–13, 1967.
439. B. Bauer, S. Nießen, and H. Hienz, "Towards an Automatic Sign Language Translation System," Proceedings of the International Workshop on Physicality and Tangibility in Interaction: Towards New Paradigms for Interaction Beyond the Desktop, Siena, Italy, 1999.
440. P. Duchnowski, D. Lum, J. Krause, M. Sexton, M. Bratakos, and L. Braida, "Development of Speechreading Supplements Based on Automatic Speech Recognition," IEEE Trans. on Biomedical Engineering, vol. 47, no. 4, pp. 487–496, 2000.
441. G. Gibert, G. Bailly, F. Elisei, D. Beautemps, and R. Brun, "Evaluation of a speech cuer: from motion capture to a concatenative text-to-cued speech system," Language Ressources and Evaluation Conference (LREC). Lisbon, Portugal, pp. 2123–2126, 2004.
442. H. Yashwanth, H. Mahendrakar, and S. David, "Automatic speech recognition using audio visual cues," India Annual Conference, 2004. Proceedings of the IEEE INDICON 2004. First, pp. 166–169, 2004.
443. X. Liu, Y. Zhao, X. Pi, L. Liang, and A. Nefian, "Audio-visual continuous speech recognition using a coupled hidden Markov model," Proceedings of the International Conference on Spoken Language Processing (ICSLP), pp. 213–216, 2002.
444. T. Kristjansson, B. Frey, and T. Huang, "Event-coupled hidden Markov models," Multimedia and Expo, IEEE International Conference on, vol. 1, 2000.
445. A. Nefian, L. Liang, X. Pi, X. Liu, and K. Murphy, "Dynamic bayesian networks for audio-visual speech recognition," EURASIP Journal on Applied Signal Processing, vol. 2002, no. 11, pp. 1274–1288, 2002.
446. I. Arsic and J. Thiran, "Mutual information eigenlips for audio-visual speech recognition," in 14th European Signal Processing Conference (EUSIPCO), ser. Lecture Notes in Computer Science. IEEE, 2006.
447. T. Butz and J. Thiran, "From Error Probability to Information Theoretic (Multi-Modal) Signal Processing," Signal Processing, vol. 85, no. 5, pp. 875–902, 2005.
448. N. Tanibata, N. Shimada, and Y. Shirai, "Extraction of hand features for recognition of sign language words," The 15th International Conference on Vision Interface May, pp. 27–29, 2002.

References

449. S. Mehdi and Y. Khan, "Sign language recognition using sensor gloves," Neural Information Processing, 2002. ICONIP'02. Proceedings of the 9th International Conference on, vol. 5, 2002.
450. T. Burger, A. Caplier, and S. Mancini, "Cued speech hand gestures recognition tool," in EUSIPCO2005, Antalya, Turkey, Sep 2005.
451. A. Caplier, L. Bonnaud, S. Malassiotis, and M. Strintzis, "Comparison of 2d and 3d analysis for automated cued speech gesture recognition," SPECOM 2004.
452. S. Tamura, K. Iwano, and S. Furui, "A Stream-Weight Optimization Method for Multi-Stream HMMS Based on Likelihood Value Normalization," Acoustics, Speech, and Signal Processing, 2005. Proceedings.(ICASSP'05). IEEE International Conference on, vol. 1, 2005. Modality Replacement Framework for Applications for the Disabled 23.
453. S. Tamura, K. Iwano, S. Furui, "A stream-weight optimization method for audio-visual speech recognition using multi-stream HMMs," Acoustics, Speech, and Signal Processing, 2004. Proceedings.(ICASSP'04). IEEE International Conference on, vol. 1, 2004.
454. E. Erzin, Y. Yemez, and A. Tekalp, "Multimodal speaker identification using an adaptive classifier cascade based on modality reliability," Multimedia, IEEE Transactions on, vol. 7, no. 5, pp. 840–852, 2005.
455. S. Argyropoulos, D. Tzovaras, and M. G. Strintzis, "Multimodal fusion for cued speech language recognition," in Proc. of the 15th European Signal Processing Conference (EUSIPCO), Poznan, Poland, Sep. 2007.
456. J. Deller Jr, J. Proakis, and J. Hansen, Discrete Time Processing of Speech Signals. Prentice Hall PTR Upper Saddle River, NJ, USA, 1993.
457. H. Cetingul, Y. Yemez, E. Erzin, and A. Tekalp, "Robust Lip-Motion Features For Speaker Identification," Acoustics, Speech, and Signal Processing, 2005. Proceedings.(ICASSP'05). IEEE International Conference on, vol. 1, 2005.
458. M. Sargin, E. Erzin, Y. Yemez, and A. Tekalp, "Lip feature extraction based on audio-visual correlation," Proc. of the European Signal Processing Conference 2005 (EUSIPCO05), 2005.
459. T. V. Raman, "Multimodal interaction design principles for multimodal interaction," in CHI 2003, Fort Lauderdale, USA, 2003, pp. 5–10.
460. C. Colwell, H. Petrier, D. Kornbrot, A. Hardwick, and S. Furner, "Haptic virtual reality for blind computer users," in Proc. of Annual ACM Conference on Assistive Technologies, 1998, pp. 92–99.
461. C. Sjostrom, "Touch access for people with disabilities," Ph.D. dissertation, CERTEC Lund University, 1999.
462. V. Scoy, I. Kawai, S. Darrah, and F. Rash, "Haptic display of mathematical functions for teaching mathematics to students with vision disabilities," in Haptic Human-Computer Interaction Workshop, 2000.
463. K. Moustakas, G. Nikolakis, D. Tzovaras, B. Deville, I. Marras, and J. Pavlek, "Multimodal tools and interfaces for the intercommunication between visually impaired and deaf-and-mute people," in Proc. of eNTERFACE 2006, Dubrovnik, Croatia, 2006, pp. 11–22.
464. S. Argyropoulos, K. Moustakas, A. A. Kaprov, O. Aran, D. Tzovaras, T. Tsakiris, and B. Kwon, "A multimodal framework for the communication of the disabled," in Proc. of eNTERFACE 2007, Istanbul, Turkey, 2007.
465. D. Gering, A. Nabavi, R. Kikinis, et al., "An integrated visualization system for surgical planning and guidance using image fusion and interventional imaging," in Proc. of the Medical image computing and computer assisted intervention (MICCAI). London, UK: SPringer-Verlag, 1999, pp. 809–818.
466. E. Keeve, T. Jansen, Z. Krol, et al., "Julius an extendable software framework for surgical planning and image-guided navigation," in Proc. 4th International Conference on Medical Image Computing and Computer-Assisted Intervention (MICCAI). Springer-Verlag, 2001.
467. K. Cleary, L. Ibanez, S. Rajan, and B. Blake, "IGSTK: a software toolkit for image-guided surgery applications," in Proc. of Computer Aided Radiology and Surgery (CARS), 2004.

468. K. Cleary, H. Y. Chung, and S. K. Mun, "OR 2020 workshop overview: operating room of the future," in Proc. Computer Aided Radiology and Surgery (CARS), 2004.
469. J. Bouchet and L. Nigay, "Icare: A component-based approach for the design and development of multimodal interfaces," in Proc.11th International Conference on Human-Computer Interaction HCI International. Vienna: Lawrence Erlbaum Associates, 2004.
470. E. H. Chi and J. T. Riedl, "An operator interaction framework for visualization systems," in Proc. IEEE Symposium on Information Visualization, 1998, pp. 63–70.
471. J. C. Spall, "Overview of the simultaneous perturbation method for efficient optimization," Hopkins APL Technical Digest, vol. 19, pp. 482–492, 1998.
472. O. Cuisenaire, "Distance transformation, fast algorithms and applications to medical image processing," Ph.D. dissertation, UCL, Louvain-la-Neuve, Belgium, Oct. 1999. [Online]. Available: http://www.tele.ucl.ac.be/PEOPLE/OC/Thesis.html.
473. Q. Noirhomme, M. Ferrant, Y. Vandermeeren, E. Olivier, B. Macq, and O. Cuisenaire, "Real-time registration and visualization of transcranial magnetic stimulation with 3D MR images," IEEE Trans. Bio-Med. Eng., Nov. 2004.
474. M. D. Craene, A. du Bois d'Aische, B. Macq, F. Kipfmueller, N. Weisenfeld, S. Haker, and S. K. Warfield, "Multi-modal non-rigid registration using a stochastic gradient approximation," in Proc. of the 2004 IEEE International Symposium on Biomedical Imaging (ISBI), Arlington, VA, USA, apr 2004, pp. 1459–1462.
475. W. E. Lorensen and H. E. Cline, "Marching cubes, a high resolution 3D surface construction algorithm," Computer Graphics, vol. 21, no. 4, pp. 163–169, July 1987.
476. P. D Haese, A. du Bois d Aische, T. Merchan, B. Macq, R. Li, and B. Dawant, "Automatic segmentation of brain structures for radiation therapy planning," in Proc. of the SPIE (vol.5032): Image Processing, San Diego, USA, May 2003, pp. 517–526.
477. S. Osher and J. A. Sethian, "Fronts propagating with curvature dependent speed: Algorithms based on Hamilton–Jacobi formulations," Journal of Computational Physics, vol. 79, pp. 12–49, 1988.
478. R. Olszewski, V. Nicolas, B. Macq, and H. Reybhler, "Acro 4D: universal analysis for four-dimensional diagnosis, 3d planning and simulation in orthognathic surgery," in Proc. Computer Assisted Radiology and Surgery (CARS), London, June 2003.
479. D. G. Trevisan, L. P. Nedel, B. Macq, J. Vanderdonckt, Detecting interaction variables in a mixed reality system for maxillofacial-guided surgery, in: SVR 2006 – SBC Symposium on Virtual Reality, Vol. 1, 2006, pp. 39–50.
480. Huff, Rafael; Dietrich, Carlos A.; Nedel, Luciana P.; Freitas, Carla M.D.S.; Comba, João L.D.; Olabarriaga, Silvia D. Erasing, Digging and Clipping in Volumetric Datasets with One or Two Hands, In: VRCIA 2006 (ACM International Conference on Virtual Reality Continuum and Its Applications), Hong Kong, China, June 2006.
481. R. Aggarwal, K. Moorthy, and A. Darzi. "Laparoscopic skills training and assessment". Br J Surg, 91(12):1549–1558, 2004.
482. F.M. Sánchez Margallo, E.J. Gómez, C. Monserrat, S. Pascual, M. Alcañiz, F. del Pozo, J. Usón Gargallo, "Sinergia: tecnologías de simulación y planificación quirúrgica en cirugía mínimamente invasiva", XXI Congreso Anual de la Sociedad Española de Ingeniería Biomédica ,CASEIB 2003. Mérida. Pp. 323–326, 84-688-3819-5, noviembre, 2003.
483. F.M. Sánchez Margallo, E.J. Gómez, J.B. Pagador, C. Monserrat, S. Pascual, M. Alcañiz, F. del Pozo, J. Usón Gargallo. "Integración de la Tecnología de Simulación Quirúrgica en el Programa de Aprendizaje de Cirugía de Mínima Invasión". Informática y Salud (Sociedad Española de Informática y la Salud). ISSN 1579-8070. n.48, pp 9–14. 2004.
484. C. Monserrat, O. López, U. Meier, M. Alcañiz, C. Juan, and V. Grau. "GeRTiSS: A Generic Multimodel Surgery Simulation", Proc. Surgery Simulation and Soft Tissue Modeling Interntaional Symposium, Juan-Les-Pinnes (Fr), Lecture Notes in Computer Science 2673, pp. 59–66, 2003.
485. C. Forest, H. Delingette, N. Ayache. "Surface Contact and Reaction Force Models for.Laparoscopic Simulation". Medical Simulation,2004.

486. J.Brown, S.Sorkin, J.C.Latombe, K. Montgomery and M. Stephanides. "Algorithmic tools for real-time microsurgery simulation". Medical Image Analysis, 6, pages 289–300, 2002.
487. U.Meier, O.Lopez, C.Monserrat, M.C.Juan, M.Alcaniz, Real-time deformable models for surgery simulation: a survey, Computer Methods and Programs in Biomedicine, 77 (2005) 183–197.
488. K. Sundaraj, "Real-Time Dynamic Simulation and 3D Interaction of Biological Tissue: Application to Medical Simulators." PhD thesis, Institut National Polytechnique de Grenoble, France, Ene 2004.
489. W. Chou and T. Wang, "Human-Computer interactive simulation for the training of minimally invasive neurosurgery," in IEEE International Conference on Systems, Man and Cybernetics, vol. 2, pp. 1110–1115, 5–8 Oct 2003.
490. J. Kim and M. A. S. Suvranu De, "Computationally Efficient Techniques for Real Time Surgical Simulation with Force Feedback," in 10th Symposium on Haptic Interfaces for Virtual Environment and Teleoperator Systems, pp. 51–57, 24–25 Mar 2002.
491. Domínguez-Quintana L., Rodríguez-Florido M.A., Ruiz-Alzola J., Soca Cabrera D. Modelado 3D de Escenarios Virtuales Realistas para SimuladoresQuirúrgicos Aplicados a la Funduplicatura de Nissen. Informática y Salud (Sociedad Española de Informática y Salud).n. 48 de la revista ISSN 1579–8070. pp 14–20. 2004.
492. N.E.Seymour, A.G.Gallagher, S.A.Roman, M.K.O'Brien, V.K.Bansal, D.K.Andersen, R.M.Satava, "Virtual reality training improves operating room performance: results of a randomized, double-blinded study", Ann Surg, 236, 458–463, 2002.
493. A.Darzi, V.Datta, S. Mackay, "The challenge of objective assessment of surgical skill", Am. J Surg, 181, 484–486, 2001.
494. E.D.Grober, S.J.Hamstra, K.R.Wanzel, R.K.Reznick, E.D.Matsumoto, R.S.Sidhu, K.A.Jarvi, "The educational impact of bench model fidelity on the acquisition of technical skill: the use of clinically relevant outcome measures", Ann. Surg, 240, 374–381, 2004.
495. R.Kneebone, "Simulation in surgical training: educational issues and practical implications", Medical Education, 37, 267–277, 2003.
496. P.Lamata, E.J.Gomez Aguilera, F.M.Sanchez-Margallo, F.Lamata Hernandez, F.del Pozo Guerrero, J.Uson Gargallo, "Study of consistency perception in laparoscopy for defining the level of fidelity in virtual reality simulation", Surgical Endoscopy, (in press) 2005.
497. P.Lamata, E.J.Gomez, F.J.Sanchez-Margallo, F.Lamata, F.Gaya, J.B.Pagador, J.Uson, and F.del Pozo, "Analysis of tissue consistency perception for laparoscopic simulator design," Proc. 18th International CARS 2004 – Computer Assisted Radiology and Surgery., International Congress Series 1268C, pp. 401–406, 2004.
498. P.Lamata, R.Aggarwal, F.Bello, F.Lamata, A.Darzi, and E.J.Gomez, "Taxonomy of didactic resources in virtual reality simulation," Proc. The Society of American Gastrointestinal Endoscopic Surgery (SAGES) Annual meeting, (in press), 2005.

Printing: Krips bv, Meppel, The Netherlands
Binding: Stürtz, Würzburg, Germany